Garlic and Other Alliums
The Lore and the Science

To my wife Judy

Garlic and Other Alliums
The Lore and the Science

Eric Block

Department of Chemistry
University at Albany, State University of New York
Albany, New York, USA

RSC Publishing

Disclaimer

This book is primarily for reference and education. It is not intended to be a substitute for a physician. The author does not advocate self-diagnosis or self-medication and urges anyone with continuing symptoms, however minor, to seek medical advice. The author documents in this book that a potent plant such as garlic, whether used as a food or medicine, externally or internally, may cause an allergic reaction, may damage sensitive tissue particularly when used with young children, or may interfere with medication.

ISBN: 978-1-84973-180-5

A catalogue record for this book is available from the British Library

Published by The Royal Society of Chemistry,
Thomas Graham House, Science Park, Milton Road,
Cambridge CB4 0WF, UK

Registered Charity Number 207890

For further information see our website at www.rsc.org

Foreword

The story of garlic and its plant relatives (the alliums) is at once extraordinary, multifaceted and unique. That tale is recounted in this book by Professor Eric Block, an unrivaled authority on every aspect of this field, with enormous care and attention to detail – from the cultural, historical, botanical, environmental, culinary and medicinal aspects of garlic and the alliums, to their fundamental science. This last area is covered authoritatively and comprehensively, as would be expected from Dr. Block, who has been at the forefront of the field of organic sulfur compounds and studies of the unusual sulfur-containing constituents of garlic and the alliums for about four decades. The scope and completeness of this volume and the expert presentation mark it as a future classic.

Yet another gift to the reader is the wonderful collection of paintings and photographs which splendidly connect the alliums to society, art, history, architecture and culture.

It is rare, indeed, to find a juxtaposition of so much interesting material – from ancient to modern – on a common plant family within the confines of a single book. There is something in this volume for everyone, and on top of that, a richness and liveliness which are delightful as well as educational. It is also a coherent, well written and most enjoyable work, clearly a labor of love. Personally, I look forward to rereading it from time to time, as well as using it as a reference resource.

Eric Block deserves our thanks and congratulations for a magnificent contribution to the literature of the endlessly fascinating

Allium family, including that special member – garlic. It will be widely read and much appreciated for many years to come.

E. J. Corey
Harvard University
Cambridge, MA

Preface

Why am I, a chemist specializing in the organic chemistry of the element sulfur, writing this book on garlic and its relatives – the alliums? Forty years ago I began laboratory work on the unusual sulfur-containing compounds found in garlic and onion. With the aid of talented students and coworkers, secrets of the often smelly and sometimes tear-provoking compounds slowly revealed themselves. The surprising complexity found in the chemistry of compounds extracted from the plants and synthesized in the laboratory served as the basis for my group's scientific publications and my students' doctoral theses. I took advantage of opportunities to talk about our work to audiences of specialists, as well as to high school students, audiences at garlic festivals and, sometimes informally, to participants at garlic-themed dinners.

To enliven my talks, I interspersed information on the history of these plants and their medicinal and culinary use with the science. On occasion I was invited to speak at professional meetings in the fields of horticulture, botany, medicine, and other areas where the science of garlic and its relatives were of interest. To prepare for these meetings, I educated myself on historical and cultural aspects of these plants, often journeying into areas far afield from chemistry. I found myself a student again, learning as much as I could from colleagues I met and those whose work I read, as well as from my international travels, which sometimes brought me to exotic produce markets that invariably included alliums. In the course of

Garlic and Other Alliums: The Lore and the Science
By Eric Block

Published by the Royal Society of Chemistry, www.rsc.org

my self-education I amassed a wealth of information on garlic and its botanical allies. The more I researched the subject, the more it became apparent just how fascinating was the interplay between the science of these plants and their incorporation into different cultures, since these plants were first noticed by our ancient ancestors.

I was hardly the first to take a fancy to the chemistry of genus *Allium* plants. Indeed, *Allium* chemistry attracted such luminaries as August von Hofmann, first Director of the Royal College of Chemistry in London, now Imperial College, and founder of the German Chemical Society; Arthur Stoll, President of the Swiss pharmaceutical company Sandoz Ltd., and a pioneering researcher on chlorophyll with Nobel Laureate Richard Willstätter; and Finnish chemist Artturi Virtanen, winner of the 1945 Nobel Prize in Chemistry, among a host of other international scientists. I found it humbling to follow in the footsteps of such scientific giants.

In my reading I learned that garlic, onion, leek, chives and other members of the genus *Allium* occupy a unique position as edible plants and herbal medicines, appreciated since the dawn of civilization. Alliums have been featured through the ages in literature, where they are both praised and reviled, as well as in architecture and the decorative arts. Planting ornamental alliums adds beauty to gardens while simultaneously protecting nearby plants against pests. Garlic pills are among the top-selling herbal supplements, while garlic-based products show considerable promise as environmentally friendly pesticides. Careful observations by early scholars, as well as recent research, has established that the remarkable and varied properties of the alliums can be well understood based on the occurrence of a number of relatively simple chemical compounds ingeniously packaged by Nature in these plants to protect them against attack by predators.

This book represents an effort to examine and document the fascinating past and present uses of these plants, sorting out fact from fiction based upon the detailed scrutiny of historic documents, as well as numerous studies conducted in laboratories around the world. Every effort has been made to explain in the clearest terms the sometimes complicated scientific concepts and results in a nonspecialist language. At the same time, I have sought to provide sufficient additional detail and full referencing to the older original sources, as well as the most recent literature, to satisfy readers who want to know more, as well as researchers in disciplines as diverse as archeology, culinary arts, medicine, ecology, pharmacology, food and plant sciences, agriculture, and organic and analytical chemistry. It is my hope that those just beginning their careers in chemistry, or related areas, will find this book useful in illustrating how scientific research is conducted, and how an understanding of organic chemistry proves very useful in many different areas of science. I am very grateful that my publisher encouraged me to include a variety of photographs and drawings illustrating the subject matter in ways not possible with words alone.

This book is also a story of a personal journey, both figurative and real. My training at Harvard University with Nobel Laureate E. J. Corey prepared me for a career in the areas of natural products and organic sulfur chemistry, the latter being the subject of my 1967 doctoral dissertation and of a 1978 monograph. My interest in *Allium* chemistry was purely serendipitous. In the late 1960s, as a young faculty member at the University of Missouri–St. Louis, looking for a subject for laboratory research, I recalled from my graduate studies the notable chemistry of dimethyl sulfoxide (DMSO; $CH_3S(O)CH_3$) as a powerful solvent, a reagent for chemical processes, and a controversial treatment for arthritis. I thought it would be of interest to explore the chemistry of a relatively little studied analog of DMSO, namely dimethyl thiosulfinate ($CH_3S(O)SCH_3$), differing from DMSO by one additional sulfur atom. The properties of this easily prepared compound proved to be quite fascinating chemically. What soon caught my attention was the fact that dimethyl thiosulfinate was formed naturally from a variety of common vegetables, including members of the *Allium*, as well as *Cruciferea* genera. Furthermore, I discovered that the knowledge I gained exploring the chemistry of dimethyl thiosulfinate was directly relevant to the chemistry of the more complicated compound allicin. Allicin ($CH_2{=}CHCH_2S(O)SCH_2CH{=}CH_2$), the active compound formed when garlic is cut or crushed, has six carbon atoms instead of the two found in dimethyl thiosulfinate.

As I continued this research at the University at Albany, State University of New York, in Albany, New York, I was amazed at how, from a stable, odorless white solid, an onion enzyme could instantly release a highly reactive three-carbon compound, which, through action of another enzyme, could rapidly reorganize its atoms forming a second, profoundly lachrymatory (tear-inducing) compound of unique molecular structure. In the absence of this second enzyme, a cascade of chemical events occur, resulting in a doubling in size of the first-formed three-carbon compound to one with six carbon atoms. This chemical event is immediately followed by a remarkable sequence of molecular reorganizations, ultimately giving stable substances with aesthetically pleasing structures – "zwiebelanes" – possessing a delicious, sweet onion flavor. Other parallel processes result in a tripling in size through combination of these same three-carbon molecules yielding new molecules with nine carbon atoms, "cepaenes," if from onion, and "ajoene" if from garlic. Both nine-carbon compounds show significant biological ability. Ajoene is currently under clinical study for the treatment of leukemia and fungal infections, while formulations containing cepaenes are being used for the healing of scars.

The novelty of the chemical transformations that occur when garlic or onion are crushed seemed to me to be a story worth communicating to a broader audience than just my fellow chemists. In 1985, while on a fellowship from the John Simon Guggenheim Foundation, I published "The Chemistry of Garlic and Onions" in *Scientific American*. In the process of preparing that article, as well as more recent *Encyclopaedia Britannica* articles, I learned how best to explain complicated chemical phenomena to a general audience. I even had the unusual opportunity to participate in three short documentary films dealing with garlic and onion. With the above experiences, I feel comfortable telling the full story of genus *Allium* plants here.

The journey associated with my *Allium* research has quite literally taken me around the world, allowing me to pursue my passions for travel, botany and photography. The first two chapters document with photographs, mostly mine, the many exquisite decorative alliums found in botanical gardens, the cultivation and marketing of edible alliums, and the incorporation of the shapes of onion, and even garlic, in architecture. The central two chapters chronicle the chemical detective work involved in understanding, at a molecular level, what happens when these plants are cut or crushed. These two chapters also take the form of a journey both through time and space, because chemists and biochemists around the world have been studying *Allium* chemistry since the earliest days of the science of chemistry. The final two chapters complete this journey, focusing on alliums in folk and complementary medicine through the ages, and on alliums in the environment, including promising recent work on the use of *Allium* extracts as environmentally benign pesticides.

I am extremely grateful to the National Science Foundation for more than 30 years of continuous, generous support for my own research program in the area of *Allium* chemistry, as well as for support from the American Chemical Society's Petroleum Research Fund and Herman Frasch Foundation, the North Atlantic Treaty Organization, United States Department of Agriculture, the National Institutes of Health, the Jack H. Berryman Institute and the American Heart Association. I am also very much indebted to the John Simon Guggenheim Foundation, to corporate sponsors of my research, particularly McCormick & Company, Inc. (United States), Societé Nationale Elf Aquitaine (France), and ECOspray Ltd. (United Kingdom), and to my hosts at Harvard University, the University of Bologna, the Weizmann Institute and Cambridge University where I conducted *Allium*-related research during sabbatical leaves. Current University at Albany funding in conjunction with my Carla Rizzo Delray Professorship is also deeply appreciated. I also thank: Wolfson College of Cambridge University for hosting me as a Visiting Fellow in 2006 and 2007, where I formally began work on the book; Dr. Tim Upson, Superintendent of the Cambridge University Botanic Gardens, for providing full access to their botanical library and encouraging me to include copies of historical botanical prints in this book; Dr. Dmitry Geltman of the St. Petersburg (Russia) Botanical Gardens for his hospitality and assistance researching the early botanical studies in St. Petersburg by Eduard Regel; and Dr. Dorothy Crawford Thompson and Dr. John Emsley, both of Cambridge University, for helpful discussions.

It was been particularly inspiring for me to get to know in person Dr. Chester J. Cavallito, whose 1945 discovery of allicin from garlic set the stage for the chemistry described in this book. It was a happy coincidence that he made his seminal discoveries in a laboratory in Rensselaer, New York, located a short distance from my own university. I warmly thank Chester Cavallito for sharing with me details of his discovery of allicin and giving me his collection of early references pertaining to garlic. Because research on genus *Allium* plants is of necessity highly collaborative, I am deeply appreciative to my many colleagues and former and current students who shared their wisdom and specialized techniques in the course of the research described in this book and in our joint publications. I also sincerely thank Professors Christopher Gardner, Rabi Musah and Frank Hauser, Dr. Murree Groom, Dr. Larry Lawson, and David Stern, all of whom carefully reviewed one or more chapters of the book, and Janet Freshwater, Katrina Harding, Rebecca Jeeves and their colleagues at the Royal Society of Chemistry for their expert editorial assistance. I particularly thank my wife Judy, my constant companion, without whose encouragement, help and love, this book, and the years of laboratory and fieldwork that preceded it, would not have been possible.

Eric Block

Contents

Garlic and Other Alliums: The Lore and the Science
By Eric Block

Published by the Royal Society of Chemistry, www.rsc.org

CHAPTER 1

Allium Botany and Cultivation, Ancient and Modern

We remember the fish which we did eat in Egypt,
the cucumbers, melons, leeks, onions and garlic!

Numbers 11:4-6

1.1 INTRODUCTION

Cultivation of leek, onion and garlic is as old as the history of the human race, and as extensive as civilization itself. References to these plants in the Bible and the Koran reflect their importance to ancient civilizations both as flavorful foods and as healing herbs. We individually have vivid personal memories of the smell of wild onions in the meadows each spring, of a steaming pot of leek and potato soup, of the aroma of sautéing onions, of a garlic-scented roast, or other olfactory recollections of *Allium* plants from our own life experiences.

Species sharing similarities are grouped together as a *genus*. So it is that leek, onion and garlic are all members of the genus *Allium*, a word said to come from the Greek αλεω, to avoid, because of its offensive smell (Boswell, 1883). One of the largest plant genera, the genus *Allium* includes 600 to 750 species. Whilst a few alliums are

Garlic and Other Alliums: The Lore and the Science
By Eric Block

Published by the Royal Society of Chemistry, www.rsc.org

of concern as invasive weeds, most are edible and have been consumed by indigenous populations for thousands of years, and some *Allium* species are cultivated as important food crops (Kiple, 2000; Peffley, 2006). Many members of this genus are popular with gardeners as hardy perennial ornamentals, though the smell of some can be off-putting. The smell is due to the presence of sulfur-containing compounds, which is a characteristic of this genus, and a focal point of this book. By analyzing the *Allium* sulfur compounds, chemotaxonomists can establish relationships between members of the genus based on similarities and differences in the types and amounts of these compounds.

It is commonly stated that as plants, alliums are both reviled and revered. In the past, they have even been forbidden. Onions were taboo for priests in ancient Egypt, orthodox Brahmins, Buddhists and Jains; aristocratic Romans shunned garlic (Peterson, 2000). Typical of those reviling garlic were late nineteenth and early twentieth century English writers, who reflected prevailing Victorian standards. Mrs. Henry Perrin in *British Flowering Plants* (Perrin, 1914) writes: "It has been truly said of this species that, if we could only divest it of its evil smell, it would rank among the most attractive of our British plants . . . The rank pungent taste and smell that pervade the stems and leaves of all the species of the large genus *Allium* results from a volatile essential oil, which is rich in sulphur."

Those favoring garlic include growers who earn their living from their garlic crop; both professional and amateur chefs who skillfully use garlic to enhance their dishes; foodies who seek out restaurants where virtually all dishes on the menu feature garlic; and the tens of thousands of attendees at garlic festivals all over the world. A relatively small but dedicated group of "alliophiles" seek out exotic alliums for their gardens, while the professionals among their lot attend conferences dedicated to the study of alliums, such as the World Onion Congress and the International Symposium on Edible Alliaceae (ISEA). Chemists can also be counted among the fans of the alliums because of the remarkable compounds and noteworthy molecular transformations that characterize all members of this genus. In fact, it was the story of the chemistry of alliums, in the context of their cultivation and their use in cooking, traditional and contemporary medicine and agriculture, that was the inspiration for writing this book.

1.2 *ALLIUM* BOTANY AND PHYTOCHEMISTRY

1.2.1 The Naming of Alliums

The genus *Allium* includes some of the most ancient cultivated crops, including *A. sativum* (garlic), *A. cepa* (onion), *A. schoenoprasum* (chives), *A. ampeloprasum* (great-headed or elephant garlic), *A. tuberosum* (Chinese or garlic chives), *A. fistulosum* (Japanese bunching onion) and *A. chinense* (rakkyo; when listing members of the genus, *Allium* is abbreviated as "*A.*"). These seven food crops can be easily distinguished by appearance. Two other edible, strong-smelling *Allium* species, *A. tricoccum* (ramp) and *A. ursinum* (bear's garlic or ramson; the latter name is derived from "rank," referring to its smell and taste) grow wild in North America and Europe, respectively, while a third, *A. victorialis* (caucas or "gyoja ninniku"), is widely consumed in Northern Japan. The shallot, *A. ascalonicum*, is a variety of the common onion and not a separate species; the separate Latin name is unjustified. The leek, *A. porrum*, and kurrat, *A. kurrat*, are cultigens (cultivated plants lacking wild counterparts) of *A. ampeloprasum* and are therefore members of the same species. *A. porrum* is the European cultivated leek. *A. kurrat*, the Middle-Eastern cultivated leek (*kurrat*, Arabic for leek), is grown for its leaves as a minor crop in the Delta region of Egypt and may be the leek mentioned in the Bible (Musselman, 2002). The scallion, green onion, spring onion and salad onion are simply different forms of *A. cepa*. Within each of the above species there may be many cultivars (human-selected horticultural varieties of domesticated crops) and subgroups with different characteristics. For example, garlic includes five distinct subgroups: *Sativum*, *Ophioscorodon*, *Longicuspis*, Subtropical and Pekinense. The *Sativum* group, originally from the Mediterranean region, was adapted by growers worldwide and constitutes the most common form of garlic.

"Sativum" in the botanical name for garlic means cultivated, consistent with the fact that wild "*A. sativum*" is unknown. The ancient name of garlic from Egyptian papyruses is *khidjana*. Garlic is called *skórodon* in ancient Greek, *shûm* in Hebrew, *ajo* in Spanish, *ail* in French, *aglio* in Italian, *Knoblauch* in German and *suan* in Chinese. The fact that the Chinese word for garlic, *suan*, is written as a single character is offered as evidence of the antiquity of the word, and hence of the ancient use of garlic in China (Kiple, 2000). Garlic was popular enough with the ancient Greeks that a section of the market in Athens was known simply as *ta skoroda*, "the garlic" (Davidson, 1999). "Garlic" itself is derived from "gar-leac" or "gar leek," "gar" meaning "spear" because of the spear-headed cloves (McLean, 1980) and "leac" an Anglo-Saxon root meaning plant or herb (Davies, 1992).

The English "leek" and German *Lauch* are derived from "leac." Leeks were so popular with the Anglo-Saxons that their word for a kitchen garden was "leek-garth" ("leac-tun"). The Greek word for leek, *prason*, is the basis for ampeloprasum ("ampelo" = vine), the *Allium* that grows in vineyards. The Latin word for leek, *porrum*, is the origin of its French name, *poireau*. "Onion" comes from the Middle English "unyun," which in turn comes from the French *oignon*. The French name was ultimately derived from the Latin *unio*, meaning one or unity, because an onion grows as a single bulb, in contrast to the multi-cloved garlic. In Latin, onion was called *cepa* or *caepa*, hence the botanical name, and the names *cebolla*, *cebola* and *cipollo*, in Spanish, Portuguese and Italian, respectively.

1.2.2 *Allium* Botany

The botany and horticulture of alliums have been extensively studied and are the subject of many excellent books (see Bibliography). *Allium* species growing in particular countries or geographic areas are to be found in *flora* of those areas, *e.g.*, the *Flora of North America* (2002), and *Wild Flowers of the United States* (Rickett, 1973), which respectively identify 96 and 79 different native species. Notable among those who specifically researched the botany of the genus *Allium* was the eminent nineteenth century botanist Eduard Regel (1815-1892; Figure 1.1; Anonymous, 1892). Scientific Director and then Director General of the Imperial Botanical Garden in St. Petersburg, Russia, Regel was instrumental in getting botanists attached to the Russian expeditions in Central and Eastern Asia. Regel introduced scores of plants to the Garden from Central Asia, described them and distributed them liberally to botanic gardens and nurseries outside

Figure 1.1 Eduard Regel (1815–1892). (Photo provided courtesy of Botanical Garden of the Komarov Institute of the Russian Academy of Sciences, St. Petersburg, Russia).

Russia. A prolific author, he was a founder of both the Russian and Swiss Horticultural Societies and the journal *Gartenflora*. He had a particular fascination with the genus *Allium* and wrote about them in two monographs (Regel, 1875, 1887), which feature more than 250 species, including a large number not previously described, the fruits of the explorations in Asia. More than 60 of the *Allium* species he helped to identify bear his name in the full name of the plant, *e.g.*, *A. giganteum* Regel and *A. rosenbachianum* Regel.

While the origin of *Allium* species remains speculative, evidence suggests that garlic and onion were first domesticated in the central Asian mountainous regions of Tajikstan, Turkmenia, Uzbekistan, and northern Iran, Afghanistan and Pakistan (Brewster, 2008), and most likely brought to the Middle East by Marco Polo and other Silk Road/spice route travelers. Recent research pinpoints the northwestern side of the Tien Shan Mountains (*e.g.*, Kyrgyzstan, Kazakhstan) as the most likely center of origin of garlic (Etoh, 2002). Extensive efforts have been made in conjunction with the European Union's 2000–2004 Garlic and Health Project to collect specimens from the region of central Asia shown in Figure 1.2 (Kik, 2004).

A. ampeloprasum (known to the Romans as *Ulpicum:* Sturtevant, 1888; Mezzabotta, 2000) is thought to be the ancestor species of leek and kurrat. There is considerable interest in the wild relatives of garlic for plant breeding and future genetic manipulation. It has been suggested that *A. longicuspis*, genetically identical to *A. sativum*, might have been cultivated by semi-nomadic hunter-gatherers more than 10 000 years ago and transported along trading routes between China and the Mediterranean. From the Mediterranean region garlic was brought to sub-Saharan Africa and the Americas by explorers and colonists. It is also thought that garlic was introduced to China by traders from Central Asia, and into Japan

Figure 1.2 Collection of garlic accessions in Central Asia in 2000 and 2001 under the E.U. Garlic and Health project. Red and blue dots indicate garlic accessions collected from market places and natural vegetations, respectively. (Reproduced with permission from: http://www.plant.wageningen-ur.nl/projects/garlicandhealth/Results.htm).

from Korea, where it was very popular (Etoh, 2002). DNA analysis indicates that neither *A. longicuspis* nor *A. tuncelianum* are ancestor species of garlic, as suggested earlier. Therefore the true ancestor species of garlic still remains unknown (Ipek, 2008).

The current taxonomic classification of alliums is:

Class	Monocotyledons
Superorder	Liliiflorae
Order	Asparagales
Family	Alliaceae
Tribe	Alliae
Genus	*Allium*

In the older literature Alliaceae were included in both the Lilliaceae and the Amaryllidaceae, but they are now regarded as a separate family. Within the genus *Allium* there are as many as 750 species, making it the largest genus of petaloid monocotyledons, excluding orchids. The taxonomy of alliums is complicated, with a proliferation of synonyms at least equal in number to the number of known species. Since important characters are generally lost in herbarium specimens, study of the living material is essential (Gregory, 1998). Extensive collections of living *Allium* species may be found at Kew Gardens (UK; 250 species; http://www.kew.org/; Mathews, 1996) and at the New York Botanical Gardens (65 species; http://www.nybg.org/). Cryopreservation appears to be a viable method for archiving small samples of living tissue from *Allium* species (Volk, 2004, 2009).

Botanists describe alliums as "low growing perennials in which the rhizomes, roots, . . . and bulbs can be important storage organs. The leaves [tubular in onions, flat in garlic] arise from the underground stem and often have long sheathing bases, which can give the appearance of a stem, as typified by leeks . . . No leaves occur on the flower stalk (the 'scape') except the single spathe that encloses the young inflorescence" (umbel; Brewster, 2008). The bulbs, which often clump, consist of the swollen base of the stem and several fleshy leaves or scales held together by a disk or hardened stem tissue called a basal plate.

There are two main sub-species of garlic, namely hardneck (stiffneck) and softneck (Volk, 2004). Hardneck garlic (*Allium sativum* ssp. *ophioscorodon*), also called ophio or top-set garlic, produces scapes or flower stalks and prefers northerly climates with cold winters. When sliced through the bulb midsection, the hardneck type typically reveals a single circle of 6 to 11 cloves around a central woody stalk. Before flowering, the scape ("top set") curls upward as it grows, looping the loop with 1 to 3 coils, like pig's tails (Figure 1.3), before straightening and then grows little seed-like bulbils. Garlic scapes are generally pinched off while still looped, three weeks before the bulbs are ready to be harvested, enhancing bulb growth (the scapes are "delicious raw or cooked," according to one cookbook). Failure to remove the scapes can reduce the yield of the bulbs by as much as 33% (Stern, 2009). Softneck garlic (*Allium sativum* ssp. *sativum*), when sliced through the bulb midsection, reveals up to 24 cloves per bulb (but sometimes much less) in several layers around a soft central stem, with large cloves around the outside and smaller cloves in the middle. Since it only develops a short scape lacking a flower top, it is sometimes called short-necked garlic. Softneck garlic generally has better storage ability than hardneck garlic and is preferred for braiding since its neck remains soft at harvest time. Hardneck garlic adapts better to colder climates than softneck garlic.

U.S. Department of Agriculture geneticist Gayle Volk used DNA analysis to classify the many different cultivars of garlic into ten major types: Rocambole (limited storage, 6 to 11 large cloves with loose, easily peeled skin), Purple Stripe (8 to 12 cloves; bright

Figure 1.3 Typical pig's tail shape of garlic scape. (Courtesy of Ted J. Meredith).

purple streaks on bulb wrappers and clove skins), Marble Purple Stripe, Glazed Purple Stripe, Porcelain (satiny white bulb wrappers with 4 to 6 large cloves), Artichoke, Silverskin, Asiatic, Turban and Creole (Figure 1.4; Volk, 2004, 2009). Two major varieties of softneck garlic are Artichoke and Silverskin. Silverskin, the type of garlic most commonly found in grocery stores due to long storage life, has three to six clove layers and fine, smooth and shiny bulb wrappers (outer skin). Artichoke strains are large bulbed with short, wide plants. It is relatively mild in flavor, having 3 to 5 clove layers with 12 to 20 cloves per bulb; a medium clove weighs about 1.8 grams.

Garlic is completely sterile and is therefore propagated asexually only from cloves (Shemesh, 2008; Kamenetsky, 2005, 2007a,b).

Figure 1.4 Examples of the diversity of garlic cultivars. (Courtesy of Gayle Volk).

In contrast to seed-propagated crops, in garlic, cloves are separated from the underground bulbs and planted each fall. By the following summer, each planted clove has grown into a garlic plant, which produces a new bulb underground. Plants such as garlic, that have been domesticated, are genetically distinct from their wild progenitors. The latter develop *via* natural selection to ensure their survival in the wild. Domesticated plants are artificially selected primarily to suit human needs and not necessarily for survival value. Domestication and cultivation accelerated selection for larger bulbs and deselection of flowering plants, thereby promoting sterility. Other factors contributing to garlic's sexual sterility are harvesting garlic bulbs in humid areas well before flowering to prevent rotting that can occur in moist soil, and removing the scapes before flowering to promote bulb growth. The sexual sterility of garlic limits development of genetic varieties, *e.g.*, those showing superior pest resistance, size, yield, quality and tolerance of temperature extremes during growth.

Fertile wild garlic has been discovered and collected at its *center of origin* in Kazakhstan and Kyrgyzstan as described above. Collection of these species is complicated by geopolitical considerations in this region of Eastern Asia due to political instability and armed conflict. The wild ancestors of garlic are the only source of germplasm for the development of new cultivars. The phytochemistry of rare alliums should be surveyed soon because some of these wild plants are endangered and on the verge of extinction. Many of the older varieties of garlic demonstrate superior resistance/tolerance to pests and disease (Kamenetsky, 2005, 2007a,b). Local genotypes from this region of Central Asia possess wide genetic diversity, including economically useful traits which were likely lost during the approximately 10 000 year selection of garlic by man. Garlic seeds obtained from these strains are smaller (about 3 mg each) and less viable than those of onions, and germination can take several months. However, large numbers of true garlic seeds from these strains can now be generated finally making garlic breeding possible. The essential role of insects in pollination of these garlic strains was established through use of fertile plants grown in insect-proof cages. While the major portion of garlic seeds resulted from cross-pollination, a certain amount of self-pollination occurs. The seed populations employed produced normal flowers and seeds as well as a large variation in all morphological and physiological qualities similar to those seen with vegetatively propagated garlic clones.

The variability seen with sexual reproducing garlic could serve as a rich source for genetic studies and breeding work, eliminating the need to preserve clones in field gene-banks throughout the world. Use of garlic seeds for breeding purposes would eliminate the main ailments associated with the current methods involving clonal propagation, such as carryover of pests from one generation to another, low propagation rates, voluminous storage of bulbs, rotting and sprouting. The use of seeds would save the costs of vegetative propagation and spare the need for virus elimination (Shemesh, 2008). While the garlic genome has been mapped and garlic seed production is now routine (Ipek, 2005), wide scale commercial production of garlic from seeds currently seems unlikely given that seed-grown garlic is neither competitive in appearance nor cost with garlic imported from China.

Alliums vary in height from 5 to 150 cm, while the bulbs can range from very small (2 to 3 mm in diameter) to rather big (8 to 10 cm). The largest onion, according to the *Guinness Book of World Records*, weighed 10 pounds 14 ounces, while the largest garlic weighed 1.19 kg (2 pounds 10 ounces; McCann, 2009). Some *Allium* species, such as *A. fistulosum*, develop thickened leaf-bases rather than forming bulbs. Most bulbous alliums increase by forming little bulbs, "offsets," or corms around the old one, as well as by seed (except for garlic, which as noted above is sterile and can only be propagated vegetatively from bulbs). Several species can form many bulbils (tiny bulbs) in the flowerhead. Various *Allium* species are used as food plants by the larvae of some *Lepidoptera* including the cabbage moth, common swift moth (on garlic), garden dart moth, large yellow underwing moth, nutmeg moth, setaceous Hebrew character moth, turnip moth and *Schinia rosea*, a moth which feeds exclusively on alliums.

Alliums are distributed widely through the temperate, warm temperate and boreal zones of the northern hemisphere at altitudes up to 3050 m from Mexico and North Africa and southern Asia northwards, but are limited to mountainous regions in tropical areas. Garlic is more tolerant of cold than the onion. *A. junceum*, found in Tibet at altitudes of 3660 to 5000 m (12 000 to 16 400 feet), is used locally as a condiment. "It grows largely on the high hills of Ladak and used by the natives. It is sold in the bazaars in the shape of balls, the whole plant being mashed up into a semi-pulp and then made into a ball as big as the fist, to be used as a condiment. The balls are strung through the middle and carried on a string" (Baker, 1874). *Allium* species also occur in Chile (*A. juncifolium*), Brazil (*A. sellovianum*) and tropical Africa (*A. spathaceum*). Only chives are found in the Arctic and in both America and Eurasia. Alliums prefer open, sunny, dry spots in relatively arid climates and are not normally found in areas of dense vegetation, although wild varieties such as *A. ursinum* thrive in woodlands and *A. triquetrum*, *A. carinatum* and *A. vineale* (crow garlic), are found as weeds in pastures (see Sections 6.2.2 and 1.5 on *A. ursinum* and *A. vineale,* respectively). Alliums are poor sunlight receptors and are not competitive with weeds or shade in general. *A. ursinum* can grow in woodlands since most of the leaf growth is completed in early spring before leaves appear on trees.

Given the difficulty of preserving dried specimens of genus *Allium* species in herbaria, it is fortunate that over the years, flora (books describing the plant species occurring in an area or time period, with the aim of allowing identification) have been published containing meticulous drawings of alliums. Among the earliest of botanical illustrations of alliums is a parchment from Northern England dating to the late twelfth century and published in 1462 (Pseudo-Apuleius, Herbarium; labeled as Ashmole 1462, fol. 31r) of "Scordeon," presumably *A. Scorodoprasum,* the giant garlic or sand leek (Figure 1.5a). A second fifteenth century example depicts the harvesting of garlic (Figure 1.5b). An exquisite series of 27 colored plates of 77 members of the genus *Allium* appear in *Flora Germanica* by Ludwig Reichenbach, Director of the Dresden Botanical Garden, published in 1848. All of these are reproduced in Appendix 2.

1.2.3 *Allium* Phytochemistry

While unique features of genus *Allium* chemistry are discussed at length in Chapters 3 and 4, connections between this chemistry and *Allium* botany are noted here. Phytochemistry – the chemistry of plants – seeks to isolate, identify and determine the properties of chemicals derived from plants, including medicinal and insecticidal

(a) (b)

Figure 1.5 (a) Twelfth century representation of giant garlic. (Courtesy of Oxford University Library). (b) Harvesting garlic, *Tacuinum sanitatis*, 15th century. (Courtesy of Bibliothèque Nationale, Paris).

activities. Alliums are all odorless until the plant cells are crushed or otherwise damaged. At this point they generate characteristic volatile, reactive, sulfur-containing chemicals. These strong-smelling and -tasting compounds are classified as *secondary metabolites,* since they are not directly involved in the normal growth, development or reproduction of the plant as is the case with *primary metabolites*. The smelly *Allium* secondary metabolites serve two important functions: they are used as defenses against predators, parasites and diseases, and, by their smell, they attract pollinators (the faintly sweet aroma of flowers of chives and other alliums are highly attractive to bees, but may be sulfur-free since *Allium* sulfur compounds seem to repel bees). This will be discussed further in Chapter 6. *Allium* secondary metabolites may also exhibit a number of protective functions for human consumers or serve as candidates for drug development, as will be discussed in Chapter 5.

Since each *Allium* species differs in the types and amounts of its sulfur compounds, the study of *Allium* sulfur compounds has been useful in taxonomic research. Only a small fraction of the known *Allium* species has been subjected to thorough phytochemical analysis. Phytochemical analysis has revealed that the ratios and amounts of the characteristic sulfur compounds vary with the part of the plant sampled (bulb, stem, leaves, flowers), with the growth stage, if the plant has been harvested, and with the storage conditions. As less common alliums are studied, it is probable that unusual, biologically active new compounds will be discovered, as recently occurred in the case of "drumstick alliums" (*e.g.*, Figure 1.7c and Chapter 4, Section 4.13).

1.2.4 Alliums as Ornamentals

Only a few *Allium* species were cultivated as ornamental plants through the early 1800s. In 1841, Mrs. Jane Loudon in *Ladies' Flower-Garden* wrote: "The name of garlic is so associated with ideas of the rank smell and taste of some of the worst kinds of Continental cookery, and the plant itself is so repugnant to most persons of refined taste, that it seems difficult to imagine the flowers of any species of the genus to be sufficiently ornamental to deserve a place in this work." But she then goes on to say: "There is perhaps no genus of bulbous plants which contains more pretty flowers than the genus *Allium*; or flowers of one genus which possess more interest from their great variety, as they are quite distinct from each other, varying widely in colour and size, though still preserving so strong a family likeness as to render it impossible to mistake them...they are well worthy of cultivation as border flowers, though too unpleasant in their smell [when the stem is "bruised or cut"] to admit of their being gathered for nosegays. They are nearly all hardy perennials, which will grow freely in any common soil; and they generally produce their flowers in great abundance. Most of the species when once planted require no further attention, but may be left in the soil for a great many years, without any bad effect being perceptible" (Loudon, 1841). When botanists under the supervision of Eduard Regel returned from their exploration of Central Asia they brought back magnificent specimens of exotic alliums which were then introduced in the Imperial Botanical Garden in St. Petersburg, Kew Gardens in

London, and elsewhere in Europe (Dadd, 1987). Recently, ornamental *Allium* species have become popular in rock gardens and perennial borders, particularly for spring and summer blooming, and as commercial cut flowers.

Allium flowers, six segments arranged in two whorls of three, can be blue, rose, violet, white or yellow. *Allium* flowers grow in umbrella-like structures, called umbels, supporting numerous tiny flowers at the top of strong stems, typically in ball or oval shapes, giving the overall appearance of a strong, leafless stem with a sphere of flowers at the top (Figures 1.6 and 1.7). The tiny flowers on the umbel, numbering 30 to 60 for ornamental alliums, are typically in the shape of a bell, cup or star. A characteristic of alliums is that each little flower has six stamens. In some cases the umbel droops, giving the flower a dramatic appearance (Platt, 2003). Most alliums produce an underground storage bulb at the end of one growing season, which flowers in the next. Decorative alliums recommended by Loudon (Figure 1.6) include *A. caeruleum* (bright blue flowers: "blue of the heavens"; also Figure 1.7a), *A. longifolium* (dark, maroon-colored flowers), *A. moly* (golden yellow flowers; also Figure 1.7b), *A. bisculum, A. neapolitanum* (white, star-shaped flowers produced in large, loosely spreading umbels), *A. triquetrum* (flat, three-sided stem, hence the name) and *A. gramineum.*

Other popular decorative alliums, notable for their resistance to deer (Platt, 2003), include the robust *A. giganteum* (Figure 1.7c), the delicate *A. cyaneum*, the deep-red *A. rosenbachianum* (Figure 1.7d), the sweet-smelling *A. ramosum* with its red-striped bell-shaped white flowers (Figure 1.7e), the hardy and edible *A. nutans* and the spectacular *A. cristophii* (Star of Persia; formerly *A. albopilosum*). With a globular flowering head approaching soccer ball size, carrying up to 80 star-shaped, pale bluish-violet florets with a metallic sheen and maintaining their shape even on dried heads, the mature head has been likened to a huge Christmas tree ornament (Figure 1.7f; Davies, 1992). *A. schubertii*, resembling an exploding firework, with up to 200 florets can be dried for indoor display or left outside to be admired when frosted. *A. canadense* (meadow leek, rose leek or Canada garlic) is a traditional American Indian food source. It was the main food supply of the American explorer Marquette in 1674 during his journey from Green Bay to what is now Chicago, a city whose name is derived from the American Indian word "cigaga-wunj," meaning "place of the wild garlic."

Figure 1.6 Illustrations from Mrs. Jane Loudon's 1841 book, *Ladies' Flower-Garden of Ornamental Bulbous Plants*, showing *A. caeruleum*, *A. longifolium*, *A. moly*, *A. bisculum*, *A. neapolitanum*, *A. triquetrum* and *A. gramineum*.

Figure 1.7 (a) *A. caeruleum*, St. Petersburg (Russia) Botanic Gardens. (b) *A. moly*, St. Petersburg (Russia) Botanic Gardens. (c) "Drumstick" alliums in a Connecticut garden. (d) *A. rosenbachianum*, Cambridge University Botanic Gardens. (e) *A. ramosum*, St. Petersburg (Russia) Botanic Gardens. (f) *A. cristophii*, Cambridge University Botanic Gardens. (Photos by Eric Block).

The Netherlands is the leading commercial producer of ornamental *Allium* bulbs, currently providing 40 species and cultivars. More than 200 species of the large subgenus *Melanocrommyum*, sometimes called "drumstick alliums," exist worldwide, growing mainly in Southwest and Central Asia. While most ornamental members of this subgenus are almost odorless or pleasantly scented, having low levels of cysteine sulfoxides (Kamenetsky, 2002), they are also characterized by a deep orange or red fluid discharge which forms when the bulbs are cut (Jedelska, 2008).

Despite their long history in traditional medicine in these locales, many of these "exotic" alliums have until recently been poorly characterized and studied, mostly due to geopolitical reasons. Novel biologically active, sulfur-containing pigments have been isolated from *Melanocrommyum* varieties; medicinal uses of these plants have been reported (see Chapters 4 and 5). Many attractive varieties of ornamental alliums are sold over the Internet.

1.2.5 Alliums as Invasive Weeds: Crow Garlic (*A. Vineale*)

Wild garlic (*A. vineale*), also known as crow garlic, is a perennial weed with a particularly noxious reputation. Folk wisdom has it that when *A. vineale* is eaten by birds it so stupefies them that they may be picked up, which is perhaps the origin of the nickname "crow garlic." *A. vineale* is sometimes confused with wild onion (*A. canadense*) in North America, with both often occurring in the same habitat. The former plant has leaves that are hollow and oval to crescent-shaped in cross section; the latter has leaves that are flat in cross section and not hollow. Wild garlic has little redeeming value despite its similarity to cultivated garlic. Wild garlic gives off a distinctive garlic odor when the leaves, bulbs or flowers are crushed, the odor and flavor being even stronger, with an unpleasant aftertaste, compared to that of cultivated alliums. Wild garlic's property of passing its excessively strong flavor to food is infamous among dairymen and wheat growers, for whom contamination by the weed causes a serious reduction in product quality. Cows eat the tender young leaves of wild garlic in spring, and the lipophilic, sulfur-containing oils bind to the milk fat, imparting an offensive garlic taste to the milk cream and butter, reducing the quality and grade of the dairy products and their market values. It is even sufficient for cows to just to inhale the odorous vapors of *A. vineale* for their milk to take on an unpleasant smell (MacDonald, 1928)!

Wild garlic is even more troublesome in small grain crops. Wild garlic produces up to 300 aerial bulblets per head. The bulblets, the most rapid and abundant form of reproduction and spread of the plant, are about the size of wheat grains and mature at the same time. These bulblets are almost impossible to separate in harvesting equipment from wheat, barley, oats or rye. The wild garlic bulblets cause the grain to have a garlic odor and flavor that is passed on to the flour and even to the bread (the origin of garlic bread?). The odor can remain in the grain even after cleaning out the bulblets. Grains tainted with wild garlic odor are subject to severe reductions in grade and price. Finally, wild garlic grows more rapidly than turf grasses, causing an unsightly spike if the weed appears too quickly above lawns. Severely infested lawns can also produce distinctive garlic odors, noticeable from quite a distance. Multiple means of reproduction make the weed particularly difficult to control once it is established (Defelice, 2003).

Wild garlic seems to evoke the ire of gardeners and farmers like almost no other weed. Consider the story of an Englishwoman determined to eradicate wild garlic from her garden. She spent years applying every new control method she could find, including dousing the area with sodium chlorate, paraquat and other herbicides every spring, without much effect. She finally defoliated the area and killed the roots of the trees, causing a mudslide that collapsed an entire woodland embankment and pathway. She eventually died, leaving behind an exposed slope covered even more heavily with wild garlic as her legacy (Banks, 1980).

In light of the above description of the adverse effects of *A. vineale* on dairy products, it is ironic that recent research suggests that extracts of garlic may prove useful in "animal agriculture" in reducing the production of methane, a major greenhouse gas affecting global warming. In particular it has been found that *Allium* extracts "have the potential to reduce rumen methanogenesis without affecting rumen fermentation adversely" (Karma, 2006; Patra, 2006).

1.3 *ALLIUM* CULTIVATION IN ANCIENT AND MODERN TIMES

1.3.1 Alliums in Ancient Egypt and the Mediterranean Basin

Garlic, onion and leek have always been popular foods in Egypt and elsewhere in the Mediterranean basin. Garlic, considered a strengthening food, ideal for workers, oarsmen and soldiers was such a popular plant with the Roman army that it was said that "one could follow the advance of the Roman legions and expansion of the empire by plotting range maps for garlic" (Parejko, 2003). The Roman legions introduced garlic to many of the peoples they conquered, especially those of Northern Europe (Hicks, 1986). Our knowledge of their cultivation, culinary usage and medicinal applications in ancient Middle Eastern civilizations, comes from the study of extensive archeological evidence. This includes pictograms and other ancient art appearing on papyrus scrolls, clay tablets and models, stone carvings, painted funerary objects such as coffin covers and mummy cases, and from burial relics including intact plants themselves. Sumerian cuneiform tablets from 2300 BCE describe the Sumerian diet as consisting of grains, legumes, onion, garlic, leek and other vegetables, as well as many varieties of fish (Moyers, 1996).

Particularly notable examples of this historical record are the Yale culinary tablets from Mesopotamia – ancient Iraq. Dating from 1600–1700 BCE, the Yale Babylonian Tablets, as they are called, represent the earliest known compilation of recipes (Bottéro, 1987, 2004). These hand-size, caramel-colored clay tablets contain in 350 lines about 40 surprisingly sophisticated recipes, such as ones for gazelle, pigeon, partridge and goat-kid stews, and a meat pie baked in an unleavened crust. The cornerstone of the Mesopotamian diet was the alliaceous plants, onion ("susikillu"), leek ("karsu") and garlic ("hazanu"). The meat pie recipe requires "pounding together leek and garlic . . . which you should squeeze" in a cloth, in order to add their essence to the dish. A typical recipe is given:

Braised turnips

Meat is not needed. Boil water. Throw fat in. [Add] onion, [an unknown spice plant], coriander, cumin, and [a legume]. Squeeze leek and garlic and spread [juice] on dish. Add onion and mint.

The insatiable passion of the ancient Mesopotamians for alliums is evident in the provisions taken by King Su-Su'en's daughter on her voyage to Ansan at the end of the third millennium BCE, items which included butter, cheese, oil, fruit and "seven talents" (35 kg!)

of garlic and a similar amount of onions. From the writings of Gimil-Marduk in *ca.* 1700 BCE we learn about dehydration of garlic: "The garlic must be dried outside. Afterwards you will send me a basket of it." Presentation of dishes was an important part of the meal, as described in the Yale tablets in the first person by the chef/writer (Bottéro, 2004):

> *When it is all cooked, I remove the pot from the fire, and before the broth cools, you rub the meat with garlic, add greens and vinegar. The broth may be eaten at a later time.*

The archeological evidence suggests that there were apparently five or six varieties of alliums, although only onion, leek and garlic can be identified, and that the plants were generally mashed or chopped together to draw more flavor from them. It is also noteworthy, that in these recipes garlic and leek are mentioned together, suggesting recognition by these ancient gourmets of complementary tastes. Furthermore, from the ancient records there seems to be no evidence for a "supernatural," "magical" or religious effect in the use of the alliums or other ingredients. They were used just for flavor (Bottéro, 2004). Additional information on historic cultivation and use of alliums comes from commentary by early Greek and Roman writers, and even from examination of contemporary Egyptian agricultural practices. The depiction of laborers eating onions appears on mastabas – rectangular, flat-roofed, mud-brick buildings with sloping sides, that were the predecessors of the Pyramids in Egypt (3000 BCE; Täckholm, 1954). Even older are clay models of garlic from the tomb of El Mahasna (3700 BCE; Moyers, 1996). These unbaked models had a globular core, on top of which nine long sausage-like rolls of clay were pressed, and the whole whitewashed, presenting a very natural appearance (Ayrton, 1911).

Of all the ancient civilizations, that of Egypt is particularly well-researched and documented and is therefore featured in this chapter, even though parallel use of alliums occurred in other ancient civilizations. The onion is mentioned as an Egyptian funeral offering and is depicted on the banquet tables of great feasts. Ancient Egyptian wall carvings and drawings, as well as several finds of dried specimens, show that both leek and onion were part of Egyptian food production as far back as the second millennium

BCE, if not earlier. For example the stela (funeral plaque) of Mentuwoser from Abydos, Egypt (Figure 1.8a; *ca.* 1955 BCE; Metropolitan Museum of Art, New York), honors an official named Mentuwoser, who sits at his funeral banquet, which includes onions in a basket and leeks, amongst other offerings. Illustrations showing how onions were planted and watered appear in numerous Egyptian tombs, from the Old Kingdom onwards, for example, the tombs of Unas (*ca.* 2420 BCE) and Pepi II (*ca.* 2200 BCE). Frequently, a priest is pictured holding onions in his hand or covering an altar with a bundle of their leaves or roots, which have been elaborately bundled and tied (Figure 1.8b; Wilkinson, 1878).

Onion bulbs were found placed in body cavities of mummies, perhaps to stimulate the dead to breathe again, while garlic was used in the embalming process (Täckholm, 1954). Excellently preserved garlic was found in Tutankhamun's tomb (1325 BCE). Dry remains of garlic were found in other eighteenth dynasty and later tombs (Figure 1.8c) and numerous carbonized garlic cloves were found in Tell ed-Der in Iraq, dating from the second millennium BCE (Zohary, 2000).

A fascinating glimpse of ancient Egyptian agricultural practices can be gleaned from archeological records of customs-house receipts and local taxes paid, as well as other written records from the time. Thus it is known that garlic growing on a sizeable scale in ancient Egypt was introduced in the third century BCE on Greek initiative and continued through the Greco-Roman period (332 BCE to 639 CE). This agricultural specialization represented a significant break with the tradition, begun in the first dynasty (*ca.* 3000 BCE), of subsistence farming.

An analysis of early writings indicates that in about 257 BCE, in a relatively large-scale operation in the rich Egyptian agricultural area of Al Fayyum, southwest of Cairo, a load of some 300 bushels of garlic bulbs, brought by boat from the port city of Alexandria and then further transported using donkeys, were split into cloves, dried and planted according to the agricultural practices of the time. These early records also hint that there may have been experimentation with new strains of garlic, possibly East African. The garlic was planted on a large scale in late November, following the recession of the Nile flood, and harvested 60 days later in January, suggesting use of a particularly rapidly growing garlic variety. There was a good market for this garlic in Alexandria,

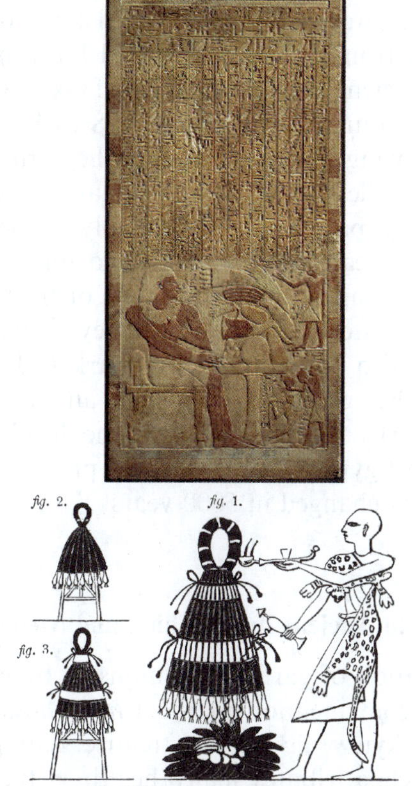

Figure 1.8 (a) Stela of Mentuwoser from Abydos, Egypt, *ca.* 1955 BCE. (Courtesy of Metropolitan Museum of Art, New York). (b) Tying onions in ancient Egypt (Wilkinson, 1878). (c) Egyptian burial relics including intact *Allium* plants. (Courtesy of Fitzwilliam Museum, Cambridge).

especially among the Greek population. From one set of tax records it is estimated that garlic formed 1.5% of the value of the food crossing a frontier in the North Al Fayyum area. Land surveys from the ancient Al Fayym village of Oxyrhyncha suggest that in the second century BCE as much as 8 to 18% of the land was devoted to growing garlic by four garlic farmers and that two specialist garlic-sellers worked in the village (Crawford, 1973).

In Egypt today, passengers on boats plying the Nile can observe endless miles of neat rows of garlic and onion plants lining the fertile river banks and can spot wagons of freshly harvested produce drawn by water buffalo and donkeys traveling the riverside paths. In the Cairo onion and garlic market, donkey-drawn carts laden with garlic, garlic-braiding men and enormous stacks of the pungent herbs can be seen and the heady aroma sampled (Figure 1.9). In Egypt some agricultural practices have apparently not substantially changed in 5000 years!

1.3.2 Alliums in Ancient India, China and Medieval Europe

Ancient cultivation of garlic and onions in India can be inferred, for example, from mention in the *Charaka-Samhita*, the oldest known Indian Ayurvedic medical treatise, dating from *ca.* 400 to 200 BCE, which attributes many health virtues to these plants (Jones and Mann, 1963) as well as from similar mention in the sixth century CE Bowers Manuscript (see Chapter 5). On the other hand, quoting a sixth century source, "onions and garlic are little used [in India] and people who eat them are ostracized" (Watters, 1904).

The *Shi-Ching* (*Book of Odes*), a Chinese classic compiled by Confucius (551–479 BCE), features garlic (*suan*) eaten by animals and people and offered to the gods for good luck, while the even older *Calendar of the Hsia* (2000 BCE) also mentions garlic (Davies, 1992), suggesting that cultivation of garlic in China occurred simultaneously to that in ancient Mesopotamia. It has been suggested that an indigenous wild garlic, *var. pekinense*, was domesticated in North China, and the cultivated garlic of China and Japan derives from it (Simoons, 1991). Other alliums of Chinese origin are Chinese chives (*A. tuberosum; chiu-ts'ai*), cultivated in China for more than 3000 years (Debin, 2005), *A. chinense*

Figure 1.9 Cairo garlic and onion market. (Photos by Eric Block).

(*A. bakeri* or rakkyo; *ch'iao*), and *A. fistulosum* ("Welsh" onion; *ts'ung*). There is some anti-*Allium* sentiment in China, believed to have been introduced by Buddhists in scholarly texts of the fourth century CE, which forbade use of "five vegetables of strong odor," including some alliums, by Buddhist clergy. This practice has continued until the present. However, there is also evidence from the Han Dynasty (206 BCE–220 CE) of putting onions and garlic on red cords hung on house doors to repel harmful insects (Simoons, 1991). In the Han Dynasty, garlic, scallions and leek were common foods.

The first Chinese agricultural book to describe allelopathy (the ecological phenomenon of plant–plant interference through release of organic compounds known as allelochemicals), *Fan Sheng Zhi Shu* (also called *Fan Sheng Chih Shu*), appeared in the first century BCE (Zeng, 2008). Allelopathy exists between crops and weeds, previous and consecutive crops, or interplanted crops. The author Fan Sheng Zhi claimed that "cucurbit [cucumber, squash, *etc.*] and leek could be interplanted to reduce the disease of cucurbit, because leek could produce special substances to inhibit pathogens of cucurbit." Leek was said to be historically recognized as a plant "that exerts a control on microbial pathogens" (Zeng, 2008). *Fen Men Suo Sui Lu* (twelfth century CE) states that planting chives, garlic or leek around flowering plants could effectively protect the latter against attack by the musk deer (*Moschus moschiferus*; Zeng, 2008).

Onions, garlic and leeks were favorites in medieval monastary gardens as they were easy to grow, hardy and, above all, strong tasting. It is said that the medieval palate couldn't get enough of these alliums, whose popularity continued throughout the Middle Ages, and that there was always a smell of garlic in English medieval houses and gardens, where it served to disguise less welcome odors. In eighteenth century France, Vinegar of Four Thieves (garlic macerated in wine), named after graverobbers using it, gained fame as an antidote to the plague (Block, 1985).

1.3.3 *Alllium* Cultivation Today

Onions, consumed either as a substantial part of the diet or as a flavoring on every continent (except Antarctica) at an annual rate of 58 million metric tons (2005), are the most widely used *Allium*, and rank third among produce consumed, after tomatoes and cabbage (Food and Agricultural Organization, 2005). During the U.S. Civil War, General Ulysses S. Grant sent an urgent message to the War Department: "I will not move my army without onions." The very next day, three train loads of onions were on their way to the front. Grant also employed the juice of onions medicinally. Today, China is by far the world's largest producer and exporter of garlic and dry onions (see Figure 1.10), as indicated by 2007 data from the Food and Agriculture Organization of the United Nations (FAO): garlic – China, 12 088 000 t (metric tons), India 645 000 t, Korea 325 000 t, Russia 254 000 t, United States 221 810 t, World 15 686 310 t; dry onions – China 20 552 000 t, Mexico 12 000 000 t, India 8 178 300 t, United States 3 602 090 t, Pakistan 2 100 000 t, Turkey 1 779 392 t, World 64 475 126 t (FAO, 2008).

Onions are divided into two categories: sweet and storage. The availability of sweet onions (*e.g.*, Bermuda, Maui, OSO Sweets, Spanish, Vidalia, Walla-Walla) is short-lived, whereas storage onions are available year round. Storage onions, which are more pungent than sweet onions, come in yellow (most common), red or purple (milder), and white (stronger, favored in Mexican cuisine). Onions are grown in most climate zones around the world, from tropical to cool temperate climates. The transition from leaf growth to bulb formation in the onion depends both on temperature and on the adaptation of the particular cultivar to day length. After harvesting, onions require a drying (curing) period of two weeks in a shady, warm place with circulating air. Ideally, onions should not be refrigerated or washed until use, and should be stored in a dark, cool (50 °F or 10 °C, or less) pantry with good air circulation and 65 to 70% relative humidity (U. C. Davis, 2008). Onions should not be stored with potatoes, since vapors from onions accelerate ripening of potatoes. Milder onions with low levels of solids are rarely stored for more than one month, while more pungent varieties with higher levels of solids can be stored longer. Exposure of onions to light accelerates chlorophyll synthesis and greening. Onions hung in the legs of old pantyhose, with excellent air circulation, can be stored for months. When selecting onions in the market, choose those which are firm, symmetrical

Figure 1.10 (a) Side by side fields of garlic (left) and onion (right) in Shandong Province, China. (b) Workers in garlic packing plant cutting, weighing and packaging freshly harvested garlic, Jinxiang, Shandong Province, China. (Photos by Eric Block).

and unblemished, with bright, dry skins that glisten. Avoid those with a spongy feel and soft necks, soft discolorations, green shoots, or a wilted, weathered or leathery look, indications of poor quality and rotting (Cavage, 1987).

Scallions, immature green onions pulled from the ground before they fully mature, while their tops are tender and green, are often consumed raw or are cut into tiny green rings as a garnish to soups, pasta and salads. They are perishable and should be refrigerated, unwashed, in perforated plastic bags for no more than five days before use. It is then essential to thoroughly wash and peel the outer layer of skin from scallions before use to avoid the risk of bacterial and parasitic contamination that may have occurred sometime during prior handling. Cipolline Borettanes are small, dry (mature) Italian pearl onions, one to three inches (2.5 to 7.5 cm) in diameter with flat, saucer-like shapes and a mild and sweet taste.

The shallot is a member of a horticultural subgroup (*Aggregatum*) of *A. cepa.* Major varieties, such as French red and Dutch white, can have clustered bulbs smaller than onions or fist-size single bulbs. Shallots were introduced to the West in the third century BCE by Alexander the Great during his conquests in Asia Minor. The name *A. ascalonicum* is based on the ancient city of Ascalon (now Ashkelon, Israel; Moyers, 1996). Shallots grow in bulbs covered with a papery brown or purple skin and filled with cloves. They have a distinctive flavor which combines that of onion and garlic, but is less overpowering than either, with a delicacy favored for sauces in French cuisine. In India, shallots are pounded with chillies and other seasonings to add pungency to curries. Shallots tend to dry up quickly and should be stored in a cool, dark place and never refrigerated. When cooking they should not be allowed to brown as this causes bitterness (Cavage, 1987; Bareham, 1995).

Today, garlic is the second most widely consumed *Allium* worldwide, at an annual level of 14 million metric tons (Food and Agricultural Organization, 2005). The garlic bulb is made up of numerous smaller bulbs, called cloves, although occasionally bulbs contain only a single clove, usually called a "round." Garlic bulbs are cleanly broken apart into cloves (called "cracking") within 24 hours of planting, when the temperature starts to turn cool (*e.g.*, in the fall in the Northern Hemisphere) to prevent premature sprouting. The cloves are planted with the pointed end of the clove up/basal plate down, at least two inches (5 cm) below the surface and seven inches (18 cm) between plants and rows. In warmer climates hardneck garlic should be stored at 7 to 10 °C (45 to 50 °F) for about three weeks before planting. If desired, garlic can be grown to maximum bulbil formation and the bulbils planted. Advantages of using the bulbils are that there are many more bulbils than cloves, and that the bulbils are not in contact with the soil, which prevents transmission of soil-borne diseases and pests. A disadvantage of using the bulbils for propagation is that it takes several years to produce decent sized bulbs from them (http://www.garlicfarm.ca/).

Garlic and other alliums are sensitive to periods of sunlight and temperature, and each plant will shut down growth when a certain point in the daylight cycle is reached. Garlic requires adequate moisture during its early growth, but then no additional moisture during the last few weeks. Garlic is harvested when the lower leaves are one-half to three-quarters brown (the green leaves start to die from the bottom up); a sample plant should be uprooted during this period to verify harvest-ready status. It is then bundled or braided and stored away from direct sunlight, *e.g.*, in a curing barn, ideally at low humidity in an area with good airflow, where it loses 18–20% of its original water content over the course of 10–20 days. Long-term, commercial storage of garlic entails refrigeration (as low as 22 °F) in a nitrogen atmosphere, to prevent sprouting.

Garlic is the strongest-flavored edible *Allium*. The fresher the garlic, the milder the taste. Fresh-from-the-ground garlic, sometimes called "green garlic," which has not been cured has a higher moisture content, is milder and is a popular item at farmers markets. Good garlic should be firm and the cloves not dried up or discolored. Garlic is relatively fragile; bruising will trigger the chemical reactions that release allicin, as discussed in Chapter 3. Garlic should not be stored in a refrigerator since the combination of low temperature, oxygen and moisture will trigger mildewing and premature sprouting. "Frost-free" refrigerators will dry out the bulb. Garlic should be stored at room temperature in a cool, ventilated location. The preferred means of storing garlic (typically for 2 to 6 weeks) is in a clay jar with holes for air circulation. Longer term storage is possible, although some cooks recommend purchasing only the amount of garlic needed for short-term use to ensure freshness.

While the large bulb of the elephant garlic (*A. ampeloprasum*) closely resembles garlic, with its fewer cloves (about four to eight) it is easier to peel. Due to its lower sulfur content, elephant garlic's flavor is milder than that of garlic. Botanically elephant garlic is the ancestor of today's leeks.

Leeks are grown in the bottom of narrow furrows, 4 to 6 inches (10 to 15 cm) deep. As the plants grow, the furrow is gradually filled with soil, which keeps the stems white (blanched). Hilling, piling sand and soil around the base of the leek with a hoe or hiller, is also used to force stem growth and blanch the stems. Leeks are at their best when they are young, tender and freshly dug. They do not dry well. Called "the asparagus of the poor" and "the king of the soup onions," leek is plentiful and cheap and is excellent in soup (*e.g.*, the author's favorite, leek and potato soup), as a filling in omelets, quiches, stews and tarts, or as a vegetable served on its own, braised and eaten hot or cold. It has a milder, sweeter taste than onion. Like scallions, leek is perishable and should be refrigerated, unwashed, in perforated plastic bags for no more than five days before use. It does not freeze well. Leeks retain sand, and therefore need to be cleaned carefully after cutting off the roots and most of the green top (Rosso, 1982).

Two different alliums share the name "chives": the readily multiplying, hardy, ordinary chive (*A. schoenoprasum*) with its bright green, tubular, grass-like stem and edible, pink, mauve or lavender flowers, and the larger, less readily multiplying, Chinese or garlic chive (*A. tuberosum*) with its white, starry flowers and flat, foot-long stems. With the exception of ordinary chives, all of the domesticated, economically important alliums in North America have come from the Near East or far eastern Asia. Chives, the tenderest and mildest *Allium*, are always given in the plural, since one chive could hardly be used. They are frequently grown on the kitchen windowsill. Cut with scissors to avoid crushing the cells and prematurely releasing the flavor, chives serve as a colorful garnish for vichyssoise, mixed with sour cream as a topping for baked potatoes, and a flavorful addition to omelets. Unfortunately, their flavor and color quickly fade with cooking. Chinese chives are larger and tougher than normal chives, with dark, muddy green leaves. Yellow Chinese chives are grown in the dark and have an earthier flavor (Rosso, 1982; Bareham, 1995).

While many of the wild species of *Allium* are edible and are consumed as food (*e.g.*, *A. nutans* is a tasty substitute for chives), only a few species are of economic significance. Of the wild species, *A. ursinum* (ramson) is used both as an ingredient in food and as a remedy for scurvy. Caution should be exercised by those collecting *A. ursinum* in the wild since it can be mistaken for the autumn crocus, *Colchicum autumnale* (sometimes called meadow saffron), which is highly toxic due to the presence of approximately 0.5% of the alkaloid colchicine in the leaves. Autumn crocus and wild garlic are quite similar, especially their leaves, and unfortunately they grow in the same areas at the same time. Several deaths attributed to cases of mistaken plant identity have been recorded (Wehner, 2006; Brvar, 2004).

The ramp, *A. tricoccum*, does not grow easily in culture and is collected from the wild. It is a perennial plant that grows in colonies along the Appalachian Mountain region of North America from northern Georgia to Quebec. It is most easily vegetatively propagated by bulbs because it has a complex multiple dormancy that makes propagation from seeds difficult and unreliable. Ramps have specific climatic requirements, preferring acid (~pH 4.5), moist, rich soil (Vasseur, 1994) and low temperature. They have a day length requirement for bulbing, which occurs only in late spring, apparently in a short growth phase, with most vegetative growth in April and May. They cease vegetative growth in June and remain quiescent until the following spring. Ramps, "vulnerable" plants because of commercial exploitation of natural populations (Nault, 1993; Vasseur, 1994), are featured in trendy restaurants in the United States.

CHAPTER 2

All Things *Allium*: Alliums in Literature, the Arts and Culture

There are spices and vegetables that you can grow
Some are under the ground, some grow tall
Though they all have their qualities, this you should know
That the garlic is best of them all

Ruthie Gordon, *The Garlic Waltz* (1980)

2.1 INTRODUCTION

Soon after beginning my research on *Allium* chemistry in 1970, I had sufficient results for publications and lectures. To enliven my talks, entitled "Chemistry in a Salad Bowl," I drew on relevant literary references. Over the years, my collection of *Allium*-related material grew substantially, reinforced by systematic searching, vastly facilitated by using the Internet. I was impressed by the existence of an entire *Allium*-based lifestyle complete with garlic festivals, garlic and onion specialty restaurants, garlic and onion in movies, songs and ballet, garlic jewelry, and garlic and onion in art and architecture. There is even an onion newspaper. However, *The Onion*, a satirical publication calling itself "America's finest news source," in fact has nothing to do with alliums other than

Garlic and Other Alliums: The Lore and the Science
By Eric Block

Published by the Royal Society of Chemistry, www.rsc.org

occasionally featuring articles on topics such as the 'U.S. Department of Breath and Human Services' warning of dangerously high levels of bad breath in America.

Literary quotations based on the lore of alliums abound. Great writers are keen observers of humanity and the world around them. In their writing they featured garlic and onion in an inventive manner, skillfully highlighting their unique, often amusing physiological effects (and their consequences!). Sometimes garlic or onions were even made the centerpiece of a tale, poem or song. The representative literary selections in this chapter, spanning more than 4000 years, are presented chronologically. Specific literary references to medicinal applications of alliums, *e.g.*, in herbals and their early Greek and Roman predecessors, are presented in Chapter 5.

In their own way, art and architecture can also tell a story or depict a concept. The works of art in this chapter go beyond the mere botanical representation of the plants, in conveying images of humble peasants selling their crop, of an unhappy cook mashing garlic, of the natural beauty of translucent onion skins and even of the direct connection between onions and health. Clever artistic use has also been made of onion and garlic shapes by architects and craftsmen.

2.2 ALLIUMS IN LITERATURE

Among the earliest quotations is that from the Bible, which describes the longing for garlic, onions and leeks felt by the Israelites as they wandered in the desert.

> *The Bible [Numbers 11:6]:* We remember the fish which we did eat in Egypt, the cucumbers, the melons, leeks, onions and garlic...
>
> Similar wording appears later in the Koran.
>
> *The Koran [2.61]:* [the people longed for] what the earth grows, of its herbs and its cucumbers and its garlic and its lentils and its onions.

Perhaps even earlier than the Bible are the stone tablets of the Sumerians of ancient Mesopotamia (modern Iraq), dating from *ca.* 2500 BCE onwards. Sumerian is the first language for which we have written evidence. Several of these early writings, made available on the Web through the Electronic Text Corpus of Sumerian Literature

(ETCSL) project at the University of Oxford (Oxford, 2006) make reference to garlic and leeks in a surprisingly modern manner.

> *Dumuzid's ("Shepherd's") Dream*: Those who come for the king are a motley crew, who know not food, who know not drink, who eat no sprinkled flour, who drink no poured water, who accept no pleasant gifts, who do not enjoy a wife's embraces, who never kiss dear little children, who never chew sharp-tasting garlic, who eat no fish, who eat no leeks.

The Greek comic dramatist Aristophanes (*ca.* 446–388 BCE) frequently uses garlic in his plays, suggesting that it played a substantial role in the diet and everyday life of ancient Greece. Thus, in *The Knights* (424 BCE) he writes of the presumed strength-giving virtues of garlic: "Now, bolt down these cloves of garlic." "Pray, what for?" "Well primed with garlic, you will have greater mettle for the fight." *The Acharnians* (425 BCE) contains the lines: "Do not go near them; they have eaten garlic" while in *Peace* (420 BCE) he writes: "Take this cuff on the head for your pains...Oh! How it stings! Master, have you got garlic in your fist, I wonder?" The latter quote suggests that Aristophanes knew that garlic could be quite irritating to open wounds. *The Wasps* (420 BCE) also refers to the use of garlic to treat afflictions of the skin, and the bizarre appearance of such plasters: "Why, you look like a garlic plaster on a boil." In another of his plays, *The Thesmophoriazusae* (411 BCE), an adulterous wife consumes "garlic early in the morning after a night of wantonness, so that our husband, who has been keeping guard upon the city wall, may be reassured by the smell and suspect nothing." Scholars discussing this text suggest that someone engaged in adultery would typically shun garlic.

Plutarch, the noted ancient Greek historian, biographer and essayist, comments on the thirst- and tear-producing effects of onion and the avoidance of the plants by priests in his *Isis and Osiris* (CE 100), regarded as a crucial source of information on Ancient Egyptian religious rites. He writes: "The priests keep themselves clear of the onion and detest it and are careful to avoid it, because it is the only plant that naturally thrives and flourishes in the waning of the moon. It is suitable for neither fasting nor festival, because in the one case it causes thirst and in the other tears for those who partake of it."

Chaucer, through the character of The Summoner in *The Canterbury Tales*, notes the pleasures of eating alliums [*The Canterbury Tales. Prologue. The Summoner.* (1387–1400)]: "Wel loved he garleek, oynons, and eek [moreover] lekes, and for to drynken strong wyn, reed as blood." An unknown contemporary of Chaucer's wrote: "If Leekes you like but do their smell dis-leeke eat Onyuns and you shall not smell the Leeke. If you of Onyuns would the scent expelle eat Garlicke and that shall drowne the Onyun's smelle."

Cervantes gives us memorable quotations on alliums from *Don Quixote de la Mancha* (1605/1615), such as the following, dealing with the effect eating garlic has on the breath; the second is quite similar to words written by Cervantes' contemporary, William Shakespeare.

> Observe too, Sancho, that these traitors were not content with changing and transforming my Dulcinea, but they transformed and changed her into a shape as mean and ill-favoured as that of the village girl yonder; and at the same time they robbed her of that which is such a peculiar property of ladies of distinction, that is to say, the sweet fragrance that comes of being always among perfumes and flowers. For I must tell thee, Sancho, that when I approached to put Dulcinea upon her hackney ... she gave me a whiff of raw garlic that made my head reel, and poisoned my very heart (Chapter 10).

> Sancho listened to him with the deepest attention, and endeavoured to fix his counsels in his memory, like one who meant to follow them and by their means bring the full promise of his government to a happy issue. Don Quixote, then, went on to say: ..."Eat not garlic nor onions, lest they find out thy boorish origin by the smell; walk slowly and speak deliberately, but not in such a way as to make it seem thou art listening to thyself, for all affectation is bad." (Chapter 43)

Several of Shakespeare's plays allude to the effects of garlic and onion on the breath and on the tear-inducing or lachrymatory effect of onion, as in the following:

> *Measure for Measure* [Act 3, Scene 2]
> The duke, I say to thee again, would eat mutton on Fridays. He's not past it yet, and I say to thee, he would mouth with a beggar, though she smelt brown bread and garlic.

Antony and Cleopatra [Act 1, Scene 2]
Indeed the tears live in an onion that should water this sorrow.

Antony and Cleopatra [Act 4, Scene 2]
What mean you, sir, to give them this discomfort? Look, they weep; and I, an ass, am onion eyed: for shame, transform us not to women.

Coriolanus [Act 4, Scene 6]
You have made good work, you and your apron-men; you, that stood so much upon the voice of occupation and the breath of garlic eaters.

Winter's Tale [Act 4, Scene IV]
Mopsa must be your mistress: marry, garlic, to mend her kissing with!

Jonathan Swift (1667–1745) in his *Verses for Fruitwomen* shares chemically reasonable secrets on avoiding problems of onion breath among lovers: both should consume onions to avoid smelling them on the other's breath or boil the onions before consuming them to destroy the odor-inducing compounds:

Come, follow me by the smell, here are delicate onions to sell; I promise to use you well. They make the blood warmer, You'll feed like a farmer; For this is every cook's opinion, no savoury dish without an onion; But, lest your kissing should be spoiled, your onions must be thoroughly boiled: Or else you may spare your mistress a share, the secret will never be known: She cannot discover the breath of her lover but think it as sweet as her own.

Humorous allusions to the lachrymatory power of onions include quotations by Benjamin Franklin in *Poor Richard's Almanac* (1734), "Onions can make ev'n heirs and widows weep" and by Nathaniel Hawthorne in the *House of Seven Gables* (1851), "As for pathos, I am as provocative of tears as an onion."

Elizabeth Barrett Browning in *Aurora Leigh* (Third Book, 1884) decries the unpleasant smell of garlic breath:

I took a master in the German tongue, I gamed a little, went to Paris twice; but, after all, this love!...you eat of love, and do as vile a thing as if you ate of garlic – which, whatever else you eat, tastes uniformly acrid, till your peach reminds you of your onion!. . .I came home uncured, convicted rather to myself of being in love...in love! That's coarse you'll say I'm talking garlic.

Guy de Maupassant in *The Rondoli Sisters* (1884) is highly critical of the odor of garlic:

> I cannot lift up the sheets of a hotel bed without a shudder of disgust. Who has occupied it the night before? Perhaps dirty, revolting people have slept in it. I begin, then, to think of all the horrible people with whom one rubs shoulders every day, people with suspicious-looking skin which makes one think of the feet and all the rest! I call to mind those who carry about with them the sickening smell of garlic or of humanity.

Thomas Hardy in *Tess of the d'Urbervilles* [Chapter 22 (1891)] describes a well-known property more likely due to wild garlic, *Allium vineale*, rather than garlic itself, of flavoring the milk of cows that have ingested small quantities of the plant, and the efforts the dairymen took to eradicate this noxious weed from the pasture:

> All having armed themselves with old pointed knives they went out together. As the inimical plant could only be present in very microscopic dimensions to have escaped ordinary observation, to find it seemed rather a hopeless attempt in the stretch of rich grass before them. However, they formed themselves into line, all assisting, owing to the importance of the search. . . With eyes fixed upon the ground they crept slowly across a strip of the field, returning a little further down in such a manner that, when they should have finished, not a single inch of the pasture but would have fallen under the eye of some one of them. It was a most tedious business, not more than half a dozen shoots of garlic being discoverable in the whole field; yet such was the herb's pungency that probably one bite of it by one cow had been sufficient to season the whole dairy's produce for the day.

The hypnotic cadence of Rudyard Kipling's poem *Mandalay* (1880) invokes the heady aroma of garlic to symbolize the romance of far-off East Asia:

> But that's all shove be'ind me – long ago an' fur away,
> An' there ain't no 'busses runnin' from the Bank to Mandalay;
> An' I'm learnin' 'ere in London what the ten-year soldier tells:
> 'If you've 'eard the East a-callin', you won't never 'eed naught else.'
> No! you won't 'eed nothin' else but them spicy garlic smells,
> An' the sunshine an' the palm-trees an' the tinkly temple-bells;
> On the road to Mandalay...

A tale from Kipling's *Second Jungle Book* [Chapter 13 (1895)] hangs on the bee-repellent properties of wild garlic as Mowgli, an abandoned "man cub" raised by wolves in the Indian jungle, lures a pack of vicious, wolf-killing dholes (Asiatic wild dogs) to their death. Mowgli "knew that the Little People [bees] hated the smell of wild garlic. So he gathered a small bundle of it ... [and] took the garlic and rubbed himself all over carefully." With the dholes in hot pursuit, Mowgli races through a bee-infested area, arousing the bees that sting the dholes while Mowgli is protected by the smell of wild garlic.

In Bram Stoker's *Dracula* (1897) garlic is used to repel vampires. The bedroom of the heroine Lucy is decked with garlic, and a garlic wreath is placed around her neck:

> The Professor's actions were certainly odd and not to be found in any pharmacopeia that I ever heard of. First he fastened up the windows and latched them securely. Next, taking a handful of the [garlic] flowers, he rubbed them all over the sashes, as though to ensure that every whiff of air that might get in would be laden with the garlic smell. Then with the wisp he rubbed all over the jamb of the door, above, below, and at each side, and round the fireplace in the same way. It all seemed grotesque to me, and presently I said, 'Well, Professor, I know you always have a reason for what you do, but this certainly puzzles me. It is well we have no sceptic here, or he would say that you were working some spell to keep out an evil spirit.'... I could see that the Professor had carried out in this room, as in the other, his purpose of using the garlic. The whole of the window sashes reeked with it, and round Lucy's neck,

over the silk handkerchief which Van Helsing made her keep on, was a rough chaplet of the same odorous flowers.

The prolific American short-story writer William Sydney Porter, better known as O. Henry (1862–1910), casts the onion as a central character in his poignant short story *The Third Ingredient*. "A stew without an onion is worse'n a matinee without candy. . .There's a little lady – a friend of mine – in my room there at the end of the hall. Both of us are out of luck; and we had just potatoes and meat between us. They're stewing now. But it ain't got any soul. There's something lacking to it. There's certain things in life that are naturally intended to fit and belong together. One. . .is beef and potatoes with onions."

Shorter quotations on garlic can also be noted here. Sculptor and author Augustus Saint-Gaudens in *Reminiscences* (1913) wrote: "what garlic is to food, insanity is to art." F. Scott Fitzgerald in *The Beautiful and Damned* (1922) wrote: "On the crowded train back to New York the seat behind was occupied by a super-respirating Latin whose last few meals had obviously been composed entirely of garlic." Carl Sanburg (1878–1967) wrote: "Life is like an onion: you peel it off one layer at a time, and sometimes you weep." English critic and editor Cyril Connolly (1903–1974) wrote: "vulgarity is the garlic in the salad of life." There is also a traditional saying: "A nickel will get you on the subway, but garlic will get you a seat."

Garlic and onions have even found their way into children's literature and ballet. *Onions and Garlic: An Old Tale* (Kimmel, 1996), based on Talmudic legend, tells of competitive merchant brothers: one good-natured, humble brother has nothing but common onions to barter, but in his travels finds an island where diamonds litter the ground and onions are unheard of. He trades his wares for 100 sacks of diamonds and returns home triumphant. His greedy brothers eagerly put together a cargo of garlic to trade for even more diamonds on this same island. Their excellent garlic is much admired. In payment, the garlic merchants receive something "more precious than diamonds," which of course turns out to be...onions!

The Adventures of the Little Onion (*Il romanzo di Cipollino;* Rodari, 1951) is a classic children's story by Italian author Gianni Rodari of the adventures of the little onion boy, Chipollino, in a land ruled by a wicked tomato-king. Translated, Rodari's story was popular in Germany, where it was known as *Zwiebelchen*, in China and particularly in Russia, where it became the basis for an animated film, a ballet and even postage stamps!

2.3 ALLIUMS IN POETRY

Given their renown, it is not surprising that both garlic and onion have been the subject of poems. One of the earliest of these, *That wicked garlic!* is *Epode 3* by the Roman lyric poet Horace (65–8 BCE; West, 1997). These satirical verses are filled with allusions to mythology and display a truculence absent from Horace's more mature writings. The numerous references to the well known properties of garlic ending with the allusions to the off-putting properties of garlic breath must have greatly amused contemporary audiences at live readings (Gowers, 1993).

If any man with impious hand has broken
  his aged father's neck,
let him eat garlic. It is worse than hemlock.
  Peasants must have guts of brass.
What is this poison seething in my chest?
  I am betrayed. These herbs
were cooked in viper's blood or else Canidia
  has touched this filthy food.
When among all the comely Argonauts, Medea saw
  and marveled at their leader Jason,
Before he went to yoke the still unbroken bulls,
  she smeared the garlic on him,
And then before she fled on serpent wings, in garlic steeped
  her gifts to take revenge on his new whore.
The stars have never sent such heat to brood
  on parched Apulia.
It was no fiercer fire that scorched the back
  of mighty Hercules.
And if you ever take in mind to try a trick like this,
  My sly Maecenas, I do pray
your lover may put up her hand against your kiss
  and lie far from you on the bed.

Poems by two contemporary American poets so accurately portray these plants that we can visualize their form in our mind's eye, sense their smell, taste, and lachrymatory properties, and even hear them sizzling in a pan. Such is the power of good poetry!

The poem *Inside the Onion* (1984) published in a book of the same title by United States Poet Laureate Howard Nemerov (1920–1991) and excerpted below aptly describes the layered design of the onion:

Slicing the sphere in planes you map inside
The secret sections filled up with the forms
That gave us mind, free-hand asymmetries
Perfecting for us the beautiful inexact
That mathematic may approximate
And clue us into but may never mate
Exactly: bulb, root, fruit of the fortunate fall
That feeds us with the weeps and utter tang...

The equally delightful poem *Chopping Garlic* from *At the White Window* (2000) by poet David Young (1936–) captures the joy of cooking with garlic:

The bulb, an oriental palace/ shrouded in gray and lavender paper,
Splits open into a heap/ of wedge-shaped packets housing
Horns, fangs, monster toenails/ all of a pungent ivory – I
Could string them into a necklace/ but I smash them flat instead,
Loving the crunch, brushing away all the confetti–clouds
Of odor bloom around me now/ as I chop, this way and that,
With my half-moon blade in the scooped wood
That will never completely lose the fragrance that oils it, smears
My fingers, wants to be in/ the pores of my skin forever ...
Trumpets and cymbals blare/ as I dump the grainy mess
Into the pan, oh, holy to the nose/ are the incense and sizzle that summon
Folks from all parts of the house/ to ask about dinner, sniffing,
While up in one end of the sky/ a crescent moon hangs crazily,
A glowing clove, a dangerous fragrance/ filling the very corners
Of some god's smiling mouth.

2.4 ALLIUMS IN FILM, SONG AND BALLET

Like Water for Chocolate by Laura Esquivel (book, Doubleday, New York, 1992; motion picture, 1992, directed by Alfonso Arau/

Miramax Films) is a "tall-tale, fairy-tale, soap opera romance, Mexican cookbook, and home-remedy handbook all rolled into one" (*San Francisco Chronicle*) with more than two million copies in print. It opens with a bit of wisdom from one of its central settings, the kitchen: to avoid tears when chopping onions, one must simply place a slice of onion on one's head. The birth of the protagonist Tita in the kitchen is then described:

> Tita was so sensitive to onions, any time they were being chopped, they say she would just cry and cry; when she was still in my great-grandmother's belly her sobs were so loud that even Nacha, the cook, who was half-deaf, could hear them easily. Once her wailing got so violent that it brought on an early labor. And before my great-grandmother could let out a word or even a whimper, Tita made her entrance into this world, prematurely, right there on the kitchen table amid the smells of simmering noodle soup, thyme, bay leaves, and cilantro, steamed milk, garlic, and of course onion.

In 1980, Les Blank produced the documentary film *Garlic is as Good as Ten Mothers* (Blank, 1980). The film has achieved cult status and in 2004 was one of 25 films selected by the U.S. Library of Congress to be added to the National Film Registry list of films to be preserved in perpetuity. One of the highlights of the film is *The Garlic Song*, written by Ruthie Gordon:

> There are spices and vegetables that you can grow
> Some are under the ground, some grow tall
> But they all have their qualities, this you should know
> That the garlic is best of them all.
>
> The Egyptians, Phoenicians, the Vikings and Greeks
> Babylonians, Danes, and Chinese
> On their voyages took enough garlic for weeks
> And their enemies died on the breeze.
>
> Since Biblical times in all parts of the earth
> It has cured countless sufferings and ills
> If we understood what the garlic is worth
> We would throw out our poisonous pills.

In Bulgaria's mountains and Russia's wide plains
People live to a hundred years old
For it's juice of the garlic that runs in their veins
Oh it's worth twice it's weight in pure gold.

With its selenium, germanium, allicin too
It can fight off all types of disease
So if you've got arthritis, TB, or the flu
Just say, "Peel me a garlic clove, please!"

Plant some cloves in your garden to keep away worms
And the other bad things that kill plants
If you're one of those people concerned about germs
You could drop one or two in your pants.

In 1973, Russian composer Karen Khachaturian, nephew of Aram Khachaturian, wrote a frequently performed three-act ballet, *Chipollino* (*The Little Onion*), based on Gianni Rodari's children's book of the same name and the subject of a popular Soviet-era animated film (Soyuzmultfilm, 1961; Randel, 1996).

2.5 ALLIUMS IN PAINTING

Just as the alliums have figured prominently in literature, so too do their images appear frequently in paintings. Three examples are given here. The first (Figure 2.1) is an oil on canvas painting by Diego Velásquez, *Kitchen Scene with Christ in the House of Martha and Mary* (*ca.* 1618; The National Gallery, London), sometimes named *A Young Woman Crushing Garlic*. Velasquez extends this unusual still life arrangement to incorporate a religious scene placed in the background. The painting relates the biblical story of two sisters, Martha and Mary, who lived in Bethany. Jesus visited them, from time to time, and rested there. Mary sat at his feet and listened to his words, while Martha was encumbered with much serving, as St. Luke put it, and complained. In the painting she seems to be pouting and unhappy with her job crushing garlic. Martha said to Jesus, "Dost thou not care that my sister hath left me to serve alone? Bid her therefore that she help me." Jesus said to her, "Martha, Martha, thou art careful and troubled about many

Figure 2.1 Velásquez, *Kitchen Scene with Christ in the House of Martha and Mary*. (Courtesy of The National Gallery, London).

things: but one thing is needful, and Mary hath chosen that good part, which shall not be taken away from her."

The second, by impressionist painter Pierre-Auguste Renoir, *The Onions* (Figure 2.2), an oil on canvas (1881), is a remarkable work, showing a group of yellow onions and a few bulbs of garlic on a table top against a shimmering background. The work is painted so skillfully that the onions' shiny papery skins seem real enough to touch.

The final painting, *Still Life with Drawing Board, Pipe and Onions*, painted by Vincent Van Gogh in Arles in 1889 (Figure 2.3) painted in Paris in 1887, is a lovely example chosen from among four still lifes featuring onions or garlic by the artist, the others being *Still Life with Ginger Jar and Onions* (1885), *Still Life with Bloaters and Garlic* (1887) and *Still Life with Red Cabbages and Onions* (1887). Symbolic of Van Gogh's health concerns, a popular book about health appears with the onions in the painting in Figure 2.3. Two other notable paintings featuring garlic can be mentioned here: the *Garlic Seller*, an oil on paper painting by French artist Jean-François Raffaëlli (*ca.* 1880; Boston Musum of Fine Arts), shows an old bearded farmer with his dog trailing behind him bringing a basket of garlic to sell in the market; John Singer Sargent's *Venetian Onion Seller* (1882; Museo

Figure 2.2 Pierre-Auguste Renoir, *The Onions*, 1881. Oil on canvas, Sterling and Francine Clark Art Institute, Williamstown, Massachusetts, USA, 1955.588 (Photo by Michael Agee). © Sterling and Francine Clark Art Institute, Williamstown, Massachusetts, USA.

Figure 2.3 Van Gogh, *Still Life with Drawing Board, Pipe and Onions*. (Courtesy of Kröller-Müller Museum, The Netherlands).

Thyssen-Bornemisza, Madrid) was painted during Sargent's stay in Venice. This painting is an experiment by Sargent marked by freer brushstrokes and a popular subject rather than a high society portrait.

Garlic has not only served as a subject for paintings but has also played a role in the process of creating the art itself. In particular, whole garlic or its juice has been used as a glue for attaching gildings of gold, silver and tin to the surfaces of paintings as well as picture frames, manuscripts and furniture. While the volatile sulfur compounds present in garlic have long since evaporated from garlic used centuries ago in the gilding process, its presence could be confirmed by identifying the characteristic pattern of amino acids present in proteins from garlic. The pattern of the 14 amino acids present in samples of gilding was sufficiently different from those of other commonly used gilding glues, animal glue and egg proteins, to prove that garlic must have been used (see Section 3.12 for more details; Bonaduce, 2006).

2.6 ALLIUMS IN ARCHITECTURE: ONION AND GARLIC DOMES

An onion dome is a type of architectural dome usually associated with Russian Orthodox churches. Such a dome is larger in diameter than the drum it is set upon and its height usually exceeds its width. These bulbous structures taper smoothly to a point, and strongly resemble the onion, after which they are named. Through analysis of old Russian icons and miniatures, Russian art and architecture historians conclude that onion domes existed in Russia as early as the thirteenth century, and therefore could not have been imported from the Orient, where onion domes did not replace spherical domes until the fifteenth century. The ubiquitous appearance of onion domes in the late thirteenth century can be explained by the general emphasis on verticality characteristic of Russian architecture from the late twelfth to early fifteenth centuries, in an effort to make churches seem taller than they were. While onion-like structures mark Dutch secular buildings of the sixteenth century, the fully developed onion dome or spire is thought to be a local architectural development in sixteenth century Prague, which was subsequently adopted in neighboring regions of Bavaria, the Austrian Empire, and Southern Germany, especially buildings in the countryside (Schindler, 1981). For example, a seventeenth century book on architecture by a Prague architect contains drawings of several perfect onion spires (Leuthner, 1677).

Each drum of a Russian church is surmounted by a special structure of metal or timber, and is lined with sheet iron or tiles, which are often brightly colored. Early examples of Russian onion domes are the blue and golden domes of the Cathedral of Dormition, (1559–1585) built by the order of Ivan the Terrible, Trinity Monastery of St. Sergius, Lavra ("Lavra" means "main and most important monastery"). The bulbous, wildly colored domes of Saint Basil's Cathedral in Moscow (Figure 2.4) have not been altered since the reign of Ivan the Terrible's son Fyodor I, in sixteenth-century Russia. Architect Bartolomeo Rastrelli used ornate baroque gilded onion domes on the east chapel at Peterhof Grand Palace, St. Petersburg (*ca.* 1750) whereas the slightly later, black-domed Vladimirskaya church in St. Petersburg (1761–1783), where Fyodor Dostovevsky was a parishioner, shows a style that straddles baroque and neoclassicism. Saint Basil's Cathedral was the inspiration for the Church of the Savior on Spilled Blood (1883–1907; Figure 2.5a), whose name derives from the location of the church on the site of the assassination of Tsar Alexander II. Jewellers' enamel was used to cover the 1000 square meter surface of the five domes.

The Church of Saint Mary Magdalene, the seat of the Russian Orthodox Church in Jerusalem, is situated on the slope of the Mount of Olives in the Garden of Gethsemane just outside the old, walled city, and is one of the most easily recognizable landmarks of Jerusalem (Figure 2.5b). This striking example of Russian architecture was built in the Muskovite style with seven gilded onion domes jutting out from a monumental Muskovite-style body that stands proudly against the sky. The Russian Tsar Alexander III and his brothers built it in 1888 as a memorial to their mother Empress Maria Alexandrovna.

The Taj Mahal, built in 1630, also shows an onion dome, sometimes referred to as a Persian dome, an *amrud* or guava dome. The marble dome that surmounts the tomb is its most spectacular feature. Its height is about the same size as the base of the building, about 35 m. Its height is accentuated because it sits on a cylindrical "drum" about 7 m high. The top of the dome is decorated with a lotus design, which serves to accentuate its height. The dome is

Figure 2.4 Saint Basil's Cathedral, Moscow, Russia. (Photo by Eric Block).

topped by a gilded finial, which mixes traditional Islamic and Hindu decorative elements. The dome shape is emphasized by four, smaller domed *chattris* (kiosks) placed at its corners. The chattri domes replicate the onion shape of main dome. Their columned

Figure 2.5 (a) The Church of the Savior on Spilled Blood, St. Petersburg, Russia. (b) Church of Saint Mary Magdalene in Jerusalem. (c) Royal Pavilion, Brighton, England. (Photos by Eric Block).

bases open through the roof of the tomb, and provide light to the interior. The chattris also are topped by gilded finials.

Perhaps the supreme example of excessive use of onion domes is to be seen in the Royal Pavilion, built for King George IV from 1815 to 1822 in the seaside resort of Brighton by British architect Thomas Nash (Figure 2.5c). With its domes, cubes, minarets and spires, the pavilion has a distinctly Indian look, borrowing heavily from the Taj Mahal.

In his Casa Batlló (Barcelona, 1905–1907) architect Antonio Gaudí transforms what had been labeled "the most boring apartment house in Barcelona" into an art nouveau masterpiece (Figure 2.6). Gaudí's imagination and invention ran wild with a design that incorporates "fluid, organic, animal and even human forms" (Gill, 2001). This house was compared by a visitor to Hansel and Gretel's cottage, greatly amusing Gaudí. Some of the most dramatic features of the roof are hidden from view, suggesting that some of the effects were intended as "visual...puns, [with] the huge

Figure 2.6 Casa Batlló (left, photo by Eric Block) with tower detail (right, Courtesy of Wikipedia), Barcelona, Spain.

ribbed humpback ... clad on one side by armour plating resembling an armadillo's" (Gill, 2001). The tower to the left of the roof is crowned by a remarkable garlic bulb, in turn topped with Gaudí's signature four-pointed transverse cross, with anagrams referring to the holy family inscribed on the tower itself [gilded initials: JHS (Jesus), JHP (Joseph) and M (Mary); Gill, 2001]. Is Gaudí slyly asking us: "You have seen many fine examples of cross-topped onion domes, but have you ever seen a cross-topped garlic dome?"

2.7 ALLIUMS EVERYWHERE: JEWELRY, NUMISMATICS, STAMPS, PORCELAIN, AND SO FORTH

The pleasing shape and color of garlic cloves has been nicely reproduced in a shell bracelet crafted by an unknown Caribbean artist (Figure 2.7).

The British Pound Sterling coin for Wales, issued in 1985 and 1990 (Figure 2.8) features the leek. The coin, designed by Leslie Durban, shows a leek in a coronet and bears the edge inscription "PLEIDIOL WYF I'M GWLAD" ("True am I to my Country"),

Figure 2.7 Garlic bracelet from Aruba (Photo by Eric Block).

Figure 2.8 British Pound Sterling coin for Wales (1985, 1990) depicting a leek.

from the chorus of the Welsh National Anthem. The leek has been recognized as the emblem of Wales since the middle of the sixteenth century. According to legend, its association with Wales can be traced back to the battle of Heathfield in 633 CE, which took place in a leek field, when St. David persuaded his countrymen to distinguish themselves from their Saxon foes by wearing a leek in their caps. Shakespeare (*Henry V*, Act 4, Scene 7) elaborates on this legend: "Your majesty says very true: if your majesties is remembered of it, the Welshmen did good service in a garden where leeks did grow, wearing leeks in their Monmouth caps; which, your majesty know, to this hour is an honourable badge of the service; and I do believe your majesty takes no scorn to wear the leek upon Saint Davy's day." Henry responds: "I wear it for a memorable honour. For I am Welsh, you know, good countryman." As an aside, one wonders if there is any connection between the statements in many old herbals that leeks "are good for the voice" (see Chapter 5) and the international reputation of the leek-loving Welsh nation for singing!

"Onion-penny" ("onion penies") in English dialect is the name for a Roman coin dug up from the ground, particularly in the town of Silchester. Onion is also the name of a giant (Schwabe, 1917; Ward, 1748).

The reverence of Russians for their onion domes goes well beyond their architecture. Chipollino, the onion boy from Rodari's children's story and Khachaturian's ballet, is the subject of Russian

Figure 2.9 Russian postage stamp from 1992 depicting the story of Chipollino.

postage stamps in 1992 (Figure 2.9) as well as in 2004 (Russian Federation, 1992, 2004), while onion-domed churches appear on the current 1000 Ruble banknote, their porcelain, their lacquer boxes and even their chocolate (Figure 2.10).

2.8 *ALLIUM* USAGE AMONG DIFFERENT CULTURES SINCE BIBLICAL TIMES: ALLIOPHILES, ALLIOPHOBES, THE EVIL EYE AND ONION LAWS

In Chapter 1, we stated that garlic and its allies are both revered (by alliophiles) and reviled (by alliophobes). Here we examine the basis for this love/hate relationship, in some of the many cultures where alliums are an important part of their cuisines.

We begin with the Jewish people, whose love affair with the garlic was such that they called themselves "garlic eaters" [*Talmud Nedarim*, 31a (Negbi, 2004)]. Discussions concerning alliums are found in the Talmud, a written record from about 500 CE, of rabbinic discussions pertaining to Jewish law, ethics, customs and history. Thus, among the ten enactments ordained by Ezra (*Talmud Baba Kama,* 82a), it is written that "garlic be eaten on Fridays" because of the duty of marriage ("to him who performs his marital duty every Friday night"). The commentary further states that: "Five things have been said of garlic: it assuages hunger, warms the body, brings joy, increases semen, and destroys intestinal parasites. There are those that say that it engenders love and dispels jealousy." The Talmud says: "the Kufri variety [of

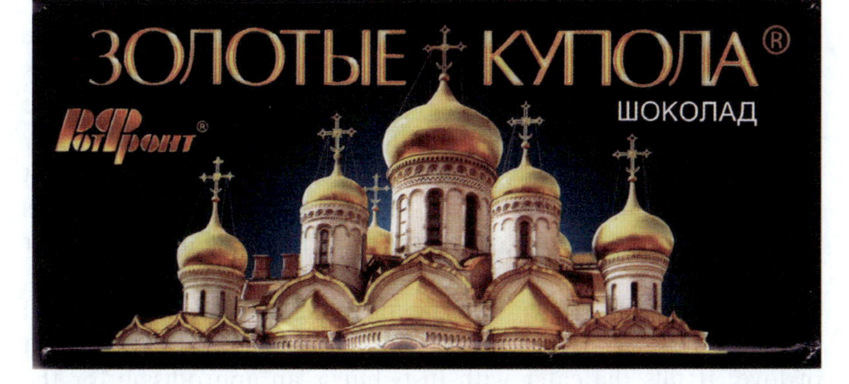

Figure 2.10 Top to bottom: Russian onion domes shown on lacquer box, Russian porcelains and a Russian chocolate bar.

onion] is good for the heart" (*Talmud Nedarim,* 26b) and also relates the story of the traveler to Jerusalem, who upon questioning whether the water was safe to drink was told: "Why worry? We have plenty of onion and garlic here." The Talmud is not completely positive about use of garlic – offensive breath disqualifies a priest and renders other garlic uses unfit for temple service (Daiches, 1936). The Talmud also relates the story of a rabbi who noticed an odor of garlic when delivering his lecture and proclaimed: "Let him who has eaten garlic go out" (Graubard, 1943; Greenspoon, 2002; Levy, 2002; Negbi, 2004).

The diet of the impoverished Ashkenazi Jews of the late nineteenth and early twentieth centuries consisted of little more than black bread, potatoes, and herring heavily seasoned with salt and garlic (Diner, 2001). The garlic-eating habits of Jews is offered as a possible explanation for the "well-recognized fact that amongst Jewish communities in this [England] and other countries the death-rate from tuberculosis is much lower than normal ... [and that] they suffer less from this disease than the surrounding population" (Minchin, 1927). It is said that garlic was so indelibly associated with Jews that the Nazis issued buttons of garlic plants to demonstrate the wearer's ardent anti-Semitism and that "the mere mention of garlic by a Nazi orator caused the crowd to howl with fury and hatred" (Miller, 2006).

Excessive use of garlic as a spice was criticized in Roman times, *e.g.*, "... a dish more stuffed with garlic than a whole gallery of Roman oarsmen," an allusion to garlic-stinking Roman galley-slaves, according to Plautus, a prominent playwright of Ancient Rome (254–184 BCE). Even handling garlic was recognized as hazardous: "If I play with garlic, my hands are bound to stink," according to Pomponius (110–32 BCE). The conspicuous absence of garlic from the fourth to fifth century CE cookbook known as "Apicius" reflects the fact that the book was directed toward the wealthiest classes, while it was the peasant class that typically cooked with garlic. The Roman attitude was contradictory in that garlic also had a reputation as an antidote to poisons or evil spirits as well as aphrodisiac properties. According to Pliny the Elder (23–79 CE), garlic mixed with coriander made a man lecherous, yet garlic was also eaten at festivals of abstinence as a physical repellant. It is suggested that the "herb is ambiguous in the context of love. It fills the eater with lust, but is an aphrodisiac for the beloved; it produces masculine vigour, but is off-putting because of its fetid smell" (Gowers, 1993).

Many cultures consider garlic and its relatives as impure foods. We have already mentioned some examples, such as the Greek poet Horace's diatribe against garlic. Another is the Moslem legend that when Satan left the Garden of Eden following the fall of man, garlic sprang up from the place where his left foot had been, and the onion, from where his right foot had been. In the Moslem tradition it is said that Mohammed disliked onions and garlic and their odor and that one should not enter a mosque after consuming alliums. The Bowers Manuscript, a fifth century Buddhist medical treatise, includes the tale that the first garlic originated in the blood of a demon. In early India, usage of garlic and other alliums was forbidden to Brahmins and other members of higher castes, an anti-allium sentiment which was also shared by Buddhists and Jains. In modern India, garlic and other alliums are considered impure and inappropriate for offering to Hindu gods (Simoons, 1998).

The view that garlic and its relatives are impure and linked to the underworld suggests that these same plants are especially suited for offerings to underworld forces, *e.g.*, evil spirits, in seeking their protection, or eliminating the evil they brought on. In ancient as well as modern Greece, garlic and onion were, and still are, considered as powerful forces protecting against spirit possession, illness and the evil eye – the belief that envy elicited by the good luck of fortunate people may result in their misfortune, such as bad luck, disease or even death. The folk wisdom about the protective nature of garlic was directly connected to its strong, pungent odor – garlic was believed to absorb the evil eye. In modern Greece a cluster of garlic bulbs may be fastened above the house door to protect against evil. Midwives in Greece use garlic in delivery rooms to protect babies against the evil eye. Garlic is tied around an infant's neck and placed beneath the mother's pillow. In some parts of Greece, merely uttering the word for garlic or exclaiming "garlic in your eyes" is thought to protect against the evil eye or against a person suspected of evil intent (Simoons, 1998).

Among the Sephardic Jews, when a neighbor came and said something complementary about her neighbors child she would often then add "May the evil eye not fall" or better "let it go to the garlic." Their houses were protected with crocheted bags with five

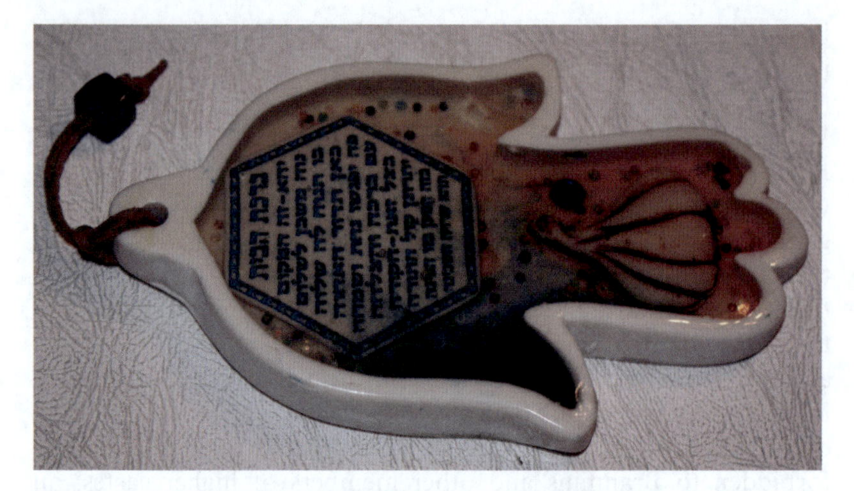

Figure 2.11 A talisman or charm to protect against the "evil eye," with garlic, a blue bead and hamsas. (Photo by Eric Block).

tiny finger-like sacks each holding a single garlic clove. The bags were placed outside a window or in a balcony just like a mezuzah. The five-fingered hand, or hamsa, was also thought to be protective among Jews as well as Arabs (see garlic and hamsa charm, Figure 2.11).

Garlic played an important role in the ancient Greek festival of Skira (or Skiraphoria), the festival of the cutting and threshing of the grain. The Skira was a special festival in May/June for the women of Athens and was one of the few days in the year when they left the seclusion of their women's chambers, assembled according to ancient custom, and abstained from sex in order to bring fertility to the land. They were said to eat garlic together so their breath reeked "to keep sex at bay," *e.g.*, making the women as repulsive to their men as possible (Dillon, 2003). The Skira is the setting for Aristophanes' comedy *Lysistrata* (411 BCE), in which the women seize the opportunity afforded by the festival, to hatch their plot to overthrow male domination.

Some humorous recent examples of views on onion usage and the consequences of their consumption are to be found in various odd laws which are still in effect in the United States. Thus, in Hartsburg, Illinois it is illegal to take onions to the local movie theater as a snack, while in Dyersburg, Tennessee, it is illegal even to enter a movie theater within four hours after having eaten raw onions. It is illegal to eat onions while attending church in Burdonville, Vermont. The pastor has the legal right to make offenders stand in a corner or leave the church until the service is finished. No one is allowed to peel onions in a Cotton Valley, Louisiana, hotel room. In Nacogdoches, Texas, there is an onions curfew for "young women". Under no circumstances are they allowed to have any raw onions after 6 p.m. (from: http://aggie-horticulture.tamu.edu/).

CHAPTER 3

Allium Chemistry 101: Historical Highlights, Fascinating Facts and Unusual Uses for Alliums; Kitchen Chemistry

> And if the boy have not a woman's gift to rain a shower of commanded tears, an onion will do well for such a shift, which in a napkin being close convey'd shall in despite enforce a watery eye.
>
> Shakespeare, *Taming of the Shrew*

> And, most dear actors, eat no onions nor garlic, for we are to utter sweet breath; and I do not doubt but to hear them say, it is a sweet comedy.
>
> Shakespeare, *Midsummer Night's Dream*

3.1 INTRODUCTION

It is appropriate to set the stage for two chapters on *Allium* chemistry with quotations from Shakespeare, poetically reminding us of the lachrymatory effect of slicing onions as well as the potential of developing halitosis (breath malodor) from ingesting alliums. Chemistry plays a central role in understanding the

Garlic and Other Alliums: The Lore and the Science
By Eric Block

Published by the Royal Society of Chemistry, www.rsc.org

remarkable characteristics of *Allium* species: their unique odors when cut, their pungent taste and tear-inducing abilities, and their biological effects both on human health and in the environment. Through chemistry we can learn how these characteristics are designed to allow the plants to survive in an often hostile world. In addition, an understanding of *Allium* chemistry is certainly helpful in the kitchen, since all cooks practice chemistry. I use a mostly historical approach in introducing *Allium* chemistry, giving simpler examples first, including the isolation, characterization and synthesis of individual chemical compounds from the plants. Several eminent chemists, including Nobel laureates or their students, played a key role in this early research. In Chapter 4, I describe methods used for separating and identifying components of mixtures of compounds found in intact and cut alliums. Once the central players in *Allium* chemistry have been identified, I discuss mechanisms of reactions – how one compound is transformed into another, spontaneously or through enzymatic action. Due to the many unique properties of compounds found in, or formed from, alliums, and because these plants are readily available, the power of new analytical methods is often tested using *Allium* plants, making them "gold standards" for analyses. Division of Chapters 3 and 4 into self-contained sections aids the reader seeking specific information with accompanying documentation. The reader is encouraged to take advantage of Web-based resources for definitions and additional background material on technical concepts or instrumental methods.

3.2 EARLY HISTORY OF GARLIC AND ONION CHEMISTRY

The manner in which I became familiar with key events in the history of *Allium* chemistry was serendipitous. While browsing through the shelves of a Paris bookstall, I chanced upon an 1856 edition of *Leçons de Chimie Générale Élémentaire* by French chemist Auguste Cahours. In the section on volatile plant oils, I found a brief description of a method for the isolation of the oil of garlic through use of an alembic (Figure 3.1), a device which separates the volatile oil from the nonvolatile plant constituents by heating in water, condensing the vapors using water-cooled copper coils and collecting the resultant liquid. Alembics, also called stills or retorts,

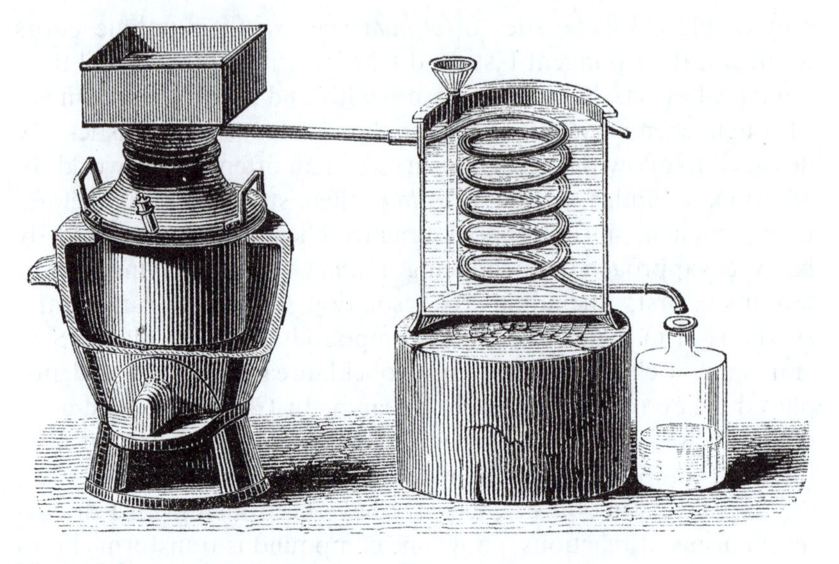

Figure 3.1 An alembic (Cahours, 1856).

were developed *ca.* 800 CE by the Arab alchemist, Jabir ibn Hayyan. Related devices are still used to make cognac. Cahours says that when using the alembic for distillation "avec l'eau les bulbes de l'ail (*Allium sativum*) . . . on obtient des huiles volatiles, douées d'une odeur forte et désagréable . . . l'essence d'ail . . . contiennent simplement du carbone, de l'hydrogene et du soufre . . . serait le monosulfure" ["with water, bulbs of garlic (*Allium sativum*) gave a volatile oil, with a strong and disagreeable odor. The essence of garlic contained carbon, hydrogen and sulfur . . . it was most likely a monosulfide"]. While it is not directly referenced in the book, Cahours is likely describing work published in 1844 by the German chemist Theodor Wertheim on the preparation and characterization of oil of *A. sativum,* described by him as "allylschwefel."

Cahours goes on to state that with colleague Hofmann, he had been able to synthesize garlic oil from "alcool acrylique," a compound they had recently discovered. The full import of Cahours and Hofmann's work became apparent in their 1857 paper in the *Philosophical Transactions of the Royal Society of London.* Hofmann was none other than August Wilhelm von Hofmann (Figure 3.2), a giant in the history of chemistry. A brilliant researcher in Germany, he was appointed by Prince Consort Albert

Figure 3.2 Professor August Wilhelm von Hofmann, Director of the Royal College of Chemistry of London and founder of the German Chemical Society. (Courtesy of http://www.encyclopedia.com/doc/1E1-HofmannA.html).

of England as first Director of the Royal College of Chemistry in London, now Imperial College. He served from 1845 to 1864, wrote a very influential book on general chemistry and was the teacher of Sir William Perkin, founder of the dye industry. Hofmann returned to Germany in 1864, founding the German Chemical Society and the German chemistry journal *Berichte*. In the 1857 paper, "Researches on a New Class of Alcohols," Hofmann and Cahours describe the first preparation of allyl alcohol, a simple unsaturated alcohol, $CH_2{=}CHCH_2OH$, and certain of its derivatives. In a footnote they explain that the earlier name given to this compound, acrylic alcohol, which relates to acrolein and acrylic acid (today a component of acrylate polymers), was changed to allyl alcohol consistent with the naming adopted in an 1844 paper by Wertheim.

In his paper Wertheim attributes garlic's appeal to "the presence of a sulfur-containing, liquid body, the so-called garlic oil. All that is known about the material is limited to some meager facts about the pure product which is obtained by steam distillation of bulbs of *Allium sativum.* Since sulfur bonding has been little investigated so far, a study of this material promises to supply useful results for science." Wertheim suggests for his garlic oil, distilling at 140 °C and analyzing for 63.33% carbon and 8.80% hydrogen, the name "allyl sulfur," which would eventually become diallyl sulfide. In this way the garlic connection to the common chemical name "allyl" was established (Wertheim, 1844).

The Hofmann and Cahours paper goes on to develop fully the chemistry of allyl alcohol (now known to be a biologically active component of garlic oil and other garlic preparations) and various derived compounds, and to describe the first synthesis of an important component of the essence of garlic, here reproduced (Hofmann, 1856; Cahours, 1856; Hofmann, 1857):

Sulfide of Allyl (Essence of Garlic)

When iodide of allyl is allowed to fall drop by drop into a concentrated alcoholic solution of protosulfide of potassium, a very energetic action ensues, the liquid becomes very hot, and an abundant crystalline deposit takes place of iodide of potassium. It is important that the iodide of allyl should only be added gradually, to avoid spirting, by which a part of the product would be lost. As soon as the action ceases, the liquid is mixed with a slight excess of sulfide of potassium; on the addition of water a light yellowish limpid oil separates, possessing a strong smell of garlic. When rectified, this liquid becomes colorless; it boils at 140 °C . . . on analysis sulphide of allyl gave 63.3% carbon and 8.9% hydrogen.

It is impressive that the above data, obtained more than 150 years ago by Wertheim and by Hofmann and Cahours, are in excellent agreement with known values for diallyl sulfide, a six-carbon, one-sulfur molecule ($C_6H_{10}S$; boiling point 138 °C; analysis: 63.10% carbon, 8.83% hydrogen), formed according to the displacement reaction shown in Equation (1).

$$2CH_2{=}CHCH_2I + K_2S \rightarrow CH_2{=}CHCH_2SCH_2CH{=}CH_2 + 2KI \quad (1)$$

In our discussion of the work of Wertheim, Cahours and Hofmann, we have touched upon most of the techniques of experimental organic chemistry that are still used today: isolation and purification of a natural product (*e.g.*, garlic oil) using separation methods (*e.g.*, distillation); determination of the formula of the natural product (*e.g.*, by elemental analysis); and confirmation of the structure by chemical synthesis, with analysis of the synthetic product to establish identity with the natural product. Of course, as will be discussed below, knowledge gained during the ensuing 150 years has allowed us to greatly refine each of the above techniques and extract valuable additional information at every step of the way.

While the work of Wertheim, Cahours and Hofmann on *Allium* chemistry occurred in the early days of development of the field of organic chemistry, there are even earlier, less specific descriptions of efforts to determine the chemical makeup of garlic and onion. For example, in *Flora Española*, a Spanish work dating from 1762, "el ajo [garlic] en la analysis chymica" is discussed, including mention of its acidic properties, smell and taste, descriptions of further investigations involving heating in a retort [alembic], and listings of various prescriptions made from 20 or 30 heads of garlic along with other traditional ingredients (Quer, 1762). On garlic, an 1814 book says: "all parts of the plant, but more especially the roots, have a strong, offensive, very penetrating and diffusible smell, and an acrimonious, almost caustic, taste. The root is full of a limpid juice, of which it furnishes almost a fourth part of its weight by expression. It also loses about half its weight by drying, but scarcely any of its smell or taste. By decoction [boiling in water] its virtues are entirely destroyed; and by distillation it furnishes a small quantity of a yellowish essential oil, heavier than water, which possesses the sensible qualities of garlic in an eminent degree. Its peculiar virtues are also in some degree extracted by alcohol and acetous [acetic] acid...The alcoholic extract was unctuous [oily] and tenacious, and precipitated metallic solutions. But the active ingredient was a thick, ropy, essential oil, according to Hagan, heavier than water, not amounting to more than 1.3 [%] of the whole, in which alone resided the smell, the taste, and all that distinguishes the garlic" (Thornton, 1814). The above description of the chemical properties of garlic and its active ingredient allicin from almost 200 years ago turns out to be remarkably accurate based on current knowledge. This same book says of onion that "it possesses in general the same properties as the garlic, but in a much weaker degree…By distillation the whole flavour of the onions passed over, but no oil could be obtained."

In 1891, the German chemist F.W. Semmler found that yellow-colored distilled oil of garlic, formed to the extent of 0.09% from distillation of garlic bulbs, when fractionally distilled at a pressure of 16 mm gave 60% diallyl disulfide, a six-carbon, two-sulfur molecule, $C_6H_{10}S_2$, or more precisely $CH_2{=}CHCH_2SSCH_2CH{=}CH_2$, with smaller amounts of diallyl trisulfide ($C_6H_{10}S_3$; $CH_2{=}CHCH_2SSSCH_2CH{=}CH_2$) and tetrasulfide ($C_6H_{10}S_4$; $CH_2{=}CHCH_2SSSSCH_2CH{=}CH_2$), collectively diallyl polysulfides, as well as allyl propyl disulfide. The diallyl sulfide reported by Wertheim was absent (Semmler, 1891). In contrast to Wertheim, Semmler used vacuum distillation (that is, distilling at 16 mm, rather than at atmospheric pressure or 760 mm mercury) to lower the temperature needed for distillation, thereby avoiding decomposition of the temperature-sensitive garlic oil components. The diallyl sulfide reported by Wertheim was most likely formed by decomposition of the diallyl polysulfides at the higher distillation temperatures he used. Semmler also characterized the distilled oil of onion as having the formula $C_6H_{12}S_2$, structurally unspecified by him, but suggested by others to be 1-propenyl propyl disulfide. For the onion oil analysis, 5000 kg of onions were steam distilled, affording only 233 g (0.005%) of oil! Today the steam-distilled oils of garlic and onion are items of commercial importance for use in food, health and agricultural products, as discussed in Chapters 5 and 6.

3.3 THE RENSSELAER CONNECTION: ISOLATION OF ALLICIN FROM CHOPPED GARLIC

It took another 50 years for the next major advance in *Allium* chemistry to come along. In 1944, chemist Chester J. Cavallito (Figure 3.3), working at Winthrop Chemical Company in

Figure 3.3 Dr. Chester J. Cavallito in his laboratory in 1947 purifying thiosulfinates using a custom-designed "molecular still". (Courtesy of C.J. Cavallito).

Rensselaer, New York, isolated a product from chopped garlic that had the formula $C_6H_{10}S_2O$, more specifically $CH_2{=}CHCH_2S(O)SCH_2CH{=}CH_2$ (**2**; Scheme 3.1), and was named allicin. Rather than steam distilling chopped garlic, Cavallito extracted the product from the garlic with ethanol, and then carefully evaporated the ethanol at temperatures below 50 °C using a vacuum pump. Thus, from four kilograms of ground garlic cloves, extracted with five liters of ethyl alcohol, followed by workup, Cavallito ultimately isolated six grams (0.15% yield) of a colorless liquid with a pungent odor that could not be distilled without decomposition. The oil had a density of $\mathbf{d}_4^{25}$ 1.1 (slightly more dense than water), was soluble to the extent of 2.5% in water,

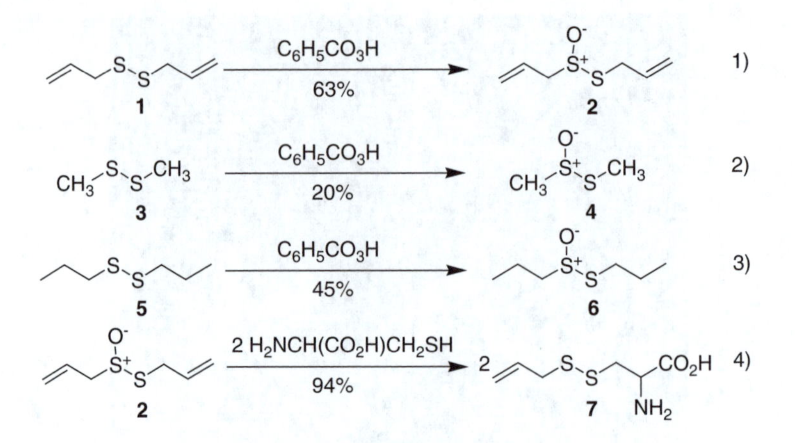

Scheme 3.1 Synthesis of thiosulfinates **2**, **4**, **6**; reaction of cysteine with allicin (**2**) giving **7**.

had a molecular weight and elemental analysis consistent with the formula $C_6H_{10}S_2O$ (calculated molecular weight: 162) and was optically inactive (the significance of the lack of optical activity for allicin will be discussed at a later point).

This product was different from that produced by vacuum distillation, diallyl disulfide (**1**) and higher polysulfides, reported by Semmler. Cavallito proved the structure of allicin by synthesis through addition of an oxygen atom to **1** using the oxidizing agent perbenzoic acid ($C_6H_5CO_3H$; Scheme 3.1). Allicin (**2**) is a *thiosulfinate* or mono-oxidized disulfide, –S(O)S– (for detailed information on organosulfur compounds see the monograph by the author; Block, 1978). Other disulfides, such as dimethyl disulfide (**3**) and di-*n*-propyl disulfide (**5**), also underwent easy oxidation to the corresponding thiosulfinates **4** and **6**, compounds that both play important roles in *Allium* chemistry. Cavallito observed that thiosulfinates **4** and **6** could be readily distilled under vacuum and appeared to be considerably more stable than allicin (**2**), and concluded: "The particular instability of the allyl compound appears to be associated with the double bond structure" (Small, 1947). In contrast to allicin, thiosulfinate **4** is completely soluble in water. Other thiosulfinates besides allicin are formed on crushing garlic. Cavallito was granted U.S. patents for his process of extracting garlic and synthesizing allicin (Cavallito, 1944a,b, 1945, 1950, 1951; Small, 1947, 1949).

Cavallito also made the very important observation that both natural and synthetic allicin showed significant antibacterial activity, comparable in some cases to penicillin. This observation provided confirmation for the very extensive medicinal literature on garlic suggesting that it "killed germs." He also demonstrated that, "like penicillin and some other antibiotic substances [allicin] . . . is inactivated by [the amino acid] cysteine" giving *S*-allylmercaptocysteine (**7**; Scheme 3.1). Allicin possesses a wide range of biological activity, as will be discussed in Chapter 5.

Additional experiments by Cavallito showed that allicin is a secondary compound formed by enzymatic action on precursors in the intact bulb. Thus, when garlic cloves were frozen in dry ice, pulverized, and extracted with acetone, "the acetone extracts upon evaporation yielded only minute quantities of residue and no sulfides [or allicin], indicating the absence of free sulfides [or allicin] in the plant. The [white] garlic powder had practically no odor, but upon addition of small quantities of water, the typical odor was detected and the antibacterial principle [allicin] could be extracted and isolated. This demonstrates that neither [allicin] nor the allyl sulfides found in 'Essential Oil of Garlic' are present as such in whole garlic. When the powder was heated to reflux for thirty minutes with a small volume of 95% ethanol, no activity could be demonstrated by addition of water to the insoluble residue. When, however, a small quantity (1 mg mL^{-1}) of fresh garlic powder was added to the alcohol insoluble fraction in water (20 mg mL^{-1}), the activity of the treated sample was shown to be equal to that of the original untreated powder. The 95% ethanol treatment has inactivated the enzyme required for cleavage of the precursor, and addition of a small quantity of fresh enzyme brought about the usual cleavage" (Cavallito, 1945).

Since a strict definition of "essential oil" is "a volatile substance contained in certain aromatic plants that imparts a distinctive odor to the plant," and since the volatile substances of *Allium* species are not found in the plant but result from disrupting the plant cells, it is preferable to refer to "distilled oil of garlic" rather than incorrectly calling it an "essential oil." Allicin accounts for 70% of all thiosulfinates found in freshly chopped or crushed garlic, amounting to approximately 0.4% of fresh weight. At room temperature, the half-life (the time period required for 50% decomposition) of pure allicin from crushed garlic is about 2.5 days; in water at 23 °C at concentrations from 0.01 to 0.1 %, the half-life is about one month, while at –70 °C allicin is indefinitely stable (Lawson, 1998). A more recent study of the stability of 0.1–0.2% solutions of allicin in water indicates half-lives of about a year at 4 °C, 32 days at 15 °C and only one day at 37 °C, and reveals that ajoene (see Chapter 4) is the main product formed when allicin decomposes (Fujisawa, 2008a,b). In general, solvents such as water that hydrogen-bond to the oxygen of allicin retard its decomposition (Vaidya, 2008).

One of the best methods for cleanly isolating allicin and other thiosulfinates from the bulk of carbohydrates and proteins present in crushed garlic involves the technique of supercritical fluid extraction (SFE), using carbon dioxide as a liquid under pressure and at a temperature no higher than 35 °C in pressurized apparatus (Rybak, 2004).

To assist in following the next sections of this chapter, the processes that occur when garlic and onion are cut are represented diagrammatically in Scheme 3.2. A more detailed discussion will be found in Chapter 4. As is indicated, the precursor compounds, designated as "alliin" in garlic and "isoalliin" in onion, are exposed to and then cleaved by the *Allium* alliinase enzymes. Enzymatic cleavage leads in both cases to very short-lived chemical species ("intermediates"), which, through different chemical processes, afford either allicin in the case of garlic or the lachrymatory factor (LF) in the case of onions. Further reactions of allicin give rise to the diallyl polysulfides found in garlic oil. In a similar manner, the LF and related compounds formed from the intermediates in onion chemistry give the characteristic polysulfides found in onion oil.

Special conditions are needed to produce and remove the maximum possible amount of allicin that can be formed from a garlic clove. It is necessary to finely chop the garlic by hand and then use a tissue grinding apparatus to ensure the maximum enzymatic yield and the uniform maceration and homogenization of the pieces of vegetable tissue. It is even necessary to chill the tissue grinder prior to and during the grinding process with an ice bath to avoid overheating the alliinase enzyme (see below). Using the above methods, elephant garlic (*Allium ampeloprasum*) was found to yield 0.073% allicin, *e.g.*, one quarter of the amount found in normal garlic.

With specialized types of mass spectrometers that can sample at atmospheric pressure rather than under high vacuum conditions, as

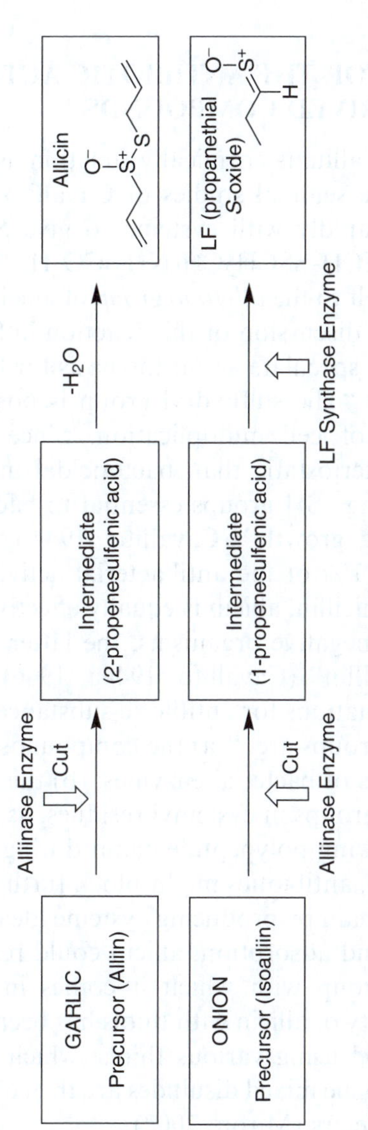

Scheme 3.2 Schematic of processes which occur on cutting garlic and onion.

is usually the case, allicin and the intermediate 2-propenesulfenic acid formed from freshly cut garlic can be directly detected. This will be discussed in detail in Section 4.5.4.1.

3.4 THE BASIS FOR THE ANTIBIOTIC ACTIVITY OF *ALLIUM*-DERIVED COMPOUNDS

To understand how alliums chemically function as antibiotics, we need to consider the seminal studies of Cavallito. He discovered that allicin reacts rapidly with cysteine to give *S*-allylmercapto-cysteine ($\mathbf{CH_2{=}CHCH_2S}SCH_2CH(NH_2)CO_2H$; **7**), that is, the cysteine attaches itself to the *allylthio group* of allicin (bold-faced in formula; see further discussion of this reaction in Section 4.5.4.2). Cavallito goes on to speculate about the basis for the antibacterial properties of garlic: "The sulfhydryl group is postulated to be a specific stimulator of cell multiplication. Since allicin is considerably more bacteriostatic than bactericidal in action, it may operate by destroying –SH groups essential to bacterial proliferation, thus inhibiting growth" (Cavallito, 1944b,c; Small, 1947). Allicin show about 1% of the antibacterial activity of penicillin. However, unlike penicillin, allicin is equally effective against Gram positive and Gram negative organisms; the latter are "practically unaffected by penicillin" (Cavallito, 1944a, 1946). Cavallito suggests that two explanations for antibiotic substances such as allicin involving the –SH groups are: "(a) the compounds may react with essential –SH groups of bacterial enzymes; (b) the compound may react with the –SH groups in cysteinyl residues, as these are joined at the end of a growing polypeptide chain during protein anabolism. In this manner, antibiotics might block further growth of the protein along that chain by producing cysteine 'dead ends.'" Based on its diffusibility and adsorption, allicin could react readily with almost any –SH group with which it comes in contact. More recently, the reactivity of allicin with thiols has been developed into an analytical method using various thiols, which gives mixed disulfides with allicin. The mixed disulfides are then easily detected by ultraviolet spectrometers (Miron, 2002).

The higher antibacterial activity and specificity of the slowly –SH-reactive antibiotic penicillin as compared with rapidly reactive allicin may be due to the loss of allicin by hydrolysis and reaction with nonessential –SH groups of structural proteins

(Cavallito, 1946). The high permeability of allicin through membranes may greatly enhance the intracellular interaction with thiols (Miron, 2000). Allicin, as well as related symmetrical thiosulfinates containing methyl or *n*-propyl groups, inhibited growth of various strains of bacteria and yeast at micromolar concentrations (Small, 1947). Cavallito also synthesized compounds known as thiosulfonates (including $CH_3SO_2SCH_3$, methyl methanethiosulfonate, with one more oxygen than a thiosulfinate), which were also found to be similar in antibacterial activity to the corresponding thiosulfinates and to react with comparable ease as thiosulfinates with cysteine (Small, 1949). Allicin reacts with thiamin giving allithiamin, more easily intestinally absorbed than thiamin itself (Fujiwara, 1958).

3.5 THE BASIS FOR THE PUNGENCY OF CUT ALLIUMS

Sliced raw garlic placed on the tongue and lips elicits a painful burning sensation. Similar effects occur with sliced raw onion, hot pepper, horseradish, ginger, mustard, wasabi and cinnamon, among other examples of spicy food. Cut raw garlic can also cause irritation and inflammation of the skin, cornea and mucosa. What is the mechanism for the above effects?

Nociceptors (noci- is derived from the Latin for "hurt") are neurons that repond to painful stimuli. When exposed to damaging stimuli, such as from heat or irritating chemicals from cut alliums, they send signals that can trigger inflammation, which causes the perception of pain. Nociceptors are silent receptors and do not sense normal stimuli. Only when activated by a threatening stimulus do they invoke a reflex. In mammals, nociceptors are found in any area of the body that can sense pain either externally (in the skin, cornea and mucosa) or internally (in the digestive tract). Chemical nociceptors employ transient receptor potential (TRP) ion-channel proteins such as TRPA1 and TRPV1, which can be activated by variety of spices commonly used in cooking. Activated TRPA1 mediates the flow of calcium ions into the endings of specialized neurons, commonly found in the mouth and skin. The ion flow excites the neurons resulting in local inflammation and pain. While many chemicals activate the ion channels by readily reversible binding, it has been found that compounds from garlic, onion, mustard and other particularly spicy foods covalently attach themselves to the thiol group of cysteine amino acid residues on the channel protein (Caterina, 2007; Brône, 2008; Salazar, 2008). A similar effect is found with environmental irritants such as acrolein, an irritant from car exhaust fumes and polluted air, as well as tear gases, which cause instant pain and irritation of the eyes, resulting in excessive tearing.

Many TRPA1 stimuli are molecules with a full or partial positive charge, called "electrophiles" (literally, *electron-lovers*). Such is the case with the onion lachrymatory factor (LF) as well as allicin from garlic. Both compounds are known to be reactive towards thiols such as cysteine (Yagami, 1980; Bautista, 2005, 2006; Hinman, 2006). A second member of the TRP family, TRPV1, has also been found to react with allicin and fresh garlic and onion extracts, as determined by studies with genetically modified, TRPA1- and TRPV1-deficient, mice (Salazar, 2008). Experiments suggest that TRPA1 is at least ten times more sensitive to garlic and allicin than is TRPV1, although the observed differences between the two ion channels seem to vary with the assay method. In these assays, diallyl disulfide and other polysulfides were active, although slightly less so than allicin; di-*n*-propyl disulfide, diallyl sulfide, alliin (the allicin precursor), as well as baked garlic (allicin decomposed) were all inactive (Bautista, 2005; Macpherson, 2005; Bandell, 2007). The reported difference in reactivity of diallyl disulfide and di-*n*-propyl disulfide (Bautista, 2005) is surprising since the difference in electrophilic character of sulfur in these two compounds should be quite small. It is likely that the reactivity of diallyl disulfide actually reflects the presence of diallyl trisulfide and higher polysulfides present in commercial samples of diallyl disulfide. Indeed, a more recent paper gives the relative concentrations required to activate TRPA1 as 254, 7.55, 0.49 and 1.47 μM for diallyl sulfide, disulfide and trisulfide and the mustard ingredient, allyl isothiocyanate ($CH_2{=}CHCH_2N{=}\mathbf{C}{=}S$), respectively (Koizumi, 2009). Thus, diallyl trisulfide is 15 times more active than diallyl disulfide and 3 times more active than allyl isothiocyanate toward TRPA1. Given the structural similarity of allyl isothiocyanate, which is readily attacked at the isothiocyanate carbon (bold-face **C** in structure) by thiols, and the onion LF ($C_2H_5CH{=}S{=}O$), it is quite likely that TRPA1 react with the onion LF in a manner similar to that postulated for allyl isothiocyanate (adding to the C=S group).

Garlic extracts have also been shown to mediate vasodilation through activation of TRPA1. It remains to be established whether this mechanism contributes to the systemic hypotensive activity of garlic *in vivo* (Bautista, 2005). TRPA1 is well conserved across the animal kingdom, with likely orthologs from human to nematode, which suggest an ancestral role for this channel, probably in sensation (García-Añoveros, 2007). The pungency of garlic and other alliums, associated with secondary metabolites such as allicin, other thiosulfinates and the onion LF, has most likely evolved as a defense mechanism to protect *Allium* bulbs against herbivorous predators, although rigorous proof of this hypothesis is lacking. The fact that many species, including European starlings, ticks, mosquitoes and nematodes, are repelled by garlic is consistent with the idea of a defensive role for the metabolites (Macpherson, 2005).

3.6 THE BASIS FOR THE STRONG ODOR OF CUT ALLIUMS

The near-universal description of alliums is that "they smell strongly when cut." The aroma is powerful, lingering and detectable even with extensive dilution. Humans are not the only primates that are very sensitive to the odor of the types of low-molecular weight, organic sulfur compounds formed by cutting alliums. Spider monkeys have been shown to be sensitive to levels of ethanethiol and propanethiol as low as 1 ppt and 1 ppb, respectively. It is argued that since "thiols . . . [and amines are] major products of the microbial degradation of proteins and thus of putrefaction processes, which are usually accompanied by the production of toxins, it seems reasonable to assume that primates should be highly sensitive to such compounds in order to avoid intoxication (Laska, 2007)." It has also been suggested that metal ions, which typically have a high affinity for sulfur and nitrogen compounds, may be involved in olfactory receptors in the form of metalloproteins (Day, 1978; Wang, 2003).

With respect to the affinity of metals for sulfur compounds, kitchen catalogues sell gadgets containing chunks of stainless steel that are reputed to remove the odor of onions and garlic upon rubbing your hands with the devices. However, an informal test of such a device conducted by a chemist at the request of National

Public Radio showed that they failed to eliminate the smell (NPR, 2006).

3.7 HOW ONION MAKES US CRY AND WHAT TO DO ABOUT IT

How does the LF induce tearing? It has been suggested that the it activates the nerve endings of "pain fibers" in the top layer of the cornea (Dostrovsky, 2002). When activated, the fibers send signals to the brain resulting in the sensation of pain, and at the same time signal the lachrymal gland to release tears. The same type of effect can occur when the eye comes into contact with a drop of shampoo or a squirt of lemon juice. The pain causes us to avoid further contact with the irritant, while the tearing helps wash it away. The molecular mechanism of "pain-sensing" presumably involves the TRPA1 and TRPV1 ion-channel proteins (Section 3.5).

There are numerous "old wives' tales" offering advice on how to prevent crying when chopping onions: lighting a match or candle (allegedly burning the LF) or holding unlit matches in your teeth (the nonsensical theory being that the sulfur in the match heads will attract the sulfurous LF); placing a wooden spoon or a piece of bread between your teeth, or breathing through your mouth. Enterprising companies market onion goggles with a foam seal, alleged to protect the eyes from the LF. Contact lens are also claimed to be helpful in minimizing tearing. However, even breathing the vapors of the LF can trigger crying since the nose is physically connected to the eyes by small ducts located at the end of the eyelids. More reasonable options include chilling the onion [thereby reducing the volatility of the LF, as well as decreasing the activity of the LF synthase (LFS)]; placing it in a pot of boiling water for 5 to 10 seconds (long enough for the skins to loosen and decrease the LFS activity); chopping it under water or in the presence of a cloud of steam from a nearby pot of boiling water (to dissolve the highly water soluble LF as it is formed), or chopping it under a kitchen hood or with a fan behind (to pull or blow the vapors of the LF away from you), and using a very sharp knife (a dull knife crushes the onion cells, mixing the LF precursor and enzymes).

Several U.S. patents advocate use of onion-derived LF to moisturize the eyes and treat the condition known as "dry eye"

(Siff, 2001a,b). Researchers measuring glucose levels in human teardrops in connection with diabetes studies induced tearing by having the subject chop onions (Taormina, 2007).

3.8 NO TEARS FROM NEW ZEALAND: GENETICALLY ENGINEERING THE TEARLESS ONION

The complexity of onion sulfur chemistry is due to the fact that *both* the onion LF and its precursor, the intermediate 1-propenesulfenic acid (Scheme 3.2), are formed when onions are cut, and both of these small molecules undergo diverse reactions. Under normal circumstances the levels of 1-propenesulfenic acid, from the action of the alliinase enzyme on the precursor isoalliin, are kept very low due to very rapid conversion to the onion LF by action of the lachrymatory factor synthase (LFS) enzyme. However, New Zealand researchers have found that it is possible to "silence" the LFS gene using molecular biology methods that result in the production of tearless onions (Eady, 2008). This discovery allows *in vivo* study of reactions associated just with 1-propenesulfenic acid, such as spontaneous self-condensation to the allicin isomer, 1-propenyl 1-propenethiosulfinate ($CH_3CH{=}CHS(O)SCH{=}CHCH_3$), which can then undergo a cascade of non-enzymatic reactions of the type predicted based on studies of the synthetically produced thiosulfinate, as will be discussed in Chapter 4 (Block, 1996).

Current "tearless" onion cultivars, *e.g.*, Vidalia, are achieved by deficient uptake and partitioning of sulfur and/or growth in sulfur-deficient soils. Such onions accumulate fewer secondary sulfur metabolites in their bulbs, leading to an onion that is "sweet" but, it is argued, with reduced "sensory and health qualities compared to more pungent high-sulfur cultivars" (Eady, 2008). Genetic engineering offers an alternative method of producing "tearless" onions without diminishing the levels of beneficial sulfur compounds. Thus, through use of the technique of "RNAi silencing," the LFS gene was suppressed in six different onion cultivars, resulting in onions with highly reduced levels of LFS activity in the leaves and particularly in the bulbs, where the LFS levels are highest. In these LFS-diminished plants, the levels of both isoalliin and alliinase were found to be within the normal range (4 to 13 mg g^{-1} dry weight for isoalliin; Eady, 2008). Analysis of leaf and bulb preparations from LFS-diminished plants by GC-MS showed that LF levels were reduced as much as thirty-fold and that levels of di-*n*-propyl disulfide were substantially reduced, while levels of various characteristic sulfur compounds were enhanced. Sensory evaluation of the modified onions revealed an aroma that was less pungent and sweeter than that given off by the non-transgenic counterparts, with no tearing or stinging of the eyes (Eady, 2008).

3.9 DETERMINING THE GEOGRAPHICAL ORIGIN OF ALLIUMS

Customs and regulatory officials as well as forensic investigators have sought laboratory methods to establish the country of origin of agricultural products, including garlic and onions. One such method involves determination of the trace metal profile of the agricultural product for comparison with an established database of trace metals profiles of the product from various countries. Through the use of inductively coupled plasma-mass spectrometry (ICP-MS) to analyze for 18 elements, it proved possible to identify the country of origin of garlic bulbs with high accuracy. U.S. Customs and Border Protection conducted a "Garlic Intervention" in 2002. When garlic was found by analysis to be from a country other than that claimed on entry documents, the importers were subject to antidumping duties of up to 376% of the value of the imported product, which often amounted to millions of dollars (Smith, 2005). Similar methods have been used to determine the geographic origin of onions (Ariyama, 2007) and Welsh onions (*Allium fistulosum*; Ariyama, 2004).

3.10 METABOLISM OF COMPOUNDS FROM ALLIUMS: GARLIC BREATH, GARLIC SWEAT, THE CASE OF THE BLACK-SPECKLED DOLLS, STINKY MILK AND AN ANCIENT FERTILITY TEST

With regard to garlic, William Woodville, M.D., writes in his 1793 book *Medical Botany*, considered the best work on medical herbs in its time in English: "Every part of the plant, but more especially the root, has a pungent acrimonious taste, and a peculiarly offensive strong smell. This odour is extremely penetrating and diffusive, for on the root being taken into the stomach, the alliaceous scent impregnates the whole system, and is discoverable in the various excretions...urine, perspiration, milk" (Woodville, 1793). A century earlier, British herbalist John Pechey observed: "If Garlick be applied to the soles of the feet, the breath will stink of it ...particles of the Garlick are mixed with the blood, and together with it, are brought to the lungs, and so are emitted by expiration" (Pechey, 1694). A modern interpretation of these perceptive observations is presented below.

3.10.1 Metabolism of *Allium* Compounds

When ingested, compounds from garlic are subject to metabolic degradation. What is known about the changes that occur to modify the original *Allium*-derived sulfur compounds? Such information is essential in understanding the effectiveness of the original compounds as potential biocides or useful drugs. In one of the earliest metabolic studies, the mold *Scopulariopsis brevicaulis* was found to convert diallyl disulfide, which accounts for more than 25% of garlic distilled oil, to 2-propenethiol and allyl methyl sulfide (Challenger, 1949). Presumably reduction to the thiol is followed by methylation (Equation (2)). In a more recent study, rats were orally administered a single 200 mg kg^{-1} dose of diallyl disulfide and concentrations of organosulfur compounds were followed over 15 days. The data confirmed the presence of 2-propenethiol, allyl methyl sulfide, sulfoxide and sulfone (Equation (3); Germain, 2002, 2003). Similarly, when dipropyl disulfide is administered as a single 200 mg kg^{-1} dose to rats, it is rapidly metabolized by the liver (eight hour half-life) to propanethiol, methyl propyl sulfide, sulfoxide and sulfone (Equation (4); Germain, 2008). Dipropyl disulfide is converted by rat liver cells to propyl propanethiosulfinate, PrS(O)SPr, mainly through oxidation by cytochrome P450 enzymes (Equation (5); Teyssier, 2000). Similarly, diallyl disulfide is converted by human liver microsomal cytochrome P-450 to allicin (Equation (6); Teyssier, 1999).

$$CH_2{=}CHCH_2SSCH_2CH{=}CH_2 \rightarrow CH_2{=}CHCH_2SH \rightarrow CH_2{=}CHCH_2SCH_3 \tag{2}$$

$$CH_2{=}CHCH_2SSCH_2CH{=}CH_2 \rightarrow CH_2{=}CHCH_2SH \rightarrow CH_2{=}CHCH_2SO_nCH_3\ (n = 0,\ 1,\ 2) \tag{3}$$

$$CH_3CH_2CH_2SSCH_2CH_2CH_3; \rightarrow CH_3CH_2CH_2SH \rightarrow CH_3CH_2CH_2SO_nCH_3 \ (n = 0, 1, 2) \quad (4)$$

$$CH_3CH_2CH_2SSCH_2CH_2CH_3 \rightarrow CH_3CH_2CH_2S(O)SCH_2CH_2CH_3 \quad (5)$$

$$CH_2{=}CHCH_2SSCH_2CH{=}CH_2 \rightarrow CH_2{=}CHCH_2S(O)SCH_2CH{=}CH_2 \quad (6)$$

3.10.2 Garlic Breath

A thorough study was performed to determine the identity and origin *(e.g.*, mouth *versus* stomach) of sulfur compounds following consumption of garlic (Suarez, 1999; also see Hasler, 1999; Blankenhorn, 1936). It was found that "immediately after garlic ingestion, transient high concentrations of methanethiol and allyl mercaptan [2-propenethiol] and lesser concentrations of allyl methyl sulfide (AMS), allyl methyl disulfide, and diallyl disulfide were observed. With the exception of AMS, all gases were present in far greater concentrations in mouth than alveolar [pulmonary] air, indicating an oral origin. Only AMS was of gut origin as evidenced by similar partial pressures in mouth, alveolar air, and urine. After 3 h, AMS was the predominant breath sulfur gas. The unique derivation of AMS from the gut is attributable to the lack of gut and liver metabolism of this gas *versus* the rapid metabolism of the other gases." It was also determined that subjects who chewed but did not swallow the garlic did not exhibit prolonged alveolar excretion of AMS, indicating the importance of gut absorption for the persistent [>30 hours after garlic consumption (Taucher, 1996)] exhalation of AMS. Furthermore, aggressive oral hygiene did not prevent the expulsion of AMS in the breath of individuals who swallowed the garlic.

From all of the above, it was concluded that breath odor after garlic ingestion initially originates from the mouth and subsequently from the gut. Other studies are in agreement with the above results concerning the identification of organosulfur compounds found in human breath after consumption of raw garlic, as well as dehydrated granular garlic (Laakso, 1989; Cai, 1995;

Taucher, 1996; Tamaki, 1999, 2008; Rosen, 2000, 2001; Lawson, 2005; Chen, 2007). It is likely that allicin and diallyl polysulfides are converted by plasma glutathione to 2-propenethiol, which is converted by the methyltransferase enzyme with *S*-adenosylmethionine (SAM) to AMS. The occurrence of acetone in breath following garlic consumption was interpreted as indicating enhanced metabolism of blood lipids (Taucher, 1996; Lawson, 2005). Trace quantities of organoselenium compounds, including dimethyl selenide (CH_3SeCH_3) and allyl methyl selenide ($CH_2{=}CHCH_2SeCH_3$), were also found in breath after consuming raw garlic (Cai, 1995). Following consumption of garlic supplements, *S*-allylcysteine is found in human plasma (Rosen, 2000) and *N*-acetyl-*S*-allycysteine in the urine (de Rooij, 1996). Bacteria in human saliva can convert odorless cysteine-*S*-conjugates, such as *S*-(1-propyl)-L-cysteine into stinky thiols, e.g., 1-propanethiol. This is one mechanism for degradation of alliums in the mouth and can explain the persistance of a sulfury taste in the mouth after consuming *Allium* species (Starkenmann, 2008). Allicin can easily pass through the skin, which explains the origin of garlic breath when fresh garlic is applied to the soles of the feet.

3.10.3 Hydrogen Sulfide: Stinky but Vital

Red blood cells convert diallyl disulfide and trisulfide to hydrogen sulfide (H_2S). While H_2S is normally a toxic, foul-smelling gas, when produced in the body in minute amounts, typically by enzymatic breakdown of cysteine, it plays a vital role in helping cells communicate with each other (Jacob, 2008a). It is also needed within blood vessels where it functions as a vasodilator, relaxing the vessel lining cells, causing the vessels to dilate (Figure 3.4). This in turn reduces blood pressure, allowing the blood to carry more oxygen to essential organs. Hydrogen sulfide from garlic (fresh is better than processed) has been shown to help prevent the severe damage that occurs in heart attacks (Chuah, 2007; Elrod, 2007; Mukherjee, 2009). Through use of a mouse model, researchers have shown that garlic extracts, by generating H_2S, can reduce tension, within blood vessels by as much as 72% (Benevides, 2007; Lefer, 2007). Formation of H_2S was demonstrated in the following manner: juice from freshly crushed garlic was mixed with a suspension of human red blood cells, such that the amount of

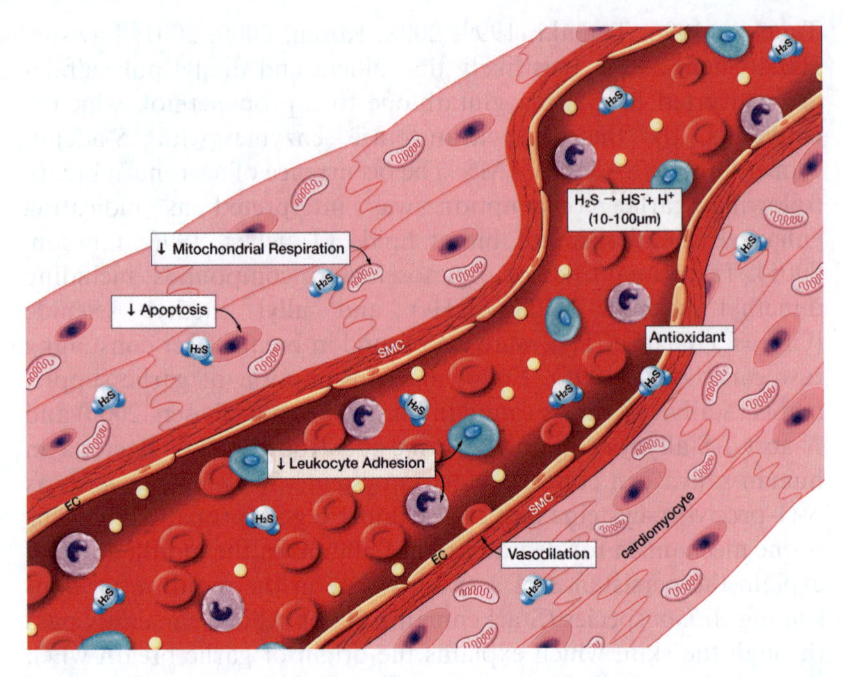

Figure 3.4 Summary of the physiological action of hydrogen sulfide protecting the cardiovascular system against disease states: antioxidant; inducer of vasodilation; inhibitor of leukocyte–endothelial cell interactions, cellular apoptosis and mitochondrial respiration. (Image reprinted from Lefer, 2007).

garlic juice in the red blood cell suspension was equivalent to two cloves of garlic in the blood volume of a typical adult human (5 L). Through the use of a very sensitive polarographic H_2S sensor, it was shown that substantial production of H_2S immediately occurred and that it occurred mainly at the surface membrane of the red blood cells. The extent of systemic H_2S generation *in vivo* after ingesting garlic is less clear since it has been shown that following garlic consumption "the concentration of H_2S in the mouth increased from about 50 to 150 ppb, whereas breath and urine concentrations remained roughly constant" (Suarez, 1999).

The abundant cellular tripeptide thiol glutathione, GSH (cytosol concentrations up to 11 mM; Valko, 2006), plays a key role in reducing the garlic-derived sulfur compounds to H_2S. Glutathione-induced formation of H_2S was greatest from diallyl trisulfide, less than half as much from diallyl disulfide, and insignificant from diallyl sulfide, allyl methyl sulfide or dipropyl disulfide (Benevides, 2007). Researchers also used intact aorta rings from rats, and under physiologically relevant levels of oxygen, exposed the rings, a model for blood vessels, to the diallyl polysulfides. Once again, production of H_2S immediately occurred. Furthermore, garlic extracts were found to reduce the tension within the vessels in a dose-dependent manner. It is significant that few plants other than garlic contain allyl-substituted sulfur compounds, and garlic and related alliums are the only ones consumed as foods (Benevides, 2007).

The levels of hydrogen sulfide normally found in human breath doubles following consumption of fresh garlic. Allyl methyl disulfide, diallyl sulfide and dimethyl disulfide are found in "garlic sweat." The odor of the sweat of garlic eaters can be particularly offensive in those rare cases where the individual is unable, due to a metabolic deficiency, to efficiently oxidize sulfides prior to excretion (see below). Through the use of a specialized type of mass spectrometer, allicin, allyl methyl sulfide and methyl vinyl sulfide were detected without any sample pretreatment in breath from healthy individuals who consumed garlic 30 minutes prior to sampling (see Section 4.5.4.3; Chen, 2007). It is also worth noting that decomposition of the onion lachrymatory factor produces H_2S, detected in onion volatiles (Block, 1992).

3.10.4 The Case of the Black-Speckled Dolls

A case that would have amused Sherlock Holmes was reported in *The Lancet,* and was the basis for a medical college valedictory address (Harris, 1986a,b). A patient who made reproduction antique china dolls complained that wherever she touched the dolls' heads when painting them, black speckles appeared after firing, requiring that those dolls' heads be rejected. It was determined that the clay was high in iron, the speckles consisted of iron sulfides, and the doll maker ate a great deal of garlic and sweated in her job. Analyses of sweat samples showed the presence of allyl methyl disulfide, diallyl sulfide and dimethyl disulfide. It was further established that the patient was a member of the small proportion of the population unable to cope with ingested sulfur compounds by the process of sulfoxidation followed by excretion of odorless oxidized sulfur compounds (Scadding, 1988). She excreted the volatile sulfides in her sweat. Thus the mystery was solved and simple solutions, abstinence from garlic or use of latex gloves, proposed.

3.10.5 Fighting Garlic Breath Naturally; Chlorophyll as a Deodorant for Garlic

Several food-based mechanisms have been proposed for neutralization of halitosis associated with *Allium* consumption (Negishi, 2002). Thiols can be removed by "enzymatic deodorization," whereby thiols bind to *ortho*-quinones. The latter are formed from natural polyphenols upon action of polyphenol oxidases and peroxidases, produced by chewing raw fruits, raw vegetables and herbs (eggplant, basil) and mushrooms (Tamaki, 2007). Natural reductase enzymes can degrade disulfides to thiols that can be trapped as above. Raw foods, which are effective in capturing disulfides, include kiwi, spinach, parsley and basil. Cooked rice, cow's milk and eggs are also effective. A physical or chemical interaction between volatile sulfur compounds and foods can involve an affinity to hydrophobic food compounds such as lipids or trapping in porous polymers contained in food. If garlic is eaten with raw foods having deodorizing activities, not only would thiols and disulfides be removed from the mouth, but also levels of allyl methyl sulfide in the gut would be reduced.

While chlorophyll is also claimed to be effective in neutralizing the odor of garlic, this is contrary to the results of an experimental and clinical study. This study assessing chlorophyll as a deodorant toward garlic and various individual sulfur compounds and other odors concluded that water-soluble chlorophyll did not remove the smell of solutions of these compounds, even after exposure for one or more months (Brocklehurst, 1953).

3.10.6 Diagnostic Value of Garlic Breath; Forensic Significance of Garlic-like Odor

Is there any useful medical information signified by the presence of garlic breath on a patient presenting symptoms of poisoning or useful forensic evidence associated with the odor of garlic in the air? Garlic breath may signify ingestion of dimethyl sulfoxide, giving dimethyl sulfide on metabolic reduction, or arsenic, tellurium or selenium compounds, whose metabolism can give monomethyl and dimethyl arsenic acid, dimethyl selenide or dimethyl telluride, all of which possess a garlic-like smell. Acute poisoning by the organophosphate pesticide chlorpyrifos ($ROP(S)(OEt)_2$; R = 3,5,6-trichloro-2-pyridyl) can result in garlic breath (Solomon, 2007). Garlic odor in the breath, sweat and urine of individuals exposed to tellurium compounds may persist for several months. Analysis of ingested specimens by inductively coupled plasma-mass spectrometry (ICP-MS) can confirm the presence of compounds of the above elements (Yarema, 2005). Mustard gas (bis(2-chloroethyl)sulfide, a sulfur-containing chemical warfare agents) is reported to have a garlic-like odor (Meyers, 2007). Phosphine (PH_3) and arsine (AsH_3) are also reputed to have *Allium*-like odors (Bentley, 2002).

3.10.7 Stinky Milk

While it has been thought that the milk of cows becomes malodorous when cows eat wild garlic, it has been demonstrated that even inhalation of the garlicky fumes from the crushed plant in the field can be transmitted to the cow's milk (MacDonald, 1928). Vacuum treatment is used to remove garlicky odors and taste from cow's milk (Granroth, 1970). Despite the adverse effects on milk odor, there are benefits from intentionally administering garlic extracts to cows and other ruminants (cattle, sheep, goats, water buffalo, deer, *etc*.). Garlic appears to reduce methane production (a major source of greenhouse gases, of concern in global warming) without adversely affecting rumen fermentation and digestion (Karma, 2006; Patra, 2006).

3.10.8 An Ancient Fertility Test

Garlic breath is the basis for a fertility test used by the ancient Greeks and Egyptians. Thus, Hippocrates describes a test to determine whether or not conception was possible: " . . . clean thoroughly a head of garlic, snip off the head and insert it into the womb; and on the following day see whether she smells of garlic through her mouth; if she smells of it, she will conceive; if not she won't" (Longrigg, 1998). A similar procedure was used in Egyptian medicine. While today we view this test as irrelevant to the question of a woman's fertility, none the less this test would likely give a positive result in most cases, because allicin, produced by cutting the garlic clove, would enter the woman's bloodstream following contact with her body and would eventually be detected in her breath.

3.11 ALLIUMS IN ART: DYEING WITH ONION SKINS AND GARLIC GLUE IN GILDING

From ancient times, the papery, dried outer skins of onions known as "tunics" have been used to provide yellow coloration for textiles and Easter eggs (Cannon, 2003; Slimestad, 2007). Onion skins are probably the best known of natural dyes and are very simple to use, with color release after only a short simmer in a pot of hot water (preferably acidified with vinegar); bonding of the dye with fibers is complete after 20 minutes (Figure 3.5). Variations in the depth of the color achieved depend mainly on the quantities of skins used,

Figure 3.5 Assorted yarns dyed with white and red onion skin. (Image reprinted from Rudkin, 2007).

the strongest colors being obtained with generous use of the skins. Thus, use of twice the weight of skins to wool gives a pinkish-brown color with no mordant (chemicals that combine with the dye and the fiber), bright orange with stannous chloride and alum (potassium or ammonium aluminum sulfate), orange-tan with potassium dichromate, and darker brown with copper sulfate. Lighter hues are obtained with a bath one fifth the strength: pale yellow with no mordant, bright yellow with alum, yellow-orange with stannous chloride, pale khaki with potassium dichromate and brown with copper sulfate. Use of the green parts of an onion imparts an undesirable smell to the wool that is difficult to remove (Cannon, 2003).

Deeper, redder shades are produced when red onion skins are added to white, while over-dyeing with indigo, woad or logwood produce attractive green shades. Onion skin can be stored in a dry, dark place for a very long time, remaining viable as a source of dye (Rudkin, 2007). Another use of onion skin color has been suggested: some cooks add yellow onion skin to broths and stews to impart an appealing golden yellow color.

An old recipe for dyeing Easter eggs involves wrapping eggs in the outer dry skins of bulbs of onions, which are tied in place with thread, and then boiling the eggs. The result is "a marbled or mottled pattern . . . in various shades of brown, orange and yellow." This traditional recipe intrigued dye chemist Arthur George Perkin, son of Sir William Perkin, who in 1896 investigated the nature of the onion skin pigments (Perkin, 1896). Perkin indicates that the brownish-orange outer skins of onion bulbs "have long been used for dyeing . . . woolen, linen and cotton materials . . . [with cotton] they give a cinnamon-brown with acetate of aluminum, a fawn with alumina and iron, a grey with iron salts, and a variety of shades with other additions." The metal salts serve as mordants. Perkin suggests that in the case of eggs, the lime of the shell acts as a mordant. It is suggested that using different mordants when dyeing with onion skins has little effect on the color achieved with silk and cotton but makes a significant difference to wool (Rudkin, 2007).

Perkin describes his experiments to verify the presence of a pigment in onion-skins and to extract it: "A piece of ordinary striped mordanted calico was dyed for about ten minutes at a boiling heat with onion skins, when the aluminum mordant became a full bright yellow, the iron mordant a dark greenish-olive. Samples of wool mordanted with chromium, aluminum, tin, and iron [salts], were also dyed with onion skins, the colors obtained being respectively brownish-olive, yellow, bright orange, and greenish-olive." Isolation of the pigment involved "boiling the onion skins (500 grams) for one hour with distilled water (9 litres). The yellow liquor thus obtained was strained through calico and allowed to cool over-night, when the impure colouring matter was deposited in the form of a pale olive precipitate ... the average yield of dry precipitate being about 1.3 percent . . . The substance was obtained as a glistening mass of yellow needles" which analyzed for $C_{15}H_{10}O_7$ and was identified as quercetin.

It is now known that quercetin (Scheme 3.3) and other pigments found in onion skins are flavonols, complex carbohydrates present in all terrestrial plants, which accumulate in the outer cell layers exposed to sunlight, protecting the plant from damaging effects of UV radiation and hydrogen peroxide (Hofenk de Graaff, 2004) as well as acting as sources of protective antifungal compounds. As it matures, the onion bulb has one to three outer dried scales of dead tissue which are, in colored cultivars, the most highly pigmented. Quercetin itself is found in the outermost, dried brown scales, while two glucoside derivatives (flavonoids), quercetin 4′-*O*-glucoside and 3,4′-*O*-diglucoside, are found in the underlying scales. The highest concentration of quercetin and its glucosides is present in the outer rings of the bulbs, decreasing significantly towards the bulb interior. Quercetin in the dried brown scales is formed by deglucosidation of these two glucosides on the border between drying and dried brown areas.

Further oxidation of quercetin in the brown scales gives 3,4-dihydroxybenzoic acid (also known as protocatechuic acid), an antifungal agent (Scheme 3.3). This oxidation is promoted by peroxidase enzymes, which show the highest activities in outer compared to inner scales. The enzymes preferentially oxidize quercetin in preference to quercetin glucosides (Takahama, 2000). It is known that browning is important for onion bulbs to resist the pathogen *Collectotrichum circinans* (Walker, 1955). Thus, if the brown scales are removed and fleshy scales are exposed to *C. circinans*, the scales are invaded and parasitized. Furthermore, cultivars of onion that do not synthesize flavonoids have non-pigmented outer dry scales, and these white bulbs are easily

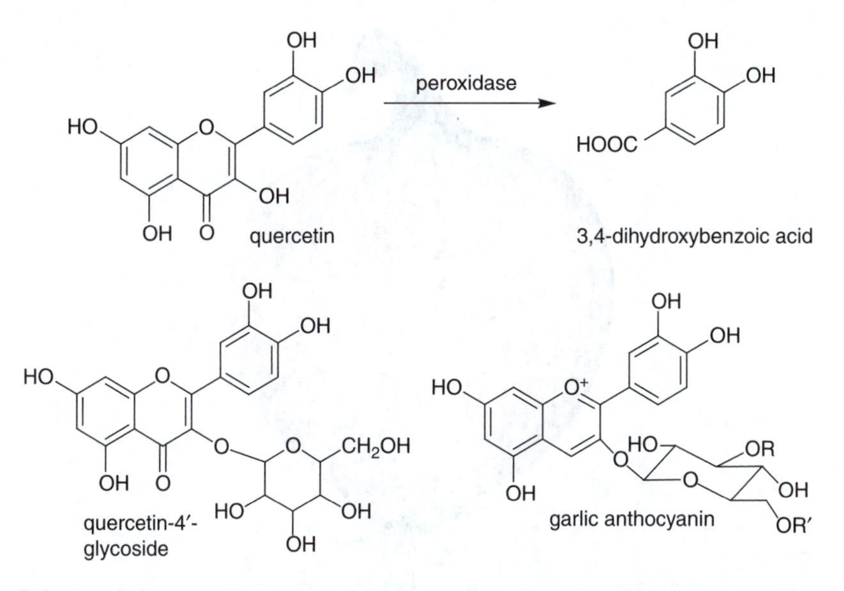

Scheme 3.3 Conversion of quercetin to 3,4-dihydroxybenzoic acid (top); structures of quercetin-4′-glycoside (bottom left) and a garlic anthocyanin (bottom right).

infected by *C. circinans* (Walker, 1955). In summary, during the process of browning and drying onion skin there appears to be a well-developed enzymatic formation of defense substances against fungal infection, involving conversion of quercetin glucosides to quercetin to 3,4-dihydroxybenzoic acid. Anthocyanin pigments structurally related to red pigments found in onion (Figure 3.6) are responsible for the red color of the inner scales of garlic; flavonols are absent in garlic cloves. (Fossen, 1997). Given the above description of the use of metal salts as mordants in dyeing with onion skins, it is not surprising that anthoxanthin pigments in onion, potatoes and other vegetables can react with metal ions from utensils used in cooking forming red, green or brown colored complexes with iron and aluminum. This is why carbon steel knives may discolor these vegetables; stainless steel knives should be used instead (McGee, 1984).

Quercetin and kampferol can be extracted from textiles of historical interest to establish the use of onion skin as dyes. Extraction, which requires the use of mild reagents such as formic acid and EDTA to avoid breaking down the characteristic glucosides, is followed by analysis by HPLC with a diode array detector or by

Figure 3.6 Halved red onion. (Courtesy of Gustoimages/Science Photo Library).

LC-MS to unambiguously identify the colorants, used to dye silk, for example (Zhang, 2005). Because of the natural antioxidants present in onion skin, compositions containing onion skin have been sold as health supplements in Japan. Garlic skin also contain antioxidants, including (*E*)-courmaric acid and (*E*)-ferulic acid [(*E*)-ArCH$=$CHCO$_2$H, where Ar$=$4-hydroxyphenyl and 4-hydroxy-3-methoxyphenyl, respectively; Ichikawa, 2003]. A substituted 4*H*-pyran-4-one, allixan, isolated from garlic (Kodera, 1989; Nishino, 1990) has been synthesized and its pharmacokinetics and accumulation on long-term storage of garlic studied. After storing garlic cloves for two years, the amount of allixin accumulated corresponded to *ca.* 1% of the dry weight of the cloves (Kodera, 2002a,b).

Another interesting use of alliums in art involves the use of a garlic-based glue in gold, silver and tin gildings in thirteenth to seventeenth century paintings, picture frames, manuscripts and furniture. Garlic has a characteristic pattern of amino acids in its proteins, and these proteins and amino acids are nonvolatile and quite stable. Through analysis of the average ratio of the constituent amino acids found in alliinase as well as other proteins using sensitive methods of amino acid analysis and statistical methods to analyze the data, it can be established that garlic rather than animal glue or egg proteins was used in centuries-old gold gildings (Bonaduce, 2006). By silylation of the amino acids formed by hydrolysis of garlic protein followed by gas chromatography-mass spectrometry (GC-MS), it was found that the amino acid composition was glutamic acid (Glu; 29%), aspartic acid (Asp; 17%), serine (Ser; 11%), alanine, glycine, valine, leucine, lysine and phenylalanine (Ala, Gly, Val, Leu, Lys and Phe; 5–6%), isoleucine, proline and tyrosine (Lle, Pro and Tyr; 2–3%), methionine and hydroxyproline (Met and Hyp; 0.5%) (Bonaduce, 2006).

3.12 ALLIUMS IN THE KITCHEN AS SPICES, HERBS AND FOODS

There is no such thing as a little garlic.

Arthur Baer

3.12.1 Introduction

The role of spices and herbs in shaping the course of history is a long and romantic story in which garlic and other alliums figure prominently. A spice may be defined as "any dried, fragrant, aromatic, or pungent vegetable or plant substance, in the whole, broken, or ground form, that contributes flavor, whose primary function in food is seasoning rather than nutrition, and that may contribute relish or piquancy to foods or beverages" (Farrell, 1990). Based on this definition, "spice" is a culinary rather than botanical term. Herbs have a more botanical definition as plants "more or less soft or succulent, mostly grown from seeds and not developing woody, persistent tissue ...[which] are usually used fresh and are considered to be a better flavouring in their fresh state" (Loewenfeld and Back, 1974). Since they are widely used as seasoning, alliums can be classified as spices and herbs despite the fact that some alliums such as onion and leek are also used as main dishes.

Humans are thought to spice foods "to take advantage of the antimicrobial actions of the plant secondary compounds that give spices their flavors... [and that] by cleansing foods of pathogens, spice users contribute to their health, survival and reproduction." For the above reasons, those living in hot climates, where the risks associated with food-borne pathogens in unrefrigerated food is greater, prefer food that is spicy (Billings, 1998; McGee, 1998). These Darwinian conclusions were drawn from an extensive study of more than 4500 meat-based recipes from 93 cookbooks from 36 countries. The authors reject the alternative hypothesis that spices are used to disguise the smell or taste of spoiled foods. An analysis of these recipes revealed that the three most widely used spices were onion, pepper and garlic (65%, 63% and 35%, respectively) and that countries with warmer climates used such spices more frequently than countries with cooler climates. Thus, onion and garlic, considered to be among the most antimicrobial spices, occur much more frequently in recipes from countries with higher mean temperatures such as Indonesia and Israel (mean temperatures 26.8 and 19.1 °C, respectively; the highest *per capita* consumption of garlic occurs in Korea and Thailand) than from countries with lower mean temperatures such as Norway and Hungary (mean temperatures 2.8 and 9.6 °C). It was hypothesized that "the benefit of spices might lie in their antimicrobial properties. The secondary metabolites and essential oils that give spice plants their distinctive flavors probably evolved to counter biotic enemies such as herbivorous insects and vertebrates, fungi, pathogens, and parasites. . .If spices kill such microorganisms or inhibit their production of toxins, spice use might reduce chances of contracting food borne illnesses or food poisoning" (Billings, 1998). A second prediction of the antimicrobial hypothesis is that "spice use should be heaviest in areas where foods spoil most quickly." This second prediction reflects the fact that the rate of bacterial growth increases dramatically with air temperature.

Mrs. Henry Perrin in *British Flowering Plants* (Perrin, 1914) writing about alliums says: "Popular as these strong flavours are, especially among the southern Latin races, it would seem that persons of refined taste have always held aloof from them." On a similar note, the earliest American cookbook (1796) mocks the use of garlic in cooking by the French: "Garlicks, tho' used by the

French, are better adapted to the uses of medicine than cookery" (Simmons, 1958).

Contemporary critics of the use of garlic in cooking complain that garlic overwhelms the taste of more subtle herbs and spices, that it is a relic from earlier times when it was needed to mask unpleasant food odors, and that garlic breath is socially offensive. A restaurateur observed, "There are lots of prejudices that people who eat and smell of garlic are second class, backward, unsophisticated. It's a class thing for many people." An Italian critic participating in an anti-garlic campaign has gone so far as to publish a guide to garlic-free restaurants. It remains to be seen how many Italians the guide will influence in a country which consumed 108 million pounds of garlic in 2006. Reports on the anti-garlic movement in Italy trumpeted, "Big Stink About Garlic" and quoted proponents saying, "What are we supposed to eat, shallots? Will that make us more elegant? More French?"

3.12.2 Onions in the Kitchen: The Effect of Cooking Temperature

The use of garlic, onion, leek, and other alliums in cooking is extensive and is well treated in general cookbooks, as well as those specializing in *Allium*-based recipes (see Bibliography). The late cookbook author, Julia Child, wrote: "It is hard to imagine a civilization without onions; in one form or another their flavor blends into almost everything in the meal except the dessert" (Child, 1966). As will be seen, there is considerable science involved in properly cutting and cooking onions and other alliums. High heat and extended cooking time bring out the worst features of onions yet they must be cooked long enough to eliminate their "bite." Strongly flavored onions can be made milder by boiling them in salted water for five minutes before proceeding with a recipe. When cutting onions (using stainless *not* carbon steel knives to avoid discolorization), larger pieces will retain more texture and more "bite" than smaller pieces, due to greater intermingling of the components causing lachrymation (tearing) when more of the onion cellular structure is disrupted. White-braised onions (glazed or lightly sautéed onions) are prepared by very slowly heating (simmering) and gently stirring the onions with liquid butter and seasonings for 40 to 50 minutes, until they are translucent and tender, without browning them. Only a very small amount of oil or butter is used for sautéing.

Brown-braised (more deeply sautéed) onions require somewhat higher temperatures and more butter or oil. The onions are tossed in the skillet for about 10 minutes to brown evenly. Simmering, even if done for just a few minutes, softens the onions and mellows the flavor, diminishing the bite of the raw onion. The oil serves both to transmit the heat and to contribute flavor. The process of browning corresponds to caramelizing (pyrolyzing and oxidizing) the onion sugars and is optimum at a temperature of about 160 °C (320 °F). The browning can be stopped when the onions are golden or continued until they turn brown. At higher temperatures the onion sugars are scorched, leading to an unpleasant bitter flavor. Because precise temperature control is important in preventing scorching, deep-frying of onions, for example to make fried onion rings, is only possible if the onions are first cloaked in a thick batter. Rapid cooking of onions or other alliums at higher temperatures allows some of the sharper flavor and texture to remain compared to slower cooking at lower temperatures (Parsons, 2001).

Quercetin and other flavonoids found in onions are antioxidants and free radical scavengers, which have anticarcinogenic and antithrombotic properties and are thought to protect against cardiovascular disease (Williamson, 1996). They have the ability to sequester copper, iron and zinc ions, which may explain their antioxidant activity (see Chapter 4.9). Quercetin 4′-*O*-glucoside and 3,4′-*O*-diglucoside account for over 85% of the total flavonoids in common varieties of onion, with another 17 flavonoids making up the remaining 15% flavonoids (Price, 1997). The total amount of flavonoids varies considerably with onion cultivars, from a low of 7 $mg\,kg^{-1}$ for white onions to 600 to 700 $mg\,kg^{-1}$ for red and gold onions, with red being slightly higher than yellow (Havey, 1999). Shallots have even higher levels of flavonoids ranging from 1000 to 1200 $mg\,kg^{-1}$ (Bonaccorsi, 2008), as do "prompt consumption onions" such as green onions. Flavonoid levels vary within the onion bulb, with the highest levels being in the outermost scales with a graduated decrease towards the center of the bulb. Postharvest field-curing of onions enhances flavonoid levels (Lee, 2008).

While onions are recognized as excellent dietary sources of flavonoids, how they are processed can determine the amount of flavonoids that are available for consumption. The majority of commercial, dehydrated onion products contain low amounts or no flavonoids, most likely due to exposure to heat during drying. Freeze-dried commercial products contain higher levels of flavonoids than the other commercial products. Losses of onion flavonoids subject to cooking are as follows: frying, 33%; sautéing, 21%; boiling, 14–20%; steaming, 14%; microwaving, 4%; and baking, 0% (Lee, 2008).

3.12.3 Garlic in the Kitchen: Crushing, Baking, Boiling, Frying, Pickling and Drying Garlic

Many cooks strongly advocate use of fresh garlic rather than garlic powder, dehydrated flakes, salt, or bottled juice in cooking, arguing that these processed forms of garlic quickly become acrid and tasteless. It has long been recognized that to get the maximum flavor benefit from a clove of garlic it has to be rigorously crushed, which efficiently mixes the flavor precursor alliin and enzyme alliinase. Crushed garlic should be allowed to "stand" for ten minutes to maximize enzymatic action before heating it (Song, 2001). The use of a garlic press, which produces crushed garlic with about three times the impact of chopped or minced garlic, has its advocates (Cavage, 1987) as well as critics (Kreitzman, 1984), with the latter arguing that a superior flavor is produced just by finely chopping and dicing garlic with a kitchen knife.

More than 300 U.S. design and regular patents have been granted for garlic presses, whose technology dates back more than 4300 years. A garlic press is designed to efficiently crush garlic cloves, typically held in a retaining pocket, by forcing them through a grate with some type of piston (Figure 3.7; Walker, 2006). Many garlic presses also have a device with a matching grid of blunt pins to clean out the holes of the grate. Garlic presses present a convenient alternative to mincing garlic with a knife, especially because a clove of garlic can be passed through a sturdy press without even removing its peel, which remains in the press while the mashed garlic is extruded. Some sources also claim that

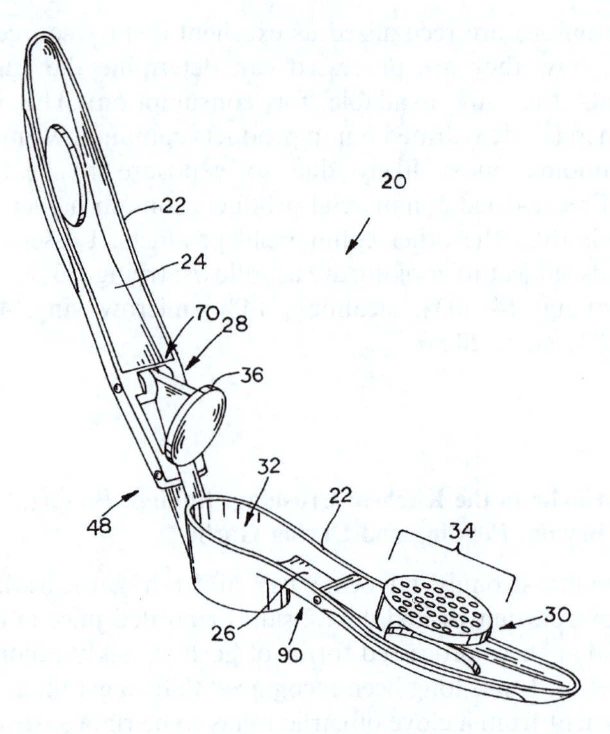

Figure 3.7 Schematic of a garlic press from a 2006 patent. (Image taken from Walker, 2006). The upper portion (labeled 24) contains a plunger (36), while the lower portion (90) contains a grate (30), which pivots (34) into the chamber (32) during crushing and back out again during cleaning.

pressing with the peel on makes cleaning the press easier. Garlic crushed by a press is generally believed to have a different flavor from minced garlic; since more cell walls are broken, more of garlic's strong flavor compounds are liberated.

The taste and aroma of baked or fried garlic is quite different from the pungent taste and powerful aroma of freshly cut raw garlic. Garlic, drizzled with olive oil and baked for 45 minutes at 400 °F (204 °C) has been described as having a sweet, mellow, nutty, caramel flavor. Chefs recommend placing "slivers of garlic cloves on meat before cooking . . . [putting] a clove of garlic on a skewer, cook it in a sauce or stew . . . drop the unpeeled cloves into boiling water. Cook for 2 minutes. Drain, peel and simmer slowly in butter about 15 minutes. Mince the blanched buttered garlic and add to sauces" (Rombauer, 1975). Laboratory study reveals additional components of garlic important for its flavor and establishes what happens to garlic when it is heated.

The best way to subtly add garlic flavor to a salad is to rub the salad bowl thoroughly with a split clove, let the bowl dry for a few minutes, and then proceed with addition of the salad ingredients. Raw crushed garlic, if sautéed until brown, becomes bitter and unpleasant; it should be carefully sautéed until it takes on a pale golden color. However, whole garlic cloves that have first been briefly placed in boiling water can be browned or even carmelized. When whole garlic cloves are roasted in their skins they become chestnut-like, sweet and buttery. Garlic can also be enjoyed when pickled. Bright red bulbs of garlic pickled with beet juice are a particular favorite in Russia (Figure 3.8). Intensely green "Laba" pickled garlic is a New Year's treat in China (see Section 4.13). Note that garlic preparations containing oil and peeled cloves *must* be acidified to pH <4.5 with vinegar or wine before storage to prevent botulism poisoning (Morse, 1990).

Garlic contains significant amounts of non-reducing water-soluble fructopolysaccharides called fructans, short-term carbohydrate

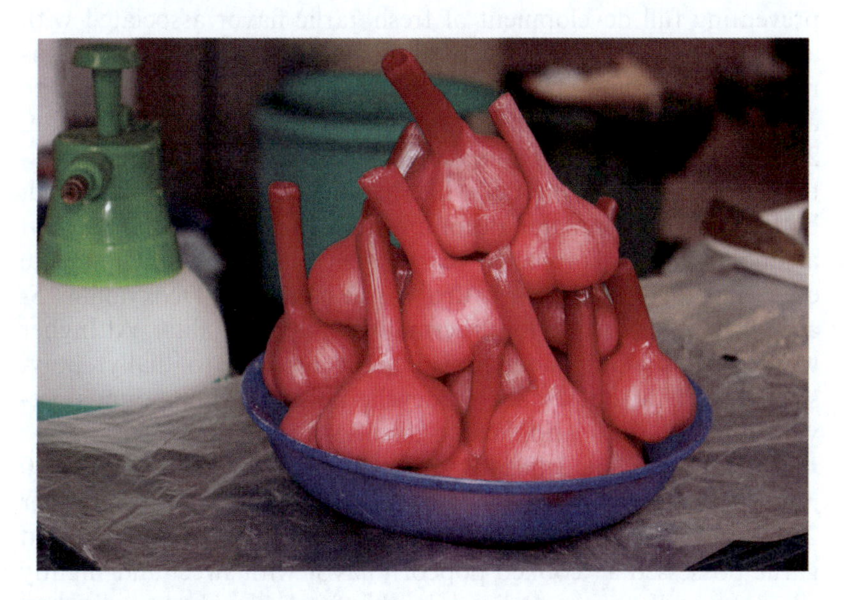

Figure 3.8 Garlic pickled with beet juice in the Kuznechny Market, St. Petersburg, Russia. (Photo by Eric Block).

storage materials that protect the plant against freezing stress. Fructans resist hydrolysis and digestion and thus represent a low caloric sweetener with effects similar to those of dietary fibers. In the colon, fructans are fermented by microflora, generating conditions which protect against *Salmonella* colonization. Fructans may also improve recovery from anemia. Starch and monosaccharides are absent in garlic; fructan and sucrose make up 96% and 4% of total nonstructural carbohydrates. Fructose and glucose are found in a 15:1 ratio. Based on the total sugar concentration in garlic cloves of 125 to 235 $mg\,g^{-1}$ fresh weight, sugar concentrations are considered to be "high" (Losso, 1997). By comparison, the alliin concentration in garlic ranges from *ca.* 5 to 10 $mg\,g^{-1}$ (Kubec, 1999).

Sugars in alliums can play an important role in flavor development through a process known as the Maillard reaction, named after the French chemist Louis Camille Maillard who investigated it in the early twentieth century. The Maillard reaction is a chemical reaction between an amino acid and a reducing sugar, usually requiring heat, and is a form of nonenzymatic browning. Drying and heating onion and garlic can promote the Maillard reaction, leading to flavor loss and color change (Cardelle-Cobas, 2005).

The alliinase enzyme from garlic is denaturated by heating, preventing full development of fresh garlic flavor associated with allicin. Thus, when drying garlic for commercial purposes it is recommended that garlic slices (optimum thickness of 10 mm) be air or vacuum dried at temperatures not exceeding 50 °C (Rahman, 2009). The effect of heat on alliin and deoxyalliin has been investigated as a way to model what happens when garlic is baked, boiled or fried. A major decomposition product of alliin is allyl alcohol, probably formed by rearrangement of alliin to a sulfenate ester that then undergoes hydrolysis (Yu, 1994a). Deoxyalliin (*S*-allylcysteine) is much more thermally stable than alliin. At higher temperatures alliin and deoxyalliin decompose to diallyl polysulfides and various cyclic and acyclic polysulfur compounds previously found on heating diallyl disulfide (Block, 1988).

Garlic cloves, blanched by immersion in boiling water for 20 minutes, were sliced and fried at 180 °C in soybean oil for another 20 minutes or baked at 180 °C for 20 to 25 minutes. The blanched garlic possessed a "cooked popcorn flavor with sweet and slightly pungent garlic character" while the fried blanched garlic had "a typical fried garlic flavor but lacked the pungency of raw garlic"

(Yu, 1994c). Pyrazines, nitrogen-containing heterocyclic compounds commonly found in cooked, roasted and toasted foods, were among the non-sulfur-containing volatiles produced by heating garlic. Thiazoles, mixed sulfur nitrogen heterocycles, were formed when alliin and glucose were heated. It is known that the Maillard reaction is a major mechanism for formation of pyrazines. Thiazoles have roasted meaty and nutty cereal and popcorn odors (Yu, 1994b).

CHAPTER 4

Chemistry in a Salad Bowl: *Allium* Chemistry and Biochemistry

Let onion atoms lurk within the bowl
and, scarce suspected, animate the whole

Sydney Smith (1771–1845)
Recipe for Salad

4.1 THE BASEL CONNECTION: ALLIIN, THE ALLICIN-PRECURSOR FROM GARLIC

Modern "*Allium* chemistry" began, as described in Chapter 3, with Cavallito's seminal discovery of allicin (**2**), his finding that diallyl disulfide (**1**) can be readily oxidized to allicin, that similar processes occur with dimethyl and dipropyl disulfides (**3**, **5**) giving methyl methanethiosulfinate (**4**) and propyl propanethiosulfinate (**6**), respectively, and that allicin readily reacts with two equivalents of cysteine giving disulfide (**7**; Scheme 3.1). The next significant event in *Allium* chemistry took place in the Basel, Switzerland laboratories of Sandoz Ltd., now a division of Novartis. Arthur Stoll (1887–1971; Figure 4.1), who was the first director of the Pharmaceutical Department and eventually President of Sandoz, as well as Professor at the University of Munich, did his doctoral work on chlorophyll chemistry with 1915 Nobel Prize winner

Garlic and Other Alliums: The Lore and the Science
By Eric Block

Published by the Royal Society of Chemistry, www.rsc.org

Photo Peter Heman, Basel

Figure 4.1 Dr Arthur Stoll (from Ruzicka, 1971).

Richard Willstätter. In 1947 to 1949, with his coworker Ewald Seebeck, Stoll turned his attention to garlic, identifying the precursor found in Cavallito's white powder as a nonprotein amino acid, *S*-allyl-L-cysteine sulfoxide (CH_2=$CHCH_2S(O)CH_2CH(NH_2)CO_2H$; **10**; Scheme 4.1). Stoll and Seebeck called this compound "alliin," using the name introduced by Carl Rundqvist, who erroneously concluded that the garlic oil precursor was a glucoside (Rundqvist, 1909).

Since this precursor readily undergoes enzymatic cleavage to allicin, extraction was carried out under conditions preventing cleavage: "Garlic bulbs are deep-frozen . . . cut into small pieces while in the frozen state . . . extracted with methanol or ethanol . . . [concentrated] at low temperature and the remaining syrupy liquid freed from fatty impurities by extraction with ethyl ether. The separation from contaminating substances . . . is completed by fractional precipitation with ethyl alcohol . . . In this way, the alliin is obtained in such a pure state that, on adding methanol to a concentrated aqueous solution, it crystallizes out in colorless needles. Final purification is effected by recrystallization from aqueous acetone" (Stoll, 1948; Stoll, 1951b). Stoll and Seebeck reported that from 1 kg of fresh bulbs of garlic, 810 mg of alliin could be obtained as fine needles, melting at 163.5 °C, and showing an optical rotation of $[\alpha]_D^{21} = +63°$. Their yield of alliin corresponds to 0.81 mg g^{-1} fresh weight (0.081%). More recent work using improved techniques shows that alliin levels of 5 to 14 mg g^{-1} fresh weight (0.5 to 1.4%) are more typical (Lawson, 1996). Seebeck was granted a U.S. Patent for alliin (Seebeck, 1953).

In addition to establishing the structure of alliin by elemental analysis, Stoll and Seebeck accomplished its synthesis from simple precursors. The synthesis starts with the chiral natural amino acid (–)-L-cysteine, which was reacted with allyl chloride giving (–)-*S*-allyl-L-cysteine (**8**; *deoxyallin*), also a natural compound. In this name the "S" indicates that the allyl group is substituted on the sulfur atom rather than elsewhere. When Stoll and Seebeck attempted to complete the synthesis of alliin in a manner similar to that used by Cavallito in his synthesis of allicin from diallyl disulfide, namely by oxidation of the sulfur to a sulfinyl (sulfoxide, S(O), S=O, or S^+–O^-) group with hydrogen peroxide, H_2O_2, they were surprised to find that although the product had the identical formula to that of alliin, its physical properties were quite different, namely a melting point of 146–8 °C and an optical rotation of $[\alpha]_D^{20} = -12°$.

To explain what had gone wrong with Stoll and Seebeck's synthesis we need to consider the stereochemistry of the molecules involved, as summarized in Scheme 4.1 While *S*-allyl-L-cysteine (**8**) shares the same stereochemical features of (–)-L-cysteine, differing only in the replacement of the hydrogen attached to the divalent sulfur by an allyl group, oxidation converts the divalent sulfur of **8** to a *trivalent* sulfur. When trivalent sulfur is surrounded by three different groups, for example, an oxygen and two differently

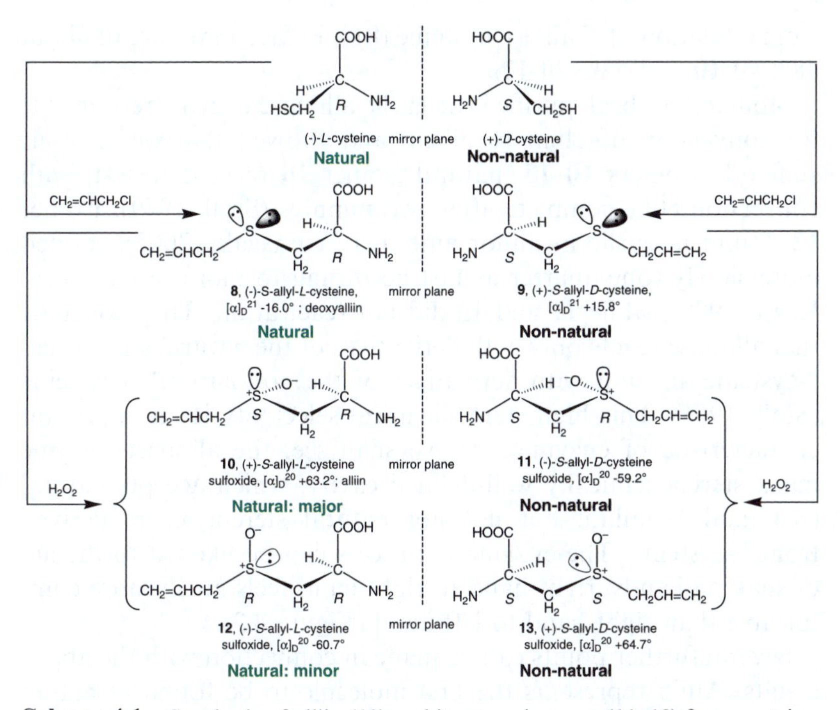

Scheme 4.1 Synthesis of alliin (**10**) and its stereoisomers (**11–13**) from cysteine. Stereochemistry at the chiral carbon and sulfur centers is described as "*R*" or "*S*" using the Cahn–Ingold–Prelog (CIP) convention, discussed below.

substituted carbon atoms, it becomes a chiral center, capable of existing in two mirror-image forms. When **8** is oxidized by a simple achiral oxidant such as hydrogen peroxide, the oxygen can be introduced from either side of the sulfur, giving a mixture of products **10** and **12**. Since the stereochemistry at the chiral carbon in **10** and **12** is identical, but that at sulfur opposite, **10** and **12** are diastereomers (non-mirror-image stereoisomers). Since diastereomeric pairs **10** and **12** are in fact different compounds rather than mirror images, they can be separated by physical and chemical methods. Similarly, hydrogen peroxide oxidation of *S*-allyl-D-cysteine (**9**), prepared from D-cysteine, gives a mixture of diastereomers **11** and **13**, and these two compounds can also be separated from each other. Of the four compounds, **10–13**, the pairs **10** and **11** are enantiomers, as are the pairs **12** and **13**. Based on the optical rotation of the pure individual compounds **10** and **12**, the product obtained by Stoll and Seebeck, with an optical rotation of –12°, from oxidation of *S*-allyl-L-cysteine (**8**) is in fact a mixture of about 48% of **10** and 52% of **12**.

Stoll and Seebeck purified the garlic alliinase enzyme responsible for conversion of alliin to allicin and showed that of the four sulfoxide isomers **10–13**, natural isomer **10** reacted fastest, with conversion 80% complete after two minutes (Stoll, 1949). Isomer **12**, found in garlic in minor amounts (Yamazaki, 2005), reacted more slowly (one-quarter as fast according to more recent work; Krest, 1999), while **11** and **13** did not react at all. They conclude that alliinase reacts only with derivatives of the natural amino acid L-cysteine and not with derivatives of the non-natural D-cysteine (Stoll, 1951). This high stereochemical selectivity is an important characteristic of enzymes. As we shall see, the alliinase enzyme has a stereochemically well-defined cavity, which accepts natural (+)- and (–)-alliin, but not non-natural stereoisomers derived from D-cysteine. The enzymatic process can be likened to the act of shaking hands: right hand to right hand feels much more comfortable than right hand to left hand (Figure 4.2).

Several further points can be made in connection with the above results. Alliin represents the first molecule to be found in nature having chirality *both* at carbon and sulfur. Stoll and Seebeck synthesized analogs of alliin with different alkyl groups on sulfur, but otherwise having the same stereochemistry at both carbon and sulfur as alliin, such as *S*-propyl-L-cysteine sulfoxide ($CH_3CH_2CH_2S(O)CH_2CH(NH_2)CO_2H$; **14**). This compound is called "propiin" and is also found naturally in *Allium* species (Stoll, 1951). On the basis of an X-ray crystallographic study of another analog methiin, (+)-*S*-methyl-L-cysteine sulfoxide ($CH_3S(O)CH_2CH(NH_2)CO_2H$; **15**; Morris, 1956), it has been determined that the absolute stereochemistry of this compound is as shown in Figure 4.3 (Hine, 1962). It is now well established that the absolute structure of alliin corresponds to **10** in Scheme 4.1. (+)-Propiin (**14**) has the same absolute configuration at sulfur as (+)-methiin (**15**; Carson, 1961b; Gaffield, 1965). Compounds **10–15** can also be named as "*S*-oxides" instead of "sulfoxides;" the latter name is used in this chapter (Bentley, 2005).

Stoll and Seebeck found that analogs of alliin, such as **14**, with the same stereochemistry at sulfur and carbon are also cleaved by alliinase (Stoll, 1951). Cleavage of **14** and **15** would be expected to give PrS(O)SPr (**6**) and MeS(O)SMe (**4**), respectively, compounds

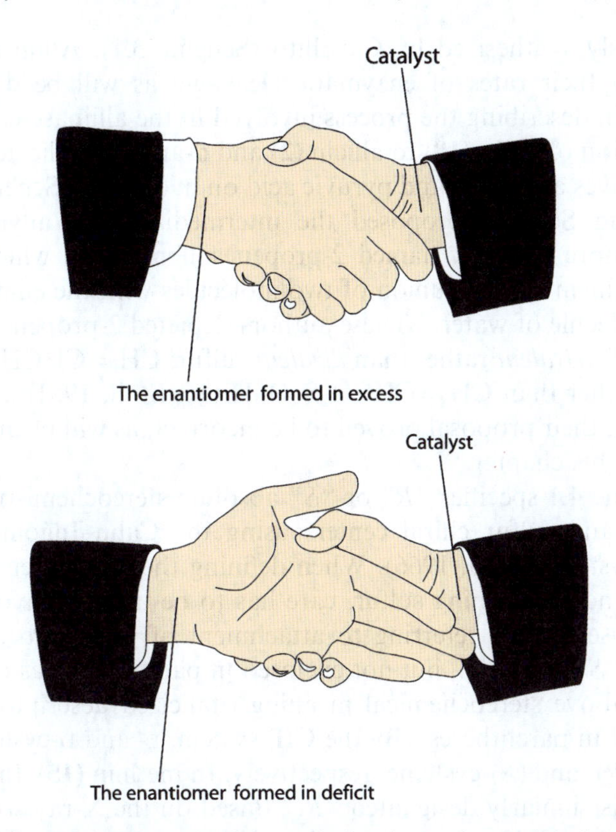

Figure 4.2 The hands on the right symbolize the catalyst and the hands on the left the products. They match better in the upper picture (the energy is lower) than in the lower picture. (Courtesy of Nobel Foundation, Nobel Prize in Chemistry 2001, Information for Public, 10 October 2001.)

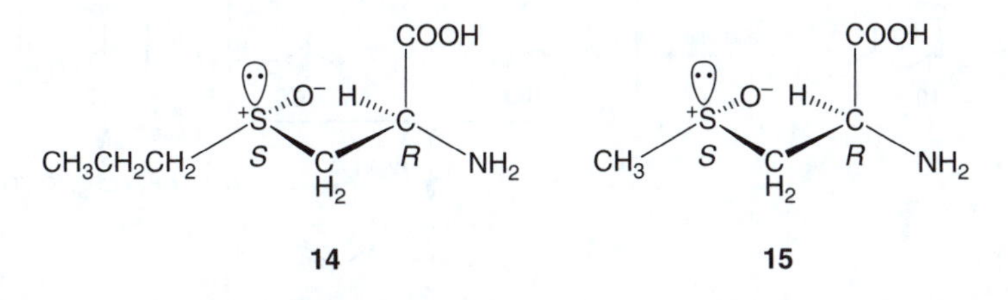

Figure 4.3 Absolute configuration of methiin, (R_CS_S)-(+)-*S*-methyl-L-cysteine sulfoxide (**15**), from the X-ray structure; configuration of propiin, (R_CS_S)-(+)-*S*-propyl-L-cysteine sulfoxide (**14**).

previously synthesized by Cavallito (Scheme 3.1). Alliin analogs differ in their rates of enzymatic cleavage, as will be discussed below. In describing the process involved in the alliinase-catalyzed conversion of alliin (**10**) to allicin (**2**) and α-aminoacrylic acid (**17**), which gives ammonia and pyruvic acid on hydrolysis (Scheme 4.2), Stoll and Seebeck proposed the intermediacy of "allylsulfenic acid", more properly named 2-propenesulfenic acid, which gives rise to allicin "by the union of two molecules with the elimination of a molecule of water." These authors depicted 2-propenesulfenic acid with *trivalent* rather than *divalent* sulfur: $CH_2{=}CHCH_2S(O)H$ (**16a**) rather than $CH_2{=}CHCH_2S{-}O{-}H$ (**16**; Stoll, 1951b). Unfortunately, their proposal proved to be incorrect, as will be discussed later in this chapter.

Scheme 4.1 specifies "*R*" or "*S*" absolute stereochemistry at all carbon and sulfur chiral centers using the Cahn–Ingold–Prelog (CIP) system (Cahn, 1966). When defining the stereochemistry of compounds containing sulfur, care has to be taken to avoid confusing use of "S" referring to attachment of groups to a sulfur center ("*S*" italicized, but not enclosed in parentheses) as opposed to the above stereochemical meaning (italicized descriptor "(*S*)" enclosed in parentheses). By the CIP system, L- and D-cysteine are called (*R*)- and (*S*)-cysteine, respectively. In methiin (**15**), the chiral carbon is similarly designated "*R*." Based on the X-ray structure, the trivalent sulfur is designated as having "*S*" stereochemistry. The full name of **15** is (R_CS_S)-(+)-*S*-methylcysteine sulfoxide. All *S*-alk(en)yl-L-cysteine sulfoxides isolated from natural sources have "*R*" carbon stereochemistry for cysteine; most have "*S*" sulfur stereochemistry.

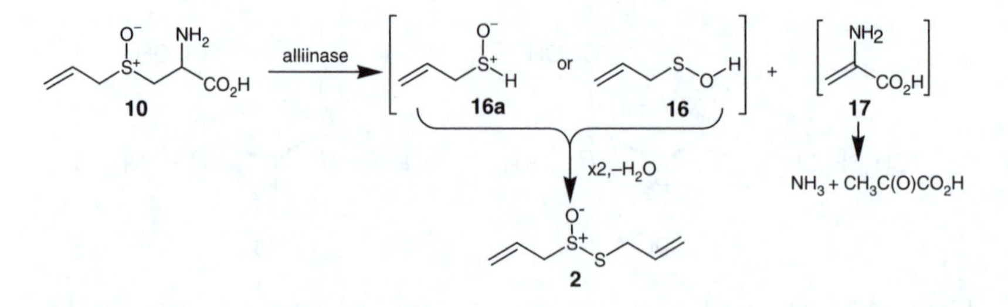

Scheme 4.2 Proposed conversion of alliin (**10**) to allicin (**2**) *via* 2-propenesulfenic acid (**16**) or **16a**.

4.2 THE HELSINKI CONNECTION: THE ONION LACHRYMATORY FACTOR (LF) AND ITS PRECURSOR, ISOALLIIN

4.2.1 Isoalliin, Precursor of the Onion Lachrymatory Factor (LF)

With key aspects of the chemistry of garlic now well understood, following the seminal studies of Cavallito and Stoll, Finnish biochemist Artturi I. Virtanen (Figure 4.4) undertook a detailed study of onion chemistry in the 1950s, seeking in particular to discover the identity of the lachrymatory factor (LF) and its precusor (LP). A brilliant scientist, Virtanen had already received the 1945 Nobel Prize in Chemistry for his work in agricultural and nutrition chemistry, when he undertook his studies on onions.

Virtanen wrote: "After some preliminary experiments it became evident that the LF would be extremely difficult to isolate and

Figure 4.4 Nobel Laureate, Artturi I. Virtanen (1895–1973; right) and colleague, Niilo Rautanen. (Photograph from the archives of Valio, Ltd, Helsinki, Finland).

characterize as such owing to its instability, probable reactivity and to the fact that it was present in onion vapours in minor amounts only together with all the other volatile constituents. Therefore it was assumed as a working hypothesis that the LF was enzymatically formed from a non-volatile and essentially stable precursor when onion is cut or crushed." To follow the isolation and purification of the precursor, Virtanen devised a very simple assay involving mixing a solution containing the onion enzyme and the solution thought to be enriched in the precursor and holding the vial "tightly, but without pressure to the eye. Depending on the concentration, the lachrymating effect was perceived after 15 to 45 seconds, usually becoming unendurable after an additional 5 seconds" (Spåre, 1963).

In his research Virtanen had access to powerful instrumental methods unavailable to earlier researchers, namely mass spectrometry (MS) and infrared (IR) spectroscopy, as well as ion exchange and paper chromatography. Mass spectrometry can be used to determine the molecular weight and elemental composition of minute samples of compounds, while infrared spectroscopy helps establish the presence of key combinations of atoms, such as sulfinyl groups, and distinguish between "$CH_3CH{=}CH{-}$" and "$CH_2{=}CHCH_2{-}$" groups. Chromatography is invaluable for separation of mixtures of amino acids, particularly those structurally related to alliin. The technique is based on differential rates of migration of compounds through a column or, by capillary action, up a piece of paper (in paper chromatography). Ion exchange chromatography involves the differential migration of ionized forms of amino acids.

In 1961, Virtanen succeeded in isolating and characterizing isoalliin (**18**; Scheme 4.3), mp 146–148 °C, optical rotation $[\alpha]_D^{25} = +74°$, as the crystalline precursor of the onion LF (Spåre, 1963). From 5 kg of fresh onions Virtanen isolated 129 mg of pure material, corresponding to a yield of 0.026 mg g^{-1} fresh weight. Virtanen was unsuccessful in synthesizing isoalliin. The subtle difference in structures between alliin (**10**) and isoalliin (**18**) needs to be emphasized: while the empirical (elemental composition) formulas for alliin and isoalliin are both $C_6H_{11}NO_3$, the C=C bond in isoalliin is directly attached to the S(O) group, while in alliin it is separated by CH_2. Attachment of the C=C bond to S(O) activates the former bond toward attack by electron-rich groups, such as

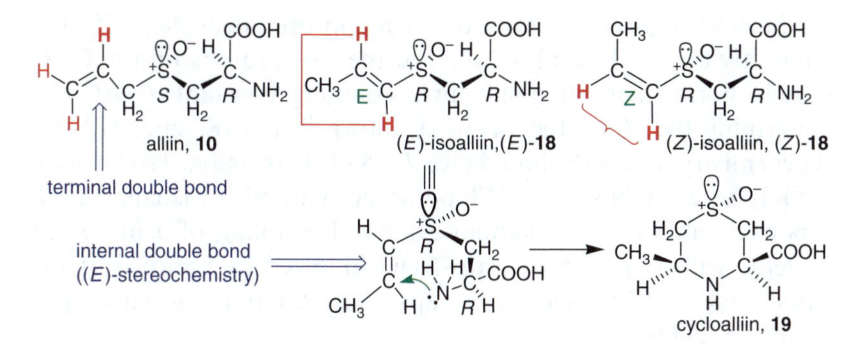

Scheme 4.3 Alliin (**10**), (*E*)- and (*Z*)-isoalliin (**18**) and formation of cycloalliin (**19**).

nitrogen, as illustrated by the spontaneous cyclization of isoalliin (**18**) to cycloalliin (**19**). Cycloalliin was previously isolated from onions (Virtanen,1959a). The full chemical name for alliin (**10**) is (R_CS_S)-(+)-*S*-2-propenylcysteine sulfoxide ("2-propenyl" is a more precise chemical name for allyl), while that for isoalliin (**18**) is (R_CR_S)-(+)-*S*-1-propenylcysteine sulfoxide. While the sulfur configuration in isoalliin, alliin and methiin are the same, the sequence rules result in an opposite descriptor for isoalliin [R_S] than for alliin and methiin [S_S].

Before discussing Virtanen's work on the LF, more needs to be said about the structure of isoalliin and how it differs from that of alliin. In 1966, USDA scientist J. F. Carson noted that in his characterization of isoalliin, Virtanen did not establish the stereochemistry of the C=C bond. With reference to Scheme 4.3, it can be seen that alliin (**10**) has a "terminal" double bond at the end of the three-carbon chain. On the other hand, isoalliin (**18**) has an internal C=C bond, which has two distinct "*E*" and "*Z*" geometric isomers. Furthermore, in isoalliin the C=C double bond is directly attached to the sulfoxide group, increasing the reactivity of the former group. Carson repeated Virtanen's work, isolating 3 g of isoalliin from 2 kg of dehydrated onion powder, corresponding to 1.5 mg g^{-1} of dry weight (0.15%). In its infrared (IR) spectrum, isoalliin show bands at 1025 and 1037 cm^{-1} for the sulfoxide group, and at 967 cm^{-1} suggesting an (*E*)-double bond (compare the –$CH{=}CH_2$ IR band for alliin at 920 cm^{-1}). The stereochemistry of the C=C bond in isoalliin was rigorously established by nuclear magnetic resonance (NMR) spectroscopy, a technique used to establish the types of protons, or other appropriate nuclei, and their connectivity. NMR analysis showed the two protons on the C=C double bond in isoalliin having a coupling constant $J = 16$ Hz, confirming the (*E*)-stereochemistry shown in (*E*)-**18**, since (*Z*)-stereochemistry in (*Z*)-**18** should give $J = 8$–9 Hz (Carson, 1963, 1966).

Only recently has a useful laboratory synthesis of isoalliin been reported, employing palladium-catalyzed coupling of a protected cysteine with (*E*)-1-bromoprop-1-ene followed by removal of the "Boc" and "Et" protecting groups and oxidation (Equation (1); Namyslo, 2006).

$$\begin{gathered}CH_3CH{=}CHBr + HSCH_2CH(NHBoc)CO_2Et + Pd(0)\\ \rightarrow CH_3CH{=}CHSCH_2CH(NHBoc)CO_2Et\\[1em] CH_3CH{=}CHSCH_2CH(NHBoc)CO_2Et + LiOH,\ \text{then acid}\\ \rightarrow CH_3CH{=}CHSCH_2CH(NH_2)CO_2H\\[1em] CH_3CH{=}CHSCH_2CH(NH_2)CO_2H + H_2O_2\\ \rightarrow CH_3CH{=}CHS(O)CH_2CH(NH_2)CO_2H\end{gathered} \quad (1)$$

4.2.2 The Onion Lachrymatory Factor (LF)

Virtanen describes the clever method he employed to generate and identify the LF using a mass spectrometer, a pure sample of isoalliin and an onion enzyme preparation that had been carefully freed from volatile, low molecular weight contaminants: "[One mg of isoalliin], the enzyme (usually 4 mg), and preferably deaerated water (usually 4 drops) were put into a small [glass] "finger" and immediately stoppered and frozen at –80 °C. At this temperature the finger was attached to the main vacuum of the mass spectrometer and vacuum pumped for 3 min. The vacuum valve was then closed and the finger warmed to 30 °C for 3 min. Then the valve to the analyzator [sic] was opened, with the pressure-guarding device disconnected, so that a relatively large sample of the volatiles entered into the mass spectrometer" (Spåre, 1963). After one minute the mass spectrum showed the presence of a compound with a pattern consistent with the formula C_3H_6SO. While numerous structures are possible for C_3H_6SO, by analogy to the earlier work of Stoll and Seebeck on enzymatic formation of allicin from alliin, Virtanen postulates that the onion LF is "propenylsulphenic acid," drawn as $CH_3CH{=}CHS(O)H$ (**20a**; Scheme 4.4). He observes that "no other aliphatic sulphenic acid has as yet been known" (Spåre, 1963). Virtanen was not the first to use mass spectrometry to obtain a mass of 90 and a formula C_3H_6SO for the onion LF. A paper with this information had already appeared in 1956 (Niegisch, 1956), although the structures proposed did not seem reasonable based on known onion chemistry.

Virtanen presents additional evidence for the $CH_3CH{=}CHS(O)H$ structure based on deuterium isotope exchange studies. When the LF was generated as described above using deuterium oxide (D_2O) instead of H_2O, the mass of the LF increased by 1 to 91, corresponding to the formula C_3H_5DSO. Here, one of the hydrogens has undergone exchange by a deuterium atom. Since a sulfenic acid structure is postulated for the LF, this should be able to exchange its acidic hydrogen (H^+) for deuterium (D^+) giving $CH_3CH{=}CHS$-(O)D (**20a**-d_1; see Scheme 4.4, with deuterium shown in red; Virtanen, 1962c, 1965).

The undeuterated LF shows a significant fragment ion at m/z 73 while the monodeuterated LF shows this same fragment ion at m/z 74. Virtanen argues, "in both instances the fragment [lost; *e.g.*, m/z 90–73 and 91–74] is OH and not OD . . . From this it can be concluded that no preformed OH exists in the molecule, because a real hydroxyl group would exchange its hydrogen for deuterium in deuterium oxide" (Spåre, 1963). That is, if the structure of the LF were $CH_3CH{=}CHS{-}O{-}H$ (**20**), exchange with D_2O would give $CH_3CH{=}CHS{-}O{-}D$ (**20**-d_1), which would be expected to fragment by loss of OD giving $CH_3CH{=}CHS^+$ of m/z 73, the same as

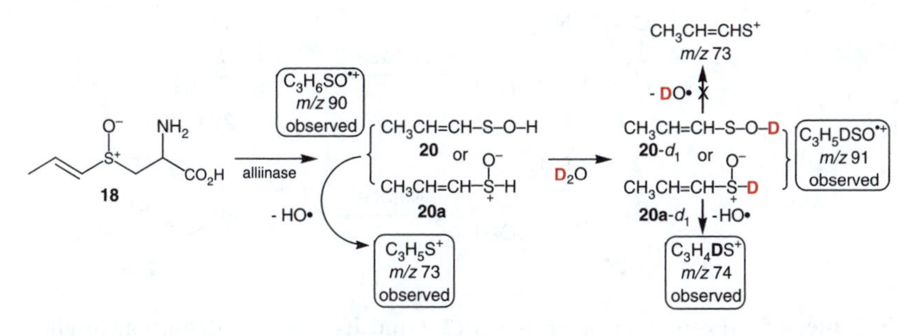

Scheme 4.4 Virtanen's proposed structure for the onion LF (**20a**) based on deuterium exchange studies.

$CH_3CH{=}CHS{-}O{-}H$ (**20**) upon losing OH, as shown in Scheme 4.4. That this does *not* occur requires that the hydrogen lost with the oxygen as OH comes from some other part of the molecule, in a manner similar to that observed for the molecule dimethyl sulfoxide ($CH_3S(O)CH_3$), which in a mass spectrometer also loses OH even though it has no OH group.

A second piece of evidence presented by Virtanen in support of his structure for the LF, $CH_3CH{=}CHS(O)H$ (**20a**), is the formation in onion vapors of propanal, $CH_3CH_2CH{=}O$ (**21**), also reported by others (Kohman, 1947), and its aldol self-condensation product, 2-methyl-2-pentenal (**22**; Scheme 4.5; Spåre, 1963). Virtanen postulated decomposition of $CH_3CH{=}CHS(O)H$ to $CH_3CH{=}CHOH$ [well known to rearrange to propanal (**21**)] and $H_2S{=}O$. The latter compound has recently been prepared and characterized as H–S–O–H, not $H_2S{=}O$ (Winnewisser, 2003). Sulfur compounds H_2S and SO_2 found in onion volatiles (Niegisch, 1956) are likely decomposition products of HSOH. Virtanen also reported that synthetic analogs of isoalliin, *S*-ethenyl-L-cysteine sulfoxide (**23**) and *S*-(buten-1-yl)-L-cysteine sulfoxide (**24**) are enzymatically cleaved to compounds presumed to be ethenesulfenic acid ($CH_2{=}CHS(O)H$, **25a**), and 1-butenesulfenic acid ($CH_3CH_2CH{=}CHS(O)H$, **26a**), respectively (Daebritz, 1964; Mueller, 1966). This latter work takes on added significance with the recent discovery of a volatile

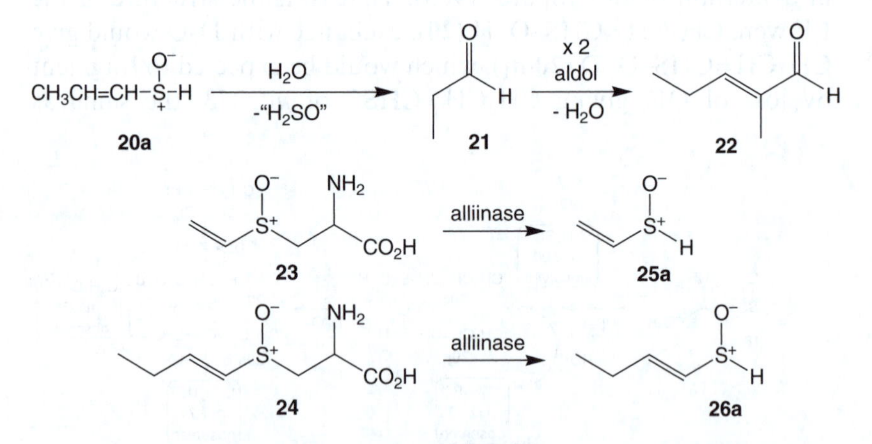

Scheme 4.5 Formation of propanal (**21**) and its aldol condensation product (**22**) from the onion LF here represented as **20a**; sulfoxides **23** and **24** their conversion to **25a** and **26a**, respectively.

lachrymatory compound with the same molecular formula (C_4H_8SO) as **26a**, and presumably derived from **24**, from the decorative plant *Allium siculum* (Block, 2009a).

In his paper on the LF, Virtanen comments on the difference between the enzymatic behavior of alliin from garlic and isoalliin from onion: "In the case of the LP [onion lachrymatory factor precursor, isoalliin] no allicin like 'double-molecule' is formed," that is, $CH_3CH{=}CHS(O)SCH{=}CHCH_3$ (1-propenyl 1-propenethiosulfinate; **27**; Spåre, 1963). This key point will also be discussed below.

While Virtanen was conducting his studies, W. F. Wilkens, a Cornell University graduate student, published his doctoral thesis on the onion LF (Wilkens, 1961). Wilkens suggests that the LF is "thiopropionaldehyde *S*-oxide," $CH_3CH_2CH{=}S{=}O$ (propanethial *S*-oxide; **28**, Scheme 4.6), depicted with a *linear* C=S=O bond. Earlier, and without spectroscopic proof, the LF was suggested to be thiopropanal ($CH_3CH_2CH{=}S$; Kohman, 1947). Evidence supporting structure **28** included: (1) the presence of strong IR absorption bands at 1113 and 1144 cm^{-1} (S=O region), together with the absence of bands attributable to S–H, O–H, or C=C; and (2) ready decomposition of the LF to propanal (**21**) [*via* 3-ethyloxathiirane (**29**)] and to a second compound with IR bands at 1140 and 1330 cm^{-1}, formulated as 2,4-diethyl-1,3-dithietane 1,3-dioxide (**30**).

In 1963 to 1964, the first syntheses of *S*-oxides of thials and thiones – termed *sulfines* – were reported, by dehydrochlorination of the corresponding sulfinyl chlorides (Sheppard, 1964). Evidence was presented for a bent C–S–O group: the methyl groups in $Me_2C{=}S^+{-}O^-$ are nonequivalent by NMR. In 1971, Brodnitz and Pascale at International Flavors and Fragrances showed that propanethial *S*-oxide (**28**) prepared from *n*-PrS(O)Cl (**31**) and

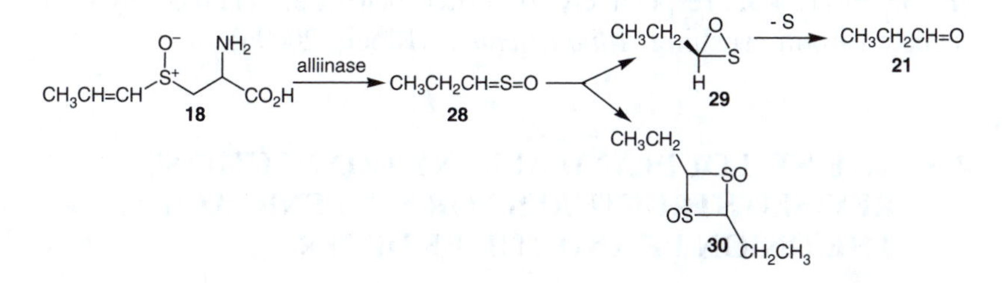

Scheme 4.6 Wilkens's proposed structure for the onion LF (**28**) and for its dimer (**30**).

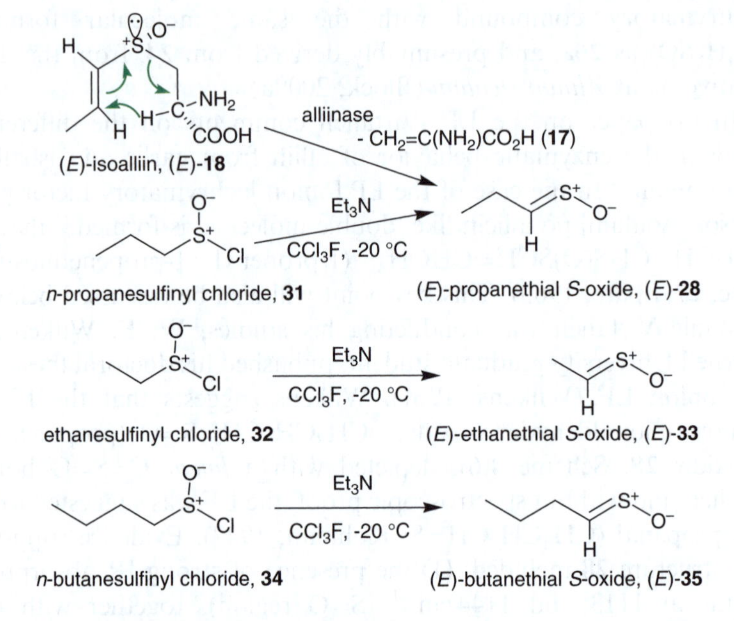

Scheme 4.7 Brodnitz's proposed structure for the onion LF, (*E*)-**28**, and its mechanism for formation.

isolated by gas chromatography (GC), was identical by IR, ^{1}H-NMR and MS to the natural onion LF (Brodnitz, 1971a). These authors proposed the mechanism for the formation of the LF, arbitrarily depicted by them as the (*E*)-isomer, (*E*)-**28**, from its known precursor **18** shown in Scheme 4.7. They also synthesized sulfines MeCH=S^+–O^- (**33**) from ethanesulfinyl chloride (**32**) and *n*-PrCH=S^+–O^- (**35**) from *n*-butanesulfinyl chloride (**34**). Based on the lachrymatory properties and mass spectra of **33** and **35**, they suggested that Virtanen's ethenesulfenic acid (**25a**; CH_2=CHS(O)H; Scheme 4.5) and 1-butenesulfenic acid (**26a**; CH_3CH_2CH=CHS(O)H) were in fact (*E*)-**33** and (*E*)-**35**, respectively. As noted below, **35** has recently been detected upon crushing *Allium siculum* (Block, 2009a).

4.3 THE ST. LOUIS AND ALBANY CONNECTIONS: REVISED STRUCTURES FOR SULFENIC ACIDS, THE ONION LF AND THE LF DIMER

So far, our discussion has focused on a historical account of the isolation and characterization, or attempted characterization, of the key compounds associated with garlic and onion chemistry, namely allicin, alliin, isoalliin, methiin and the onion LF. Based on the accumulated knowledge of chemical bonding and chemical reactions, it is also possible to address questions of *mechanism*, that is, how do chemical reactions occur? More specifically, we would like to understand the detailed step-by-step sequence of bond breaking and bond making involved in a transformation such as the conversion of alliin to allicin. While it is clear that this reaction is initiated by the alliinase enzyme, Stoll and Seebeck imply that the enzyme serves only to convert alliin to a compound they write as $CH_2{=}CHCH_2S(O)H$ (**16a;** 2-propenesulfenic acid) which then self-condenses giving allicin. Virtanen postulates that, by analogy, the onion alliinase enzyme converts isoalliin to $CH_3CH{=}CHS(O)H$ (**20a**), which he names "propenylsulphenic acid" or, more accurately, 1-propenesulfenic acid. Virtanen states that this latter compound represents the first isolated sulfenic acid, and he indicates his surprise that, unlike Stoll and Seebeck's 2-propenesulfenic acid, his "sulfenic acid" apparently does not give an allicin-like self-condensation product.

In addressing the above questions of mechanism, I draw on the personal experiences that first attracted me to the chemistry of garlic and onion. Despite the identification in 1944 of allicin as the active agent of garlic, when I began my research in 1970 at the University of Missouri–St. Louis, little systematic information existed on the chemistry of the class of compounds known as alkyl alkanethiosulfinates (RS(O)SR), of which allicin is a member, perhaps because of their reputation as malodorous, unstable substances (Moore, 1966). I began my work with the parent member of this family, methyl methanethiosulfinate ($CH_3S(O)SCH_3$; **4**; Scheme 3.1), first synthesized by Cavallito in 1947 and later by Moore and O'Connor (Small, 1947; Moore, 1966). True to its reputation, **4**, prepared by careful oxidation of dimethyl disulfide (**3;** Murray, 1971), is an unpleasant smelling, reactive and unstable skin irritant, with the nasty property of readily attaching itself to hair and clothing. Care has to be taken in the oxidation process since overoxidation gives **40** (Scheme 4.8).

Fortunately, **4** proved to have a rich chemistry. Our first discovery was that **4** readily decomposed to methanesulfenic acid (**36**) and thioformaldehyde (**37**), as well as disproportionated to **3** and **40**, at temperatures below 100 °C (Block, 1972). The curved green

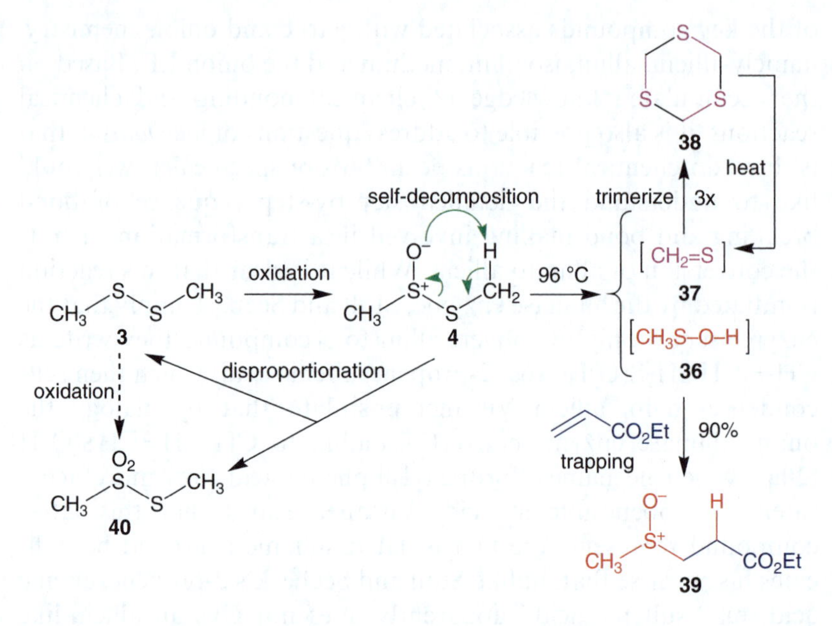

Scheme 4.8 Self-decomposition of methyl methanethiosulfinate (**4**) to methanesulfenic acid (**36**) and thioformaldehyde (**37**), trapping of **36** with ethyl acrylate giving **39**, trimerization of **37** to 1,3,5-trithiane (**38**) and disproportionation of **4** to **3** and methyl methanethiosulfonate (**40**). **4**: SO IR band at 1093 cm^{-1}; **40**: SO_2 IR bands at 1141 and 1343 cm^{-1}.

arrows associated with **4** are devices – "electron-pushing" – widely used by organic chemists to explain reaction mechanisms. Here the oxygen of **4** uses its negative charge to pull off hydrogen in the form of a proton from a methyl group. The two electrons the proton leaves behind create the carbon–sulfur bond of thioformaldehyde (CH_2=S; **37**). As this process is occurring, the sulfur–sulfur bond of **4** must break to maintain the proper number of electrons on each atom. The two electrons of the S–S bond become lone-pair electrons on the methanesulfenic acid sulfur. All of the processes shown by the curved green arrow are postulated to occur simultaneously in a single smooth step. While it is generally not possible to "see" a reaction occurring, the proposed mechanism is consistent with the expected ease of making and breaking the various bonds and with theoretical calculations, which predict a low energy barrier of 21.3 kcal mol^{-1} for this process (Cubbage, 2001).

Conversion of disulfide **3** to thiosulfinate **4** results in a substantial weakening of the S–S bond from ~70 kcal mol^{-1} to 34.8 kcal mol^{-1} (Okada, 2008), which should greatly facilitate the reaction. The brackets enclosing **36** and **37** indicate that neither can be isolated, both being highly reactive compounds. We *infer* the formation of **36** by conducting the decomposition of **4** in the presence of a large excess of ethyl acrylate ($CH_2{=}CHCO_2Et$) whereupon adduct **39** is formed in excellent yield. This "trapping experiment" confirms the presence of **36**, not otherwise isolable. The presence of **37** is inferred by isolation of its trimer, 1,3,5-trithiane (**38**). This particular experiment does not distinguish between the two possible structures for methanesulfenic acid, $CH_3S(O)H$ and $CH_3S{-}O{-}H$, but informs us that thiosulfinate **4** easily decomposes to **36** and **37**.

The highly reactive molecule thioformaldehyde ($CH_2{=}S$; **37**) is itself notable. Its structure was established by the technique of microwave spectroscopy (Johnson, 1970). Microwave spectroscopy is a technique for studying the absorption and emission of radiation in the microwave region of the electromagnetic spectrum by molecules in the gas phase as they undergo various types of rotational processes governed by the rules of quantum mechanics. The technique, which allows the determination of molecular structure with great accuracy and is complementary to X-ray crystallography in the solid phase, became possible due to the development of microwave technology for radar during World War II. Microwave spectroscopic techniques can be applied to the detection and identification of molecules in deep interstellar space using radioastronomy. Once the various rotational bands of thioformaldehyde were identified in the laboratory, it was discovered a short time later that this same molecule occurs in interstellar space (Sinclair, 1973), one of more than 125 molecules so far identified!

Since thioformaldehyde (**37**) is a highly reactive molecule, how is it possible that it exists as such in interstellar space? The high reactivity of **37** is illustrated by the great ease with which it combines with other molecules of itself, for example *trimerizing* to 1,3,5-trithiane (**38**). However, since interstellar space is a vacuum, mostly devoid of matter, molecules of **37** are too widely separated for reactions to occur, and the temperature is too low for molecules of **37** to undergo self-decomposition. Through use of a technique called flash-vacuum pyrolysis (FVP), conditions responsible for

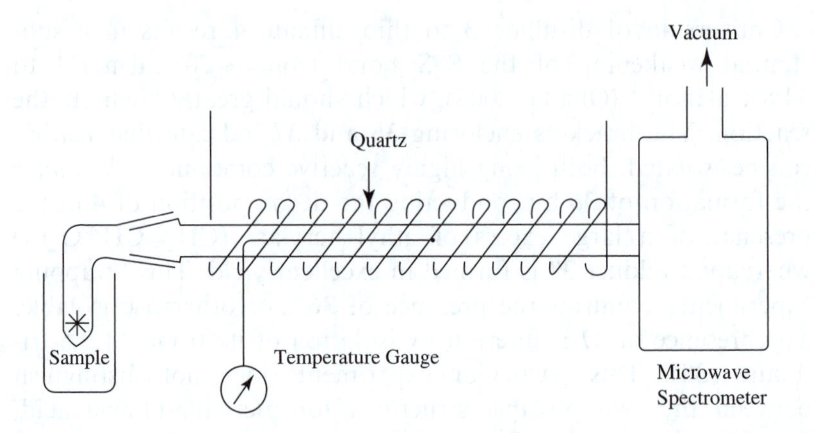

Figure 4.5 Schematic of flash vacuum pyrolysis apparatus for microwave spectrometer.

stabilization of **37** in interstellar space can be simulated in the laboratory. At the same time, the FVP device pictured in Figure 4.5 permits easy generation of compounds such as **37** through brief ("flash") exposure of suitable precursors, such as **38**, to high temperature under vacuum, allowing heat-induced decomposition ("pyrolysis"). In essence, the precursors are passed through the very hot part of the system in less than one hundreth of a second, just long enough to form the desired compound, such as **37**, but not long enough for it to decompose. The FVP device itself is simple in design, consisting of a sample receptacle, a section of high temperature-tolerant quartz glass, electrically heated like a toaster with nickel–chromium ("Nichrome") resistance wire, with an imbedded temperature-measuring device ("pyrometer"). In the device shown in Figure 4.5, the heated tube is directly connected to the chamber of a microwave spectrometer maintained under high vacuum. Other instruments, such as a mass spectrometer, could also be used here to monitor the reaction.

Initial studies done with FVP-microwave instrumentation in collaboration with Robert Penn at the University of Missouri–Saint Louis, in 1977, verified that 1,3,5-trithiane, when subjected to high temperatures, gave rise to the characteristic microwave spectrum of **37** previously measured by others. Next it was found that FVP of methyl methanethiosulfinate (**4**) at 250 °C gave the microwave spectrum of **37** along with that of methanesulfenic acid

(**36**) of structure MeS–O–H, with *divalent* sulfur. The isomeric structure MeS(O)H, with trivalent sulfur, was predicted to have a very different microwave spectrum than what was observed. Through use of a second, connected or "tandem" tube furnace heated to 750 °C, it was shown that **36** lost water giving **37**. If a small quantity of deuterium oxide, D_2O, was introduced into the microwave spectrometer, it was found that the acidic hydrogen of **36** readily exchanged with deuterium giving MeS–O–D (**36**-d_1; Penn, 1978). When condensed at –178 °C using liquid nitrogen, **36** could not be recovered on warming into the microwave spectrometer under high vacuum, the major products being methyl methanethiosulfinate (**4**) and water. Even in the gas phase at 0.1 Torr (1 Torr = 1/760 atmosphere) at 25 °C, **36** only has a half-life of about one minute, which is shorter than the reported eight minute half-life of thioformaldehyde (**37**; Johnson, 1971). Similar results were obtained using a variety of other thermal sources of **36**, such as *tert*-butyl methyl sulfoxide (**41**; Scheme 4.9) which, beginning at 250 °C, decomposes to **36** and isobutylene (**42**), whose microwave spectrum is known.

The unique characteristics of sulfenic acids merit discussion because of their important role in *Allium* chemistry. Methanesulfenic acid can be contrasted with the two other oxyacids formally derived from oxidation of methanethiol (CH_3SH), namely methanesulfinic acid (CH_3SO_2H) and methanesulfonic acid (CH_3SO_3H). The three acids are shown as ball and stick models in Figure 4.6. Methanesulfonic acid, an important commercial

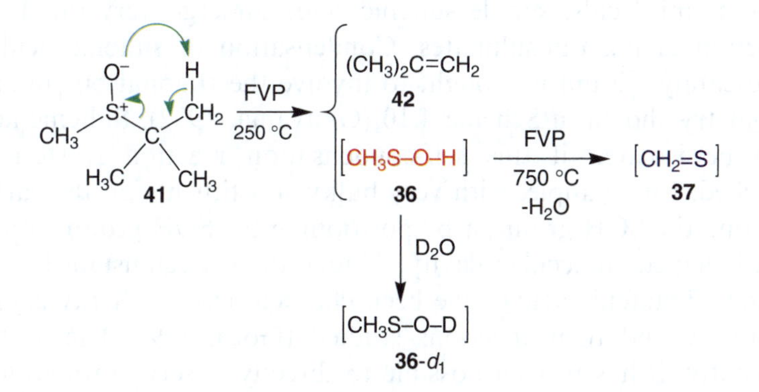

Scheme 4.9 Generation of methanesulfenic acid (**36**) and isobutylene (**42**) by FVP of *t*-butyl methyl sulfoxide (**41**).

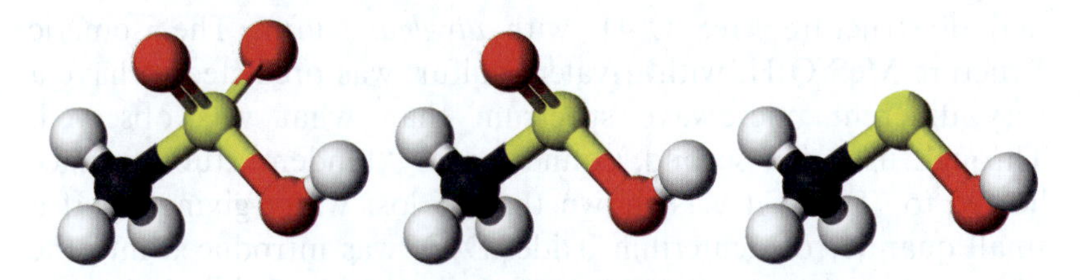

Figure 4.6 Left to right: methanesulfonic acid, methanesulfinic acid and methanesulfenic acid (red = oxygen, yellow = sulfur, black = carbon, white = hydrogen). (Courtesy of Wikipedia).

chemical, is a strong acid with $pK_a = 1.75$, while methanesulfinic acid is a stable research reagent with $pK_a = 3.13$. Methanesulfenic acid is weak acid, with $pK_a = 10–11$ estimated from values for related compounds (Okuyama, 1992), formed as a very short-lived intermediate by the action of alliinase on methiin or from various precursors by FVP. By way of comparison, methanesulfenic acid is weaker than both acetic acid and phenol ($pK_a = 4.8$ and 9.9, respectively) and only slightly stronger as an acid than hydrogen peroxide ($pK_a = 11.7$).

Other researchers have confirmed, both experimentally and computationally, that methanesulfenic acid has divalent rather than trivalent sulfur (Lacombe, 1996; Alkorta, 2004a,b; Turecek, 1988). It has a staggered or "skewed" three-dimensional structure ($\angle$CSOH = 94°) like hydrogen peroxide, H–O–O–H (90.3°) and H–S–O–H (91.3°; Winnewisser, 2003), minimizing repulsion between the adjacent lone-pair electrons. In contrast to sulfonic and sulfinic acids, simple sulfenic acids undergo very rapid self-reaction giving thiosulfinates. Condensation of sulfenic acids is acid-catalyzed and is thought to involve the trigonal bipyramidal geometry shown in Scheme 4.10 (Okuyama, 1992). Sulfenic acids can be isolated if this self-condensation reaction is sterically blocked, for example, with very bulky substituents on the carbon bearing the SOH group or by positioning the SOH group within a bowl-shaped molecular cavity. Under these circumstances, such hindered sulfenic acids have been characterized by X-ray crystallography and their reactions studied (Goto, 1997; Ishii, 1996). Recently, it has proven possible to directly observe formation of 2-propenesulfenic acid and pyruvic acid by crushing garlic and immediately monitoring volatile acids produced using a technique

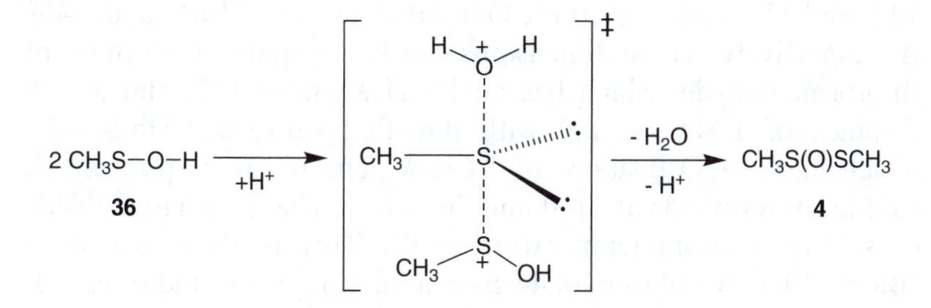

Scheme 4.10 Transition state for self-condensation of methanesulfenic acid (**36**) giving **4**.

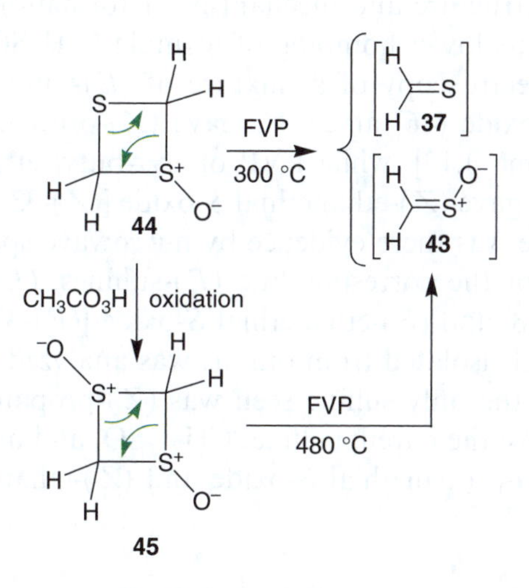

Scheme 4.11 Generation of sulfine (**43**) by FVP of 1,3-dithietane 1-oxide (**44**) and 1,3-dioxide (**45**).

called direct analysis in real time (DART) mass spectrometry (Block, 2009).

The above studies help define the nature of the parent alkanesulfenic acid molecule **36**. A second set of experiments done with Robert Penn defined the stucture of the parent sulfine, *e.g.*, thioformaldehyde *S*-oxide ($CH_2{=}S^+{-}O^-$; **43**). Thus, pyrolysis of 1,3-dithietane *S*-oxide (**44**; Scheme 4.11), using the same FVP-microwave spectroscopic apparatus as above, showed that beginning at 300 °C, **44** cleanly decomposed to thioformaldehyde (**37**) and a new molecule, identified as thioformaldehyde *S*-oxide or sulfine (**43**), having a bent shape, with HCH and CSO angles of

122° and 115° and C–S and S–O bond distances of 1.610 and 1.469 Å, respectively. These dimensions can be compared with those of thioformaldehyde, which has an HCH angle of 117° and a C–S distance of 1.611 Å, and with dimethyl sulfoxide, $Me_2S^+–O^-$, which has a S–O distance of 1.678 Å. The infrared spectrum of sulfine shows bands at 1170 and 760 cm^{-1}. The gas phase half-life of sulfine at room temperature at 0.1 Torr is about one hour (Block, 1976). Oxidation of **44** gives a mixture of *cis*- and *trans*-1,3-dithietane 1,3-dioxide (**45**), which affords pure sulfine (**43**) on FVP at 480 °C (Block, 1982).

A third set of studies was undertaken with Robert Penn to establish the structure and mechanism of formation of the onion LF as well as its lower homolog of formula C_2H_4SO. Thus, FVP-microwave spectroscopy of a mixture of (*E*)- and (*Z*)-*tert*-butyl propenyl sulfoxide (**46**) at 250 °C gave (*Z*)-propanethial *S*-oxide [(*Z*)-**28**; Scheme 4.12], while FVP of *tert*-butyl ethenyl sulfoxide (**47**) at 250 °C gave (*Z*)-ethanethial *S*-oxide [(*Z*)-**33**; Scheme 4.13]. In neither case was there evidence by microwave spectroscopy for the presence of the corresponding (*E*)-sulfines, (*E*)-propanethial *S*-oxide [(*E*)-**28**] and (*E*)-ethanethial *S*-oxide [(*E*)-**33**].

When the LF isolated from onions was analyzed by microwave spectroscopy, the only sulfine seen was (*Z*)-propanethial *S*-oxide [(*Z*)-**28**]. Unlike the parent sulfine, $CH_2{=}SO$, and methanesulfenic acid, both (*Z*)-propanethial *S*-oxide and (*Z*)-ethanethial *S*-oxide

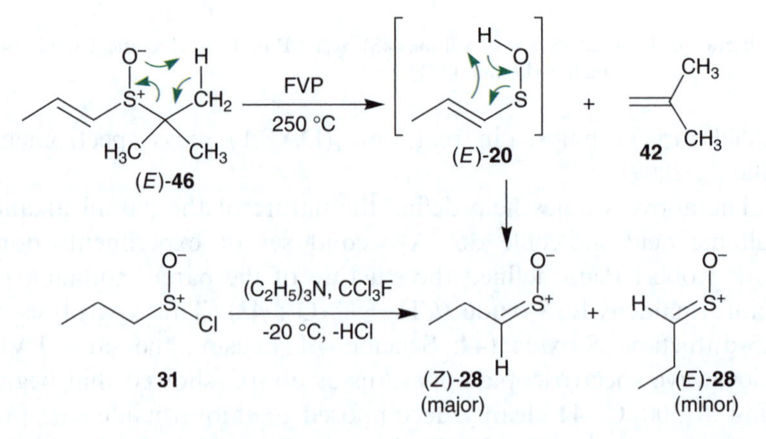

Scheme 4.12 Generation of the onion LF, (*Z*)-propanethial *S*-oxide [(*Z*)-**28**] by FVP methods.

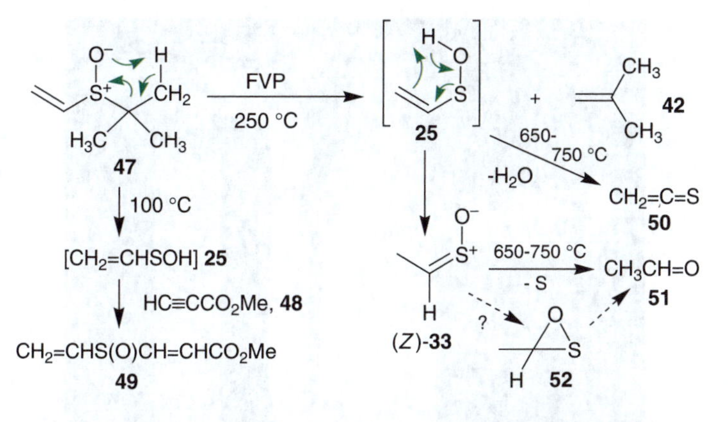

Scheme 4.13 Generation of (Z)-ethanethial S-oxide [(Z)-**33**] by FVP methods.

are stable in the gas phase at room temperature and 0.05 Torr. At 100 °C, sulfoxide (**47**) decomposed in the presence of methyl propiolate (**48**) as a sulfenic acid trap giving adduct **49**. This result requires the intermediacy of ethenesulfenic acid ($CH_2{=}CHSOH$; **25**), as an initial decomposition product of **47**. At FVP temperatures of 650 to 750 °C, precursor **47** decomposed giving thioketene ($CH_2{=}C{=}S$; **50**), and acetaldehyde ($CH_3CH{=}O$; **51**). The formation of **50** can be rationalized by dehydration of ethenesulfenic acid, much like the high temperature process observed with methanesulfenic acid undergoing dehydration to thioformaldehyde. Acetaldehyde (**51**) is likely derived from loss of sulfur from (Z)-ethanethial *S*-oxide [(Z)-**33**] perhaps *via* an oxathiirane (**52**).

Additional studies were done using NMR spectroscopy. Freshly peeled and quartered white onions were frozen in dry ice, crushed with a hammer, converted into a powder in a food processor while still cold and immediately extracted with a highly volatile solvent (Freon-11; $CFCl_3$). The solvent was separated from the onion material, dried and concentrated at –78 °C. The residual oil was then "trap-to-trap" distilled at –20 °C under high vacuum (see Figure 4.7) giving the LF as a colorless, intensely lachrymatory liquid found to be of high purity by proton NMR spectroscopy. The latter analysis showed, among other features, a low-field triplet signal at $\delta = 8.19$ ppm, with a coupling constant $J = 7.81$ Hz, corresponding to the **HC**=SO proton in (Z)-propanethial *S*-oxide [(Z)-**28**; the *triplet* results from interaction of the **HC**=SO proton with the two hydrogens of the neighboring CH_2 group]. However,

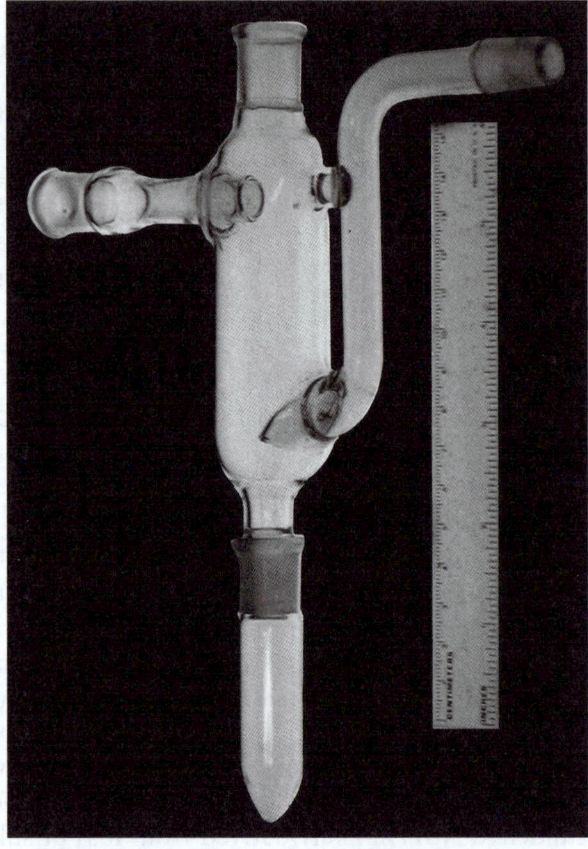

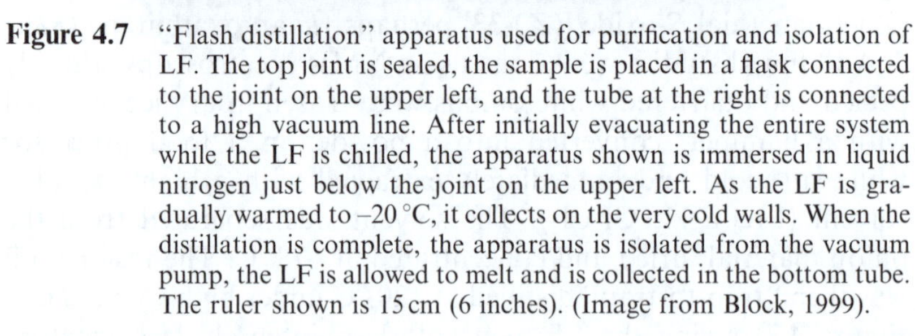

Figure 4.7 "Flash distillation" apparatus used for purification and isolation of LF. The top joint is sealed, the sample is placed in a flask connected to the joint on the upper left, and the tube at the right is connected to a high vacuum line. After initially evacuating the entire system while the LF is chilled, the apparatus shown is immersed in liquid nitrogen just below the joint on the upper left. As the LF is gradually warmed to –20 °C, it collects on the very cold walls. When the distillation is complete, the apparatus is isolated from the vacuum pump, the LF is allowed to melt and is collected in the bottom tube. The ruler shown is 15 cm (6 inches). (Image from Block, 1999).

in addition, a much smaller triplet appeared at still lower field, $\delta = 8.87$ ppm, with a coupling constant $J = 8.79$ Hz ppm. This smaller triplet, with its slightly different coupling constant, and amounting to only 5% of the area of the major triplet, was identified as belonging to (*E*)-propanethial *S*-oxide [(*E*)-**28**]. Another small triplet, appearing at δ = 9.79 ppm, with a small coupling constant $J = 1.4$ Hz, was identified as being due to propanal (**21**), a known decomposition product of the onion LF. After 5 h at 30 °C, the signals for the LF decreased in intensity as the signals for propanal grew.

Similar observations were made on propanethial *S*-oxide samples prepared by FVP of precursor **46** or by dehydrochlorination of propanesulfinyl chloride (**31**; Scheme 4.14; Block, 1980). In addition, it was possible to observe the major and minor **HC=SO** protons in (*Z*)-ethanethial *S*-oxide [(*Z*)-**33**] and (*E*)-ethanethial *S*-oxide [(*E*)-**33**] at δ = 8.31 ppm ($J = 7.33$ Hz) and δ = 8.88 ppm ($J = 8.79$ Hz), with a *Z* : *E* ratio of 97 : 3. Here the **HC=SO** protons appear as quartets due to interaction with the three identical hydrogens of the attached CH_3 group. Other alkanethial *S*-oxides gave similar results. Concentrations of (*E*)-**28** and (*E*)-**33** are too low to allow detection by microwave spectroscopy. One of the more interesting sulfines prepared is 2,2-dimethylpropanethial *S*-oxide, $(CH_3)_3CCH{=}SO$, which shows **HC=SO** singlets (since there are no neighboring hydrogens) at δ = 7.62 and 9.00 ppm for the (*Z*)- and (*E*)-isomers, respectively. The *Z* : *E* ratio is 75 : 25, which indicates that as the hydrogens on the carbon adjacent to the sufine function are replaced by methyl groups, the preference for the (*Z*)-isomer is diminished. In addition, quite remarkably, 2,2-dimethylpropanethial *S*-oxide is *devoid of lachrymatory activity*! This observation suggests that the bulky *t*-butyl group on 2,2-dimethylpropanethial *S*-oxide sterically inhibits interaction

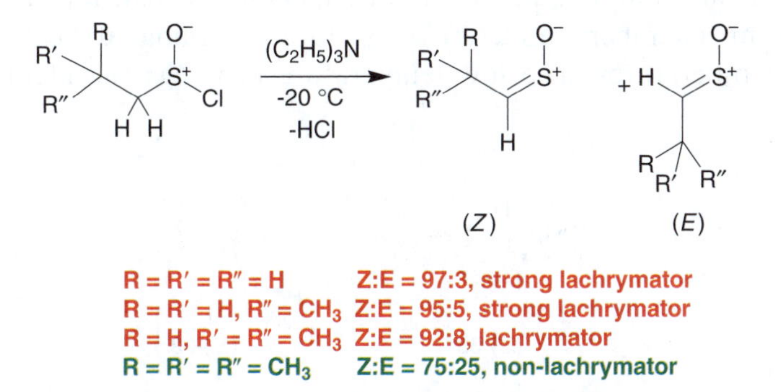

Scheme 4.14 Variation in *Z* : *E* ratio and lachrymatory potency for a series of sulfines.

with lachrymator-sensing protein thiol groups (see Scheme 3.5; Day, 1978).

In general, (*E*)-isomers are *more stable* than (*Z*)-isomers. Why is this not the case with the onion LF and its homologs? To answer this question, Dieter Cremer performed computational studies to examine the relative stability of (*E*)- and (Z)-ethanethial *S*-oxide [(*E*)- and (Z)-**33**]. He found that the (*Z*)-isomer is more stable by 1.7 kcal mol^{-1} than the (*E*)-isomer (Scheme 4.15). Several types of attractive OH interactions are thought to be responsible for making the (*Z*)-isomer more stable than the (*E*)-isomer. Such effects should be more strongly opposed by steric effects upon replacing the hydrogens with methyl groups, as indicated by the diminished *Z* : *E* ratio noted above in $(CH_3)_3CCH{=}SO$. This type of attractive interaction in (*Z*)- or "syn" isomers is well known and is termed the "syn-effect" (Block, 1981).

The above described microwave studies were performed at the University of Missouri–St. Louis, in the period 1976 through 1979 using "Stark-modulation" microwave absorption spectroscopy techniques. In the 1980s, now at the University at Albany, State University of New York, I began a collaboration with Charles and Jennifer Gillies using the newer technique of "pulsed-beam Fourier transform (FT) microwave spectroscopy" to extend and complete some of our earlier microwave studies. The equipment, at the Rensselaer Polytechnic Institute (Troy, New York), offers a number of advantages over absorption microwave methods for the study of short-lived chemical species, such as the onion LF. The pulsed-beam equipment uses a solenoid valve to generate "pulses" of gaseous samples, which, upon supersonic expansion into the vacuum chamber, undergo very rapid cooling to extremely low temperatures. The low temperature prolongs the lifetime of

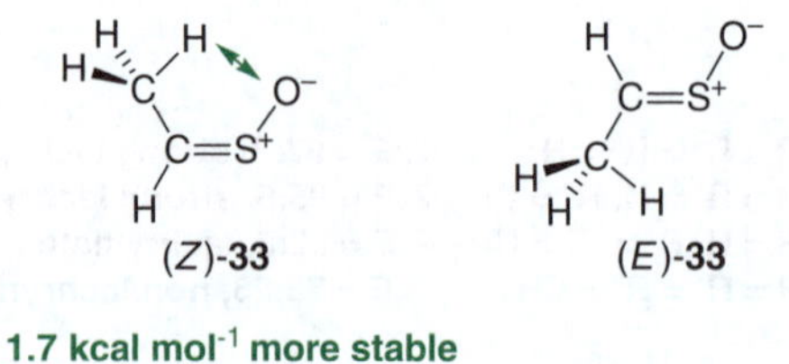

Scheme 4.15 Calculated relative stability of (*Z*)-**33** and (*E*)-**33**.

short-lived compounds and improves signal resolution, which is further enhanced computationally by Fourier transform techniques.

The newer instrumentation allowed us to repeat the experiments of Virtanen involving exposure of 1-propenesulfenic acid to D_2O to identify the ultimate location of deuterium labeling following rearrangement. Thus, pulsed-beam microwave analysis of freshly prepared onions, chopped and crushed in the presence of D_2O, led to the unequivocal identification of (Z)-**28**-d_1, where a single deuterium is attached to the second carbon atom with the observed conformation as shown in Scheme 4.16. Very similar results were obtained when precursor **46** (Scheme 4.12) was pyrolyzed in the presence of a gas flow containing D_2O. These results are consistent with the mechanism indicated by the green arrows in Scheme 4.16 for incorporation of one deuterium into the LF, a so-called 1,4-rearrangement, since the hydrogen moves from atom 1 (oxygen) to atom 4 (carbon). It also becomes clear why "OH" rather

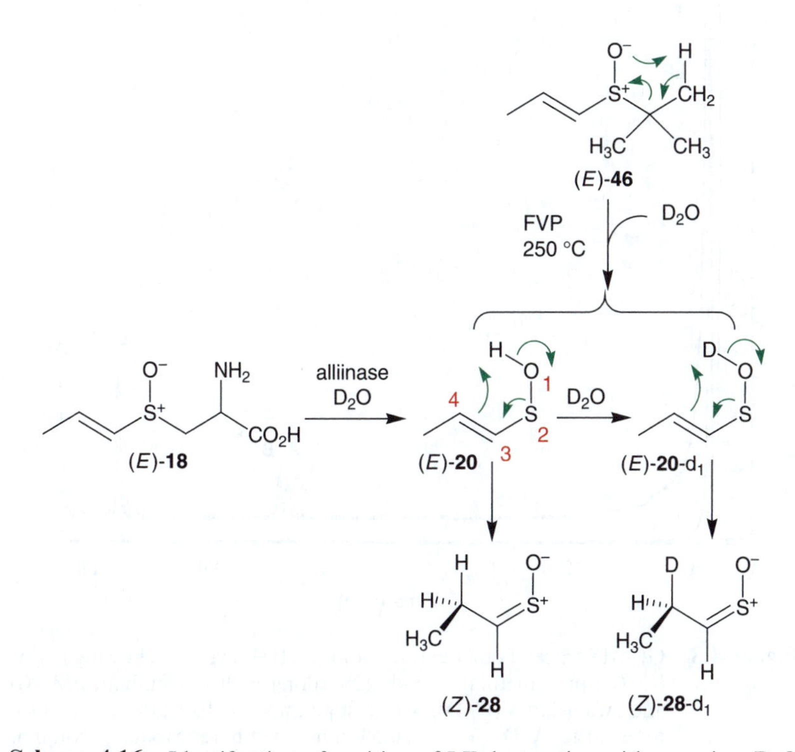

Scheme 4.16 Identification of position of LF deuteration with an onion–D_2O mixture.

than "OD" is lost upon the fragmentation Virtanen observed in his mass spectrometer. Since the C–D bond is known to be stronger than the C–H bond – a phenomenon called the "deuterium isotope effect" – it is more favorable for the oxygen to pull off one of the hydrogen atoms losing OH rather than pulling off the single deuterium atom, losing OD.

When freshly obtained juice from an onion is extracted with ether, and the concentrated extract is analyzed by gas chromatography-mass spectrometry (GC-MS) under mild conditions, the onion LF rapidly elutes as the major product (Figure 4.8). As the temperature of the GC column is increased, small peaks elute corresponding to thiosulfinates and zwiebelanes (peaks **A** to **D**; discussed below; Block, 1992c). Based on the above, a simple protocol was developed for rapidily quantitating the LF level in onion samples (Schmidt, 1996; Mondy, 2002). Thus, onions are

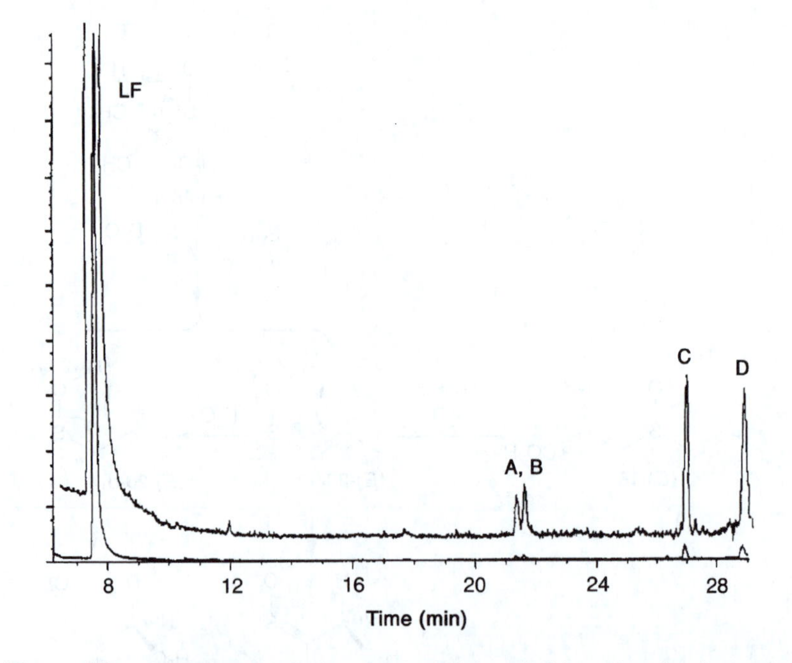

Figure 4.8 GC-MS trace of ether extract of juice of white onion showing major LF (*Z*)-propanethial S-oxide (**28**) along with thiosulfinates (**A**, **B**) and zwiebelanes (**C**, **D**), with a superimposed 10× attenuated trace to enlarge **A–D** (30 m × 0.53 mm wide-bore capillary column; heated from 0 °C, at a rate of 5 °C min^{-1}, to 200 °C). (Image from Block, 1992c).

crushed and the juice is collected in a beaker and maintained at room temperature for a maximum of two minutes to allow for completion of enzymatic conversion. The juice is then saturated with sodium chloride and extracted with an equal volume of methylene chloride containing an inert internal standard. The extract is centrifuged, concentrated under a flow of compressed air, cooled to 0 °C and immediately injected into a GC equipped with a short (5 m) column and low column and injector temperatures (~60 °C).

With this method it was found that the juice of mature onions contains an average of 2 to 10 mM of LF, while the juice of the most pungent onions has a concentration of 22 mM of LF (5–7% of RS(O)SR′ is also formed; Schmidt, 1996). These LF levels correspond to 0.02 to 0.2% of onion fresh weight and are consistent with an average level of pyruvate in the above samples of onion juice of 12 mM. As previously discussed, pyruvate is enzymatically released from isoalliin, present at levels of up to 20 μmol g^{-1} of fresh onion, when the LF precursor (*E*)-1-propenesulfenic acid (**20**) is formed. Onion pungency is typically determined by analysis of pyruvate in onion juice (Abayomi, 2006; Marcos, 2004; Anthon, 2003). Other GC studies show that the LF isomers (*Z*)- and (*E*)-propanethial *S*-oxide (**28**) can be cleanly separated as peaks with a *Z* : *E* ratio of 95 : 5 (Arnault, 2000). As discussed in Section 4.5.4, with specialized types of mass spectrometers that can sample at atmospheric pressure, the onion LF can be directly detected on the surface of freshly cut onion (Ratcliffe, 2007; Block, 2009).

The energetics associated with the 1,4-rearrangements (shown in Schemes 4.12, 4.13 and 4.16) can be probed using computational methods. The results, summarized in Figure 4.9, are surprising: while (*Z*)-ethanethial *S*-oxide, (*Z*)-**33**, is more stable by 3 kcal mol^{-1} than ethenesulfenic acid (**25**), the activation barrier for the conversion is higher than anticipated, namely 33 kcal mol^{-1} (Turecek, 1990). This barrier is higher than that calculated (21.3 kcal mol^{-1}) for the decomposition of methyl methanethiosulfinate (**4**) to methanesulfenic acid (**36**) in Scheme 4.8.

At least part of this barrier reflects the unfavorable planar transition state geometry that forces the electron pairs on sulfur and oxygen to be parallel, leading to electron–electron repulsions, rather than perpendicular, which minimizes such repulsions and, in fact, results in stabilizing electronic interactions. While the high

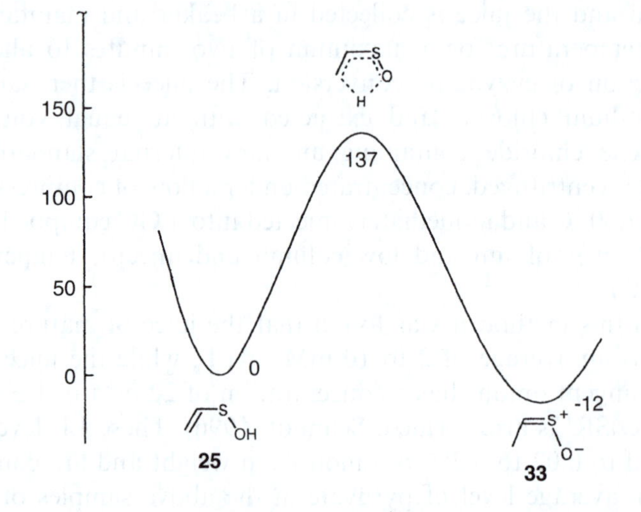

Figure 4.9 Reaction coordinates (in kJ mol^{-1}) for conversion of ethenesulfenic acid (**25**) to the more stable (*Z*)-ethanethial *S*-oxide (**33**) *via* a 137 kJ mol^{-1} (33 kcal mol^{-1}) planar transition state. (Image from Turecek, 1990).

temperatures used for FVP would provide more than sufficient energy for the 1,4-rearrangements, it is puzzling how the rearrangement to the LF at or below room temperature can still compete so effectively with self- or cross-condensation giving thiosulfinates, RS(O)SR. The next section, focusing on the enzymes associated with formation and transformation of *Allium* sulfenic acids, will provide an answer to this question.

First, another onion-related puzzle will be discussed. As noted above, Wilkens proposed that upon standing, two molecules of the onion LF combine, forming a compound he claimed was a 1,3-dithietane 1,3-dioxide. Wilkens includes the infrared spectrum in his report, and it clearly shows the presence of a sulfone group, –SO_2–, rather than two sulfoxide groups, as expected from his structure. Wilkins rationalized the inconsistency between the IR data for **30** which showed "absorption bands typical of the sulfone stretching modes" and the proposed disulfoxide structure **30** as due to "the proximity of the antipodal sulfinyl groups in this strained-ring structure . . . postulated to induce a pseudo-sulfone infrared absorption, which would be of less intensity than that of a true sulfone" (Wilkens, 1964). However, as described in Scheme 4.11, *cis-/trans*-1,3-dithietane 1,3-oxide (**45**), prepared as a thermal source of sulfine, is a nicely crystalline solid showing completely normal sulfoxide bands in its infrared spectrum (Block, 1982).

In an effort to repeat Wilkens's work, purified (*Z*)-propanethial *S*-oxide [(*Z*)-**28**], either isolated from onions or synthesized from propanesulfinyl chloride as described above, was dissolved in twice its volume of dry benzene and kept it in the dark at 5 °C, for 7 days. The resultant slightly yellow, nonlachrymatory solution was concentrated under vacuum, affording a practically colorless, clear liquid with a strong onion-like odor. Analysis by GC-MS confirmed the dimer structure, showing a single compound with *m*/*z* 180, precisely twice that of the onion LF (*m*/*z* 90), and an exact mass that confirmed the molecular formula, $C_6H_{12}O_2S_2$. Through analysis of its proton, carbon-13 and oxygen-17 NMR spectra, as well as through degradation to independently synthesized *d*,*l*-hexane-3,4-dithiol (**55**; Scheme 4.17), the structure of the LF dimer was proven to be *trans*-3,4-diethyl-1,2-dithietane 1,1-dioxide (**54**; Block, 1980).

Compound **54** was the first known example of a saturated 1,2-dithietane derivative. I proposed that the dimerization of the onion LF [(*Z*)-**28**] proceeded by way of a five-membered ring intermediate **53**, which rearranges to **54**. The proposed initial dimerization is similar to known reactions involving 1,3-dipolar cycloadditions. The dimerization mechanism for the parent sulfine **43** has been modeled using density functional theory and has been shown to initially afford a five-membered ring with an activation berrier of 12.3 kcal mol^{-1} (Arnaud, 1999). Several other examples of the dimerization of thial *S*-oxides to 1,2-dithietane 1,1-dioxides have

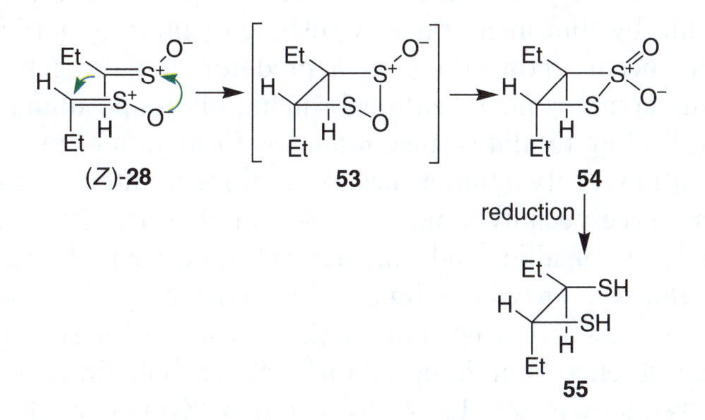

Scheme 4.17 Formation and structure of the dimer **54** of the onion LF [(*Z*)-**28**].

since been reported (Block, 1988a; Hasserodt, 1995; Baudin, 1996). It should be noted that the *dimerization* of the LF is different from the *self-condensation* sought for 1-propenesulfenic acid, which should give a product of *m*/*z* 178, due to loss of water, as occurs in the conversion of 2-propenesulfenic acid (*m*/*z* 90) to allicin (*m*/*z* 178). While the spontaneous dimerization of the onion LF is notable, even more remarkable, related processes affording "zwiebelanes," will be described below. Propanethial *S*-oxide **28** is not the only thial *S*-oxide formed from an *Allium* species. The author and coworkers recently discovered that the lachrymator (*Z*)-butanethial *S*-oxide ((*Z*)-**35**) is produced in significant quantities upon crushing the leaves, stems or bulbs of *Allium siculum* (Block, 2009a). This sulfine was identified by GC-MS and NMR analysis of a plant extract (^{1}H NMR: $\delta = 8.12$ (t, 94%) and 8.59 (t, 6%)), with the data in excellent agreement with those for a known standard (Scheme 4.7), as well as using DART mass spectrometry (see below) with the freshly cut plant.

4.4 A TALE OF TWO ENZYMES: ALLIINASE, A CELLULAR ASSEMBLY LINE FOR ALLICIN FORMATION; LF SYNTHASE, MAKING A SLOW REACTION FAST

4.4.1 Allicin from Alliin and why Allicin from Garlic is Racemic

In garlic cells, alliin is physically separated from the alliinase enzyme: the enzyme resides in microcompartments separated from alliin by thin membranes. Crushing or injuring garlic bulbs, as would occur upon attack by a predator or pathogen, breaks down the compartmentalization bringing alliin and alliinase into contact, leading to almost instantaneous formation of allicin from condensation of two molecules of 2-propenesulfenic acid. The conversion is especially rapid ($>97\%$ complete after 30 s) since the cellular levels of alliin and alliinase are almost equal. It is noteworthy that the enzyme alliinase accounts for up to 10% of the total clove protein. Garlic alliinase (EC 4.4.1.4) belongs to a family of glycoproteins containing about 6% carbohydrate and has been fully characterized by the technique of X-ray crystallography as a dimer with two equal subunits, each of 448 amino acids, with a total molecular mass of 103 000 Daltons (Shimon, 2007; Kuettner, 2002).

Alliinases from garlic, onion and *A. tuberosum* are localized in vascular bundle sheath cells, which surround the veins, while alliin and related cysteine derivatives are concentrated in the abundant storage mesophyll cells (Ellmore, 1994; Yamazaki, 2002; Manabe, 1998; Lancaster, 1981). Since alliinase contributes to the chemical defense of the plant by producing thiosulfinates and other irritating and repellent sulfur compounds, the strategic specific localization in vascular bundle sheath cells may be helpful in protecting against the invasion of microorganisms and small herbivores moving through the veins to mesophyll cells.

Based on X-ray studies by scientists at the Weizmann Institute in Rehovot, Israel, we have a good understanding of what happens when alliin and alliinase interact. First, alliinase is activated by a cofactor known as pyridoxal-5′-phosphate or "PLP". When not in use, alliinase exists as a complex known as an internal aldimine, involving the 251st amino acid (lysine 251). Due to its specific three-dimensional shape, alliin fits precisely into an enzymatic cavity. Upon contact, alliin substitutes for lysine 251 forming an external aldimine. Once alliin is activated as an aldimine, it loses a proton to a basic site in the enzyme. Synchronized with loss of a proton, the alliin carbon–sulfur bond on the amino acid side undergoes cleavage (an α,β-elimination reaction) releasing 2-propenesulfenic acid and leaving behind a PLP-bound aminoacryl intermediate. Alliinase is called a C–S lyase enzyme, "lyase" meaning, "to cleave." The immediate product of cleavage, 2-propenesulfenic acid (sometimes incorrectly named "allylsulfenic acid"), is drawn as $CH_2{=}CHCH_2S{-}O{-}H$, consistent with structural studies on other sulfenic acids. This sulfenic acid is called an "intermediate" since it has never been isolated, even at very low temperature. It very rapidly self-condenses with loss of water to form allicin, as shown in Scheme 4.18. Recently, it has proven possible to directly observe the transient formation of 2-propenesulfenic acid using special mass spectrometric methods, discussed below (Block, 2009).

The PLP-bound aminoacrylate undergoes rapid hydrolysis giving pyruvate and ammonia as lysine 251 is reinstalled. The system functions like a very efficient assembly line with alliin going in one end and allicin, ammonia and pyruvate coming out at the other

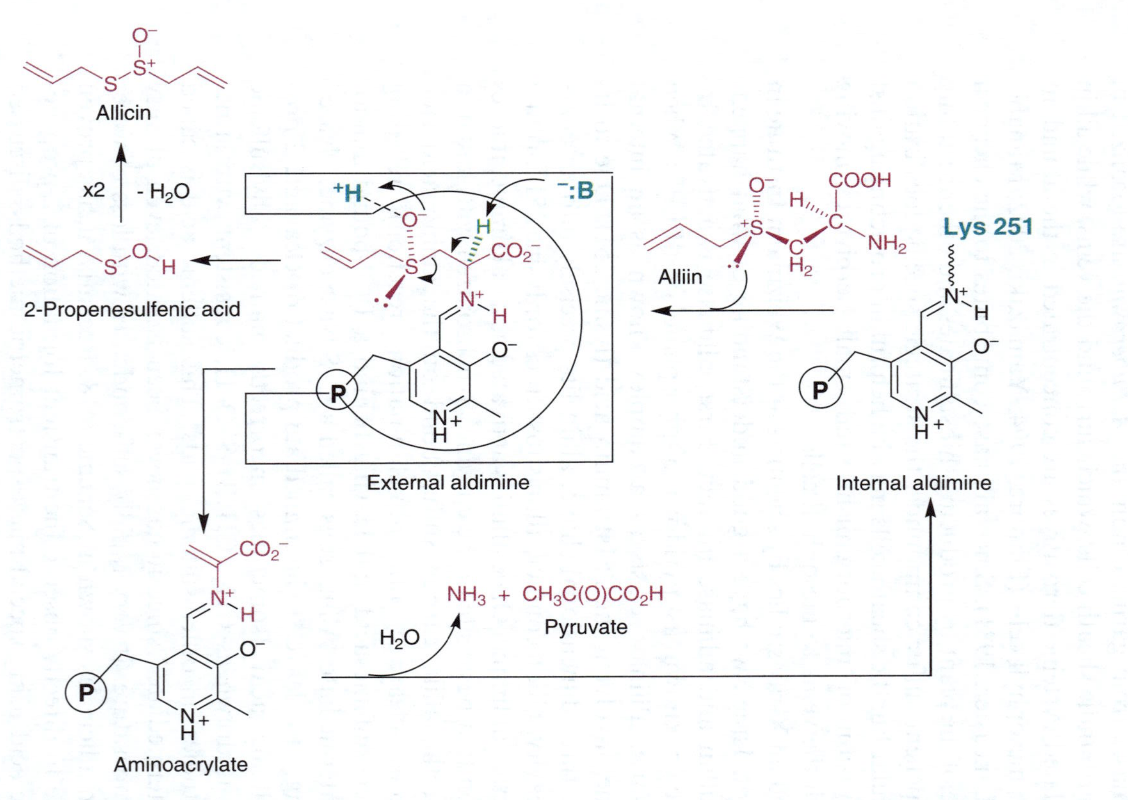

Scheme 4.18 Alliinase reaction with alliin: the key intermediates in the reaction pathway are labeled. In its resting state, in the absence of substrate, the enzyme forms a covalent, Schiff base with Lys251, the internal aldimine. When the substrate alliin is present, it binds covalently to the PLP to form the external aldimine. The C–S bond is cleaved, releasing 2-propenesulfenic acid. Two 2-propenesulfenic acid molecules join to form allicin, leaving aminoacrylate bound to the PLP. In the final step, this aminoacrylate is lysed from the PLP and hydrolyzed, yielding pyruvic acid and ammonia. The PLP is then free to reform the internal aldimine with Lys251.

end, while the alliinase remains unchanged. The overall process is so rapid that a burst of allicin is produced upon mixing precursor and enzyme. If we are keeping score of the numbers and types of atoms involved in going from alliin to allicin, a useful exercise when considering some of the other compounds found in garlic and onion preparations, we go from the three carbons, one sulfur and one oxygen supplied by alliin to the six carbons, two sulfurs and one oxygen present in allicin, the loss of one oxygen accounted for by formation of a molecule of water.

With this explanation we can better appreciate the significance of Cavallito's observations. Enzymes can generally only function in the presence of water. By freezing garlic cloves in dry ice and then pulverizing them with acetone, water is removed and the alliinase enzyme and precursor compound alliin both remain intact in the form of a white powder. On addition of water at room temperature to this powder, the enzymatic reaction can then occur, producing allicin as described above. However, if the powder is first boiled with ethyl alcohol, the alliinase enzyme is destroyed (denatured), so that even with the addition of water the odor of allicin is not detected. Boiling with water does not affect the more chemically robust precursor alliin, so that the addition of a small amount of the original white powder, not heated with ethyl alcohol, leads to allicin formation. The above explains why a garlic bulb has little odor compared to the odor upon crushing. In Chapter 5, we will learn the full significance of the sensitivity of the alliinase enzyme toward denaturation in the manufacture of garlic-based health products.

From the mechanism of conversion of alliin to allicin, shown in Scheme 4.18, we can also understand Cavallito's observation that allicin isolated from garlic is optically inactive, despite having a chiral trivalent sulfur atom bonded to three different groups, oxygen, sulfur and carbon. Indeed, a variety of experimental evidence supports the chiral nature of sulfinyl sulfur in thiosulfinates, including the observations that various thiosulfinates can be separated into their enantiomers on chiral HPLC columns (Bauer, 1991) and that chiral thiosulfinates bearing tertiary butyl groups, to preclude decomposition reactions causing racemizaton, are easily prepared and are useful as reagents (Cogan, 1998; Blum, 2003; Colonna, 2005). However, optical activity has never been detected in thiosulfinates from alliums. As shown in Scheme 4.18, cleavage of alliin by alliinase first gives sulfenic acid (CH_2=$CHCH_2S$–O–H), which is achiral. Therefore, the "handedness" present in alliin is lost in this intermediate, which will combine with itself forming only racemic allicin. It is assumed that combination of two molecules of sulfenic acid occurs *without* enzymatic involvement, which is reasonable, since chemically generated sulfenic acids condense with each other extremely rapidly.

Figure 4.10 shows the relative reactivity of synthetic *S*-alk(en)-ylcysteine sulfoxides and natural (+)-alliin toward alliinase from garlic (Stoll, 1951). Table 4.1 compares the reactivity of these same substrates toward cleavage by alliinases from garlic and *A. ursinum*

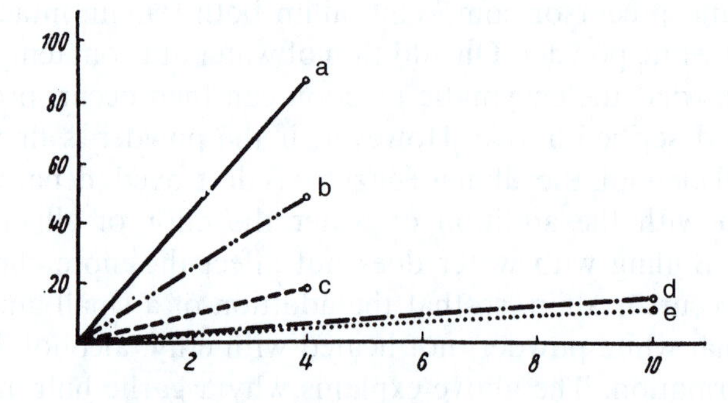

Figure 4.10 Relative reactivity of *S*-alk(en)ylcysteine sulfoxides toward alliinase from *A. sativum*. Abscissa: minutes, ordinate: cleavage in percent. Curves (top to bottom) are for (a) (+)-alliin from garlic and (b) synthetic (±)-alliin, (c) (±)-ethiin, (d) (±)-propiin and (e) (±)-butiin. (Image from Stoll, 1951).

Table 4.1 Reactivity of *S*-alk(en)ylcysteine sulfoxides toward crude alliinase enzymes from *A. sativum* and *A. ursinum* relative to (±)-alliin activity = 100 (Schmitt, 2005).

	A. sativum	*A. ursinum*
(±)-Alliin	100	100
(+)-Alliin	122	128
(−)-Alliin	20	57
(+)-Isoalliin	123	125
(+)-Methiin	7	27
(±)-Ethiin	24	58
(+)-Propiin	8	35
(±)-Butiin	13	31

(Schmitt, 2005). In commenting on the data in Figure 4.10, Stoll notes the "remarkable differences between the velocity of enzymatic cleavage of alliin and that of closely related compounds," which are typically cleaved much more slowly (Stoll, 1951). The *A. ursinum* alliinase is seen to be less selective than the garlic alliinase. In his 1951 paper Stoll observes: "reaction [of alliin] with alliinase takes place extremely rapidly, a fact which is in agreement with the instantaneous appearance of the typical odor on crushing garlic." The optimum temperature for activity of alliinase is 37 °C; 50% of a sample of alliin is split by alliinase after 1 minute at room temperature, while 20% of the alliin is split after 2 minutes at 0 °C (Stoll, 1951). More recent work shows that the rate of alliin cleavage both at 0 °C and at room temperature is at least ten times faster than reported by Stoll (Lawson, 1992; Block, 2009). The ability of alliinase to function even at low temperatures presumably reflects the need for garlic to defend itself against predator attack at low temperatures. Due to the presence of microbes such as *Bacillus subtilis* and *Escherichia coli* which reside in our intestinal tract, there is modest alliinase activity in our body (Murkami, 1960). This bacterial alliinase activity amounts to about 3% of the activity of garlic alliinase based on human studies involving alliinase-deactivated garlic supplements (Lawson, 2006).

Garlic alliinase monomer contains ten cysteine residues. Eight of these form four disulfide bridges while two are free thiols. Since neither enzyme activity nor protein structure are affected by chemical modification of the two free thiols, the thiol groups can be used as chemical handles to attach alliinase to low- or high-molecular weight molecules, for example biotin and its streptavidin complex, while still enabling efficient enzymatic production of allicin (Weiner, 2009). Alliinase can be chemically attached to an immobilized support in a glass tube, and a solution of alliin passed through the column (Miron, 2006). Under these circumstances a pure aqueous solution of allicin can be produced without continuously replenishing the alliinase. It has been shown that, despite the presence of free thiol groups in alliinase, direct contact of alliinase with its product allicin does not inactivate the enzyme, which further supports the observation that modification of the free thiol groups in alliinase is not important for enzymatic activity (Weiner, 2009).

Since, as already shown by Cavallito, and illustrated in Scheme 3.1, allicin reacts rapidly with free thiol groups (RSH), such as glutathione found in blood plasma, and easily penetrates biological membranes, it disappears from the circulation shortly after injection. For this reason the potent antibiotic and cytotoxic effects of allicin have been limited to *in vitro* applications. A new technique developed by the Weizmann researchers circumvents this limitation in antitumor therapy by *in situ* generation of allicin on the targeted cells. Thus, the alliinase enzyme is first targeted to a tumor, accomplished by conjugating the enzyme to a carrier, a monoclonal antibody (mAb) specific to a tumor-associated surface antigen. Alliin is then introduced into the circulation, which results in the formation of cytotoxic allicin only at the site of alliinase localization on the tumor cell surface. This strategy for drug delivery is based on the concept of antibody-directed enzyme prodrug therapy (ADEPT; Miron, 2003). This targeted allicin generation approach has been used to kill human lymphocytic leukemia tumor cells *in vivo* (Arditti, 2005).

4.4.2 The LF Synthase Enzyme

In 2002, a noteworthy paper appeared in *Nature* announcing the discovery of a new onion enzyme, "lachrymatory factor synthase" or "LFS," responsible for converting 1-propenesulfenic acid into (*Z*)-propanethial *S*-oxide, the onion LF (Imai, 2002). Imai and coworkers at House Foods in Tokyo observed that addition of crude garlic alliinase to isoalliin led to enzymatic cleavage of isoalliin, yet no LF was formed, in contrast to what was observed using crude onion alliinase. They concluded that the onion alliinase must contain an additional component. Chromatography of the crude onion alliinase led to the isolation of the previously unknown LFS enzyme. The LF is formed only when all three components – purified alliinase, LFS and isoalliin – are present. While we have shown that (*E*)-1-propenesulfenic acid spontaneously rearranges to the LF, our experiments were done at elevated temperatures in the gas phase. When (*E*)-1-propenesulfenic acid is generated by the onion alliinase in aqueous solution from isoalliin, it is likely (without intervention of the second LFS enzyme) that strong hydrogen bonding would hamper a rearrangement already predicted to be slow because of the unfavorable transition state geometry (Figure 4.10). Presumably, in the onion enzyme systems, (*E*)-1-propenesulfenic acid is rapidly shuttled from its point of origin in the alliinase enzyme to the LFS enzyme to limit escape of the sulfenic acid into the medium where it could self-condense (Scheme 4.19).

The LFS may enforce planarity with a U-shaped conformation on the sulfenic acid. Such type of control by the LFS enzyme is necessary to explain the specific deuterium labeling results found by the microwave spectroscopic studies involving crushing onions in D_2O. If we accept the premise that the LF is a key defensive weapon in the onion's fight for survival, then Nature is much more intimately involved in the steps in onion chemistry than was previously thought.

4.5 THE AROMA AND TASTE OF ALLIUMS: A MULTITUDE OF FLAVOR PRECURSORS

4.5.1 Analysis by Paper and Thin Layer Chromatography

The aroma and taste of genus *Allium* plants, perhaps their most defining characteristics, can be attributed to various sulfur-containing substances. To summarize what has already been discussed, cutting or crushing a garlic clove mixes the alliinase enzyme with the precursor compound alliin, triggering very rapid formation of the thiosulfinate allicin. Allicin is formed by self-condensation of the highly reactive 2-propenesulfenic acid. Distillation of chopped fresh garlic gives the distilled oil of garlic originally investigated in the nineteenth century by Wertheim and Semmler and reported to contain diallyl sulfide, diallyl disulfide and related polysulfides, all of which have garlicky odors. Similar polysulfide mixtures are formed when allicin is kept at room temperature for several days. Both allicin and the diallyl polysulfides contribute to what our nose and tongue tell us is garlic. In a similar manner, cutting onion allows the alliinase enzyme to mix with the precursor isoalliin forming (*E*)-1-propenesulfenic acid that, in turn, is immediately converted into (*Z*)-propanethial *S*-oxide, the onion LF, by the lachrymatory factor synthase (LFS) enzyme, which is thought to be in close proximity to the onion alliinase.

The compounds described previously are not the only contributors to the "garlic and onion experience." In the 1950s, using the technique of paper chromatography, Kyoto University (Japan) chemists examined the sulfur compounds present before, and after, enzymatic cleavage in *A. sativum* (garlic), *A. chinense* (*A. Bakeri*;

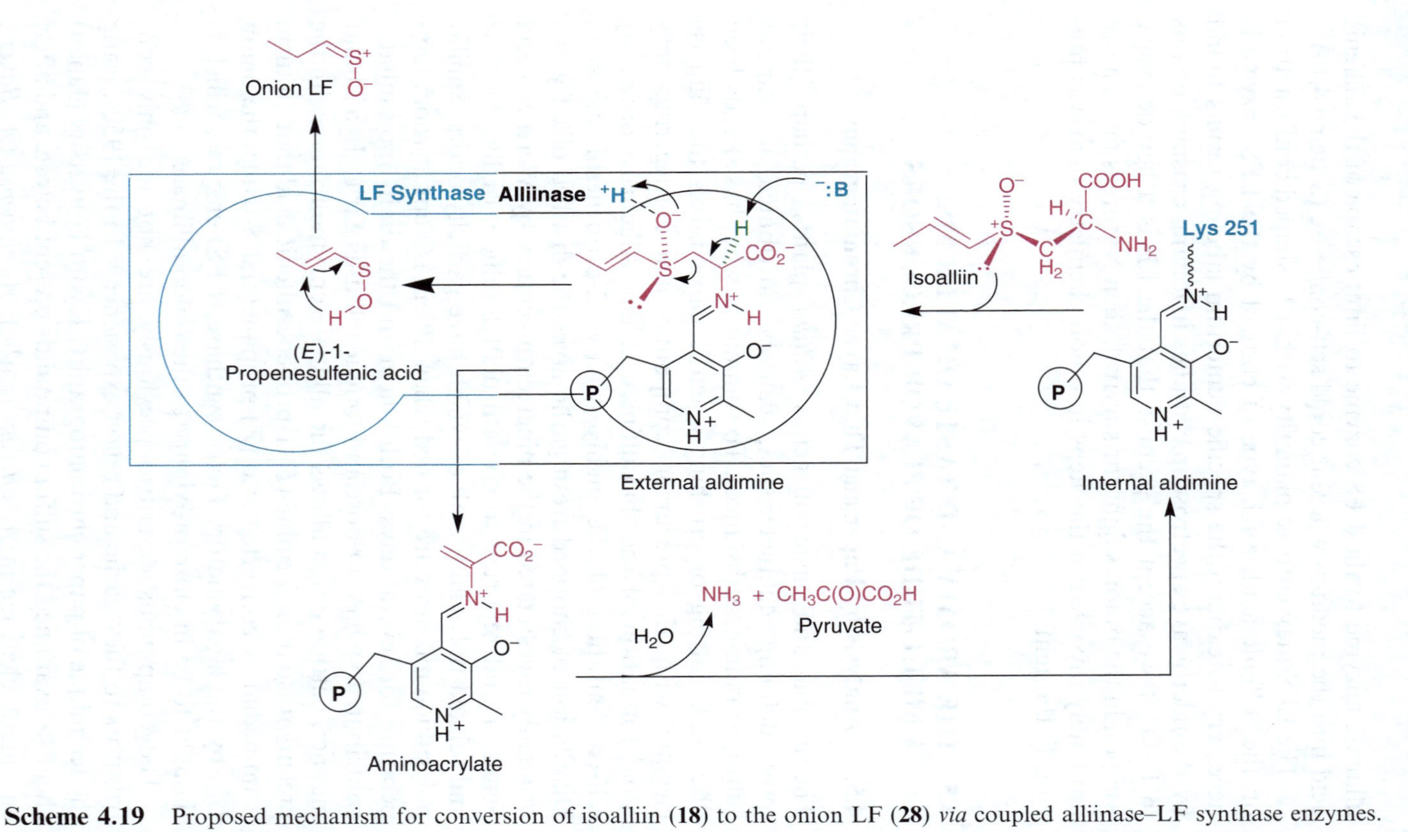

Scheme 4.19 Proposed mechanism for conversion of isoalliin (**18**) to the onion LF (**28**) *via* coupled alliinase–LF synthase enzymes.

rakkyo), *A. tuberosum* (Chinese chives), *A. victorialis* (caucas), *A. cepa* (onion), *A. fistulosum* (Japanese bunching onion) and *A. schoenoprasum* (chives). Prior to enzymatic cleavage, methiin was detected in all alliums examined, with the highest concentration being found in *A. chinense* and *A. tuberosum*. Alliin was found only in *A. sativum, A. tuberosum* and *A. victorialis*. Propiin was found in *A. cepa, A. fistulosum* and *A. schoenoprasum*, with traces present in *A. sativum* (Fujiwara, 1958). At about the same time Virtanen also reported the presence of methiin and propiin in onion (Virtanen, 1959b). Thin layer chromatography (TLC), using glass plates thinly coated with silica gel, was used to identify isoalliin in onion extracts (Granroth, 1968).

Paper chromatography of cut and ground alliums was revealing: garlic was found to produce allicin (AllS(O)SAll; All = allyl) as the major component, MeS(O)SMe (minor component; Me = methyl), MeS(O)SAll (intermediate amounts), MeS(O)SPr (minor amounts; Pr = *n*-propyl) and AllS(O)SPr (intermediate amounts), among other compounds; *A. victorialis* afforded major amounts of MeS(O)SMe and lesser amounts of allicin and MeS(O)SAll; both *A. cepa* and *A. fistulosum* generated PrS(O)SPr and minor amounts of MeS(O)SMe; *A. chinense* gave major amounts of MeS(O)SMe, intermediate amounts of MeS(O)SAll, and minor amounts of MeS(O)SPr, allicin, AllS(O)SPr and PrS(O)SPr (Fujiwara, 1955); *A. rosenbachianum* produced MeS(O)SMe (Matsukawa, 1953). Several researchers reported that onion alliinase converts methiin and propiin to the corresponding thiosulfinates, MeS(O)SMe and PrS(O)SPr (Schwimmer, 1960; Tsuno, 1960). In the above studies the presence of the particular thiosulfinates was inferred based on relative chromatographic migration values. Migration is quantified as R_F values or "ratio of fronts," that is, the distance the substance moves up the paper divided by the distance the solvent moves (solvent front). Synthetic standards, as well as the reaction product of each thiosulfinate, extracted from its spot, with cysteine (see Scheme 3.1 for reaction of allicin with cysteine) were used to calibrate the method. Colorless compounds can be visualized with a chemical spray or other methods.

The above results are helpful in appreciating the complexity of flavorants from alliums. Let us represent in generalized form the *Allium* flavor precursors alliin, isoalliin, methiin, and propiin, called *S*-alk(en)yl-L-cysteine sulfoxides (ACSOs), as **A**S(O)$CH_2CH(NH_2)CO_2H$, **B**S(O)$CH_2CH(NH_2)CO_2H$, **C**S(O)$CH_2CH(NH_2)CO_2H$

and **D**$S(O)CH_2CH(NH_2)CO_2H$. If both **A**$S(O)CH_2CH(NH_2)$-CO_2H and **B**$S(O)CH_2CH(NH_2)CO_2H$ are present, then following exposure of the mixture to alliinase, **A**SOH and **B**SOH will form and immediately combine giving **A**S(O)S**A**, **A**S(O)S**B**, **A**SS(O)**B** and **B**S(O)S**B**, that is, *four* different thiosulfinates from *two* precursor sulfenic acids and cysteine sulfoxides. If **A**$S(O)CH_2CH$ $(NH_2)CO_2H$, **B**$S(O)CH_2CH(NH_2)CO_2H$ and **C**$S(O)CH_2CH$ $(NH_2)CO_2H$ are all present, then **A**SOH, **B**SOH and **C**SOH will form and immediately combine, giving **A**S(O)S**A**, **A**S(O)S**B**, **A**SS(O)**B**, **A**S(O)S**C**, **A**SS(O)**C**, **B**S(O)S**B**, **B**S(O)S**C**, **B**SS(O)**C** and **C**S(O)S**C**, that is, *nine* different thiosulfinates from *three* precursor sulfenic acids and cysteine sulfoxides. If all four ACSOs were present, which rarely occurs, alliinase cleavage would give *sixteen* different thiosulfinates.

Thus, the total possible number of unique thiosulfinates RS(O)SR′ is the *square* of the number of ACSOs. The ratios of the various thiosulfinates would ultimately depend on the ratios of the ACSOs, their relative rates of cleavage by alliinase (which are substantially different!), room temperature stability of the thiosulfinates, and other factors to be discussed below. While we have only described four ACSOs, in fact *seven* simple ACSOs are now known (Figure 4.11); the rarer ACSOs (Kubec 2000, 2002, 2009) will be discussed below. In addition, (+)-*S*-(*E*)-3-pentenyl-L-cysteine sulfoxide has been isolated from the seeds of a red onion (Dini, 2008) while a cysteine sulfoxide, containing a pyrrole group attached to sulfur, has been identified in extracts of "drumstick" alliums. The latter sulfoxide will be discussed below (Jedelská, 2008). Three-quaters of the sulfur in *Allium* species occurs in the form of ACSOs and their γ-glutamyl derivatives (Lundegardh, 2008). In *Allium* species ACSOs occur naturally mainly as the (+)-L-enantiomers. Some of the ACSOs in Figure 4.11, as well as others of novel structures, are also found in other plant species, sometimes in the absence of the corresponding alliinase enzyme.

4.5.2 Analysis by High Performance Liquid Chromatography (HPLC) and LC-MS

Paper chromatography has been replaced by the room temperature separation method of high performance liquid chromatography

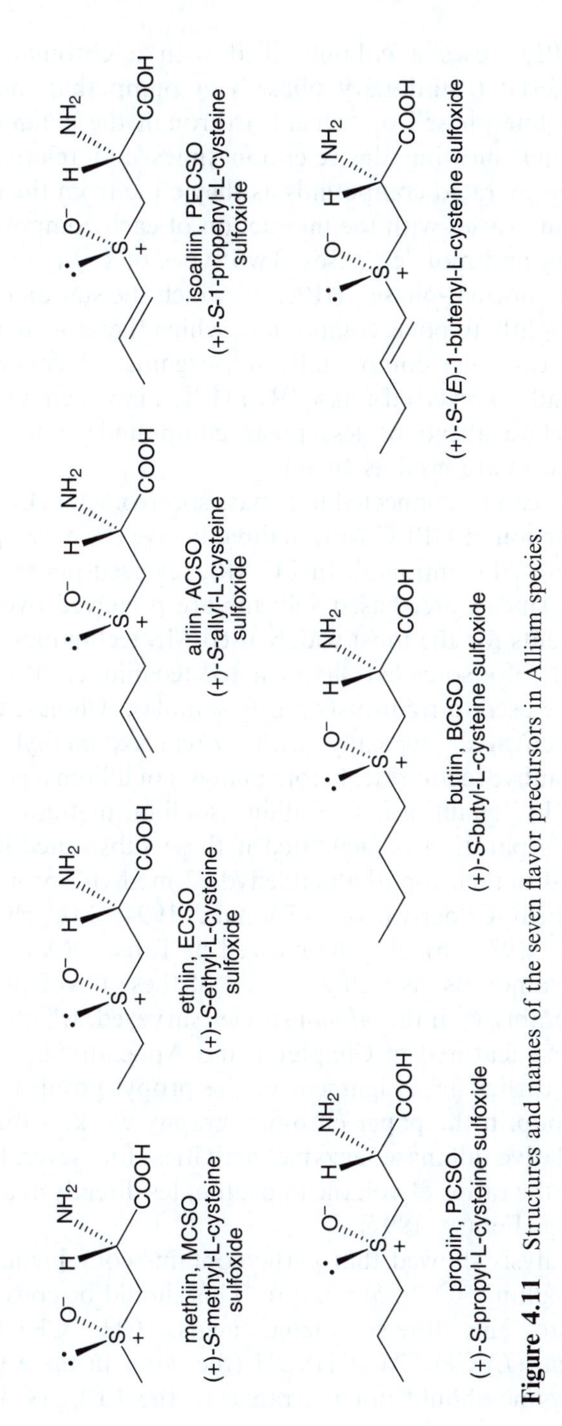

Figure 4.11 Structures and names of the seven flavor precursors in Allium species.

(HPLC). HPLC uses a column filled with a chromatographic packing material ("stationary phase"), a pump that moves the solvent ("mobile phase" or "eluent") through the column and a detector/plotter showing the retention times and relative abundance of the separated compounds as they elute from the column. Retention time varies with the interaction of each compound with the stationary and mobile phases. Two types of stationary phases are used: (1) "normal-phase" HPLC in which the stationary phase binds more tightly to polar compounds, which therefore elute more slowly than less polar compounds, and organic solvents are used as eluents; and (2) reversed-phase (RP) HPLC in which more polar compound elute ahead of less polar compounds, and aqueous solvent mixtures are used as eluents.

The HPLC can be connected to a mass spectrometer (LC-MS – a shortened version of HPLC-MS) to directly give the mass spectrum of each separated compound. In LC-MS, reversed-phase columns are favored, since water-based solvents are preferred over purely organic solvents for the most widely used MS techniques. LC-MS is an example of a so-called "hyphenated technique." Preparative HPLC can be used to rigorously purify samples. Alliinase enzymes can be deactivated, typically with *O*-(carboxymethyl) hydroxylamine hemihydrochloride, before analysis of alliums, permitting accurate HPLC quantitation of alliin, isoalliin, methiin and propiin. HPLC separation is facilitated if these substituted cysteines are converted to their *o*-phthaldialdehyde/2-methyl-2-propanethiol (OPA) or FMOC derivatives (Thomas, 1994; Mütsch-Eckner, 1992; Ziegler, 1989a,b). As summarized in Table 1 (Appendix 1), the absolute amounts, as well as ratios, of these flavor precursors vary considerably with the *Allium* species surveyed, which includes many of those featured in Chapter 1 and Appendix 2. This table corrects the earlier misassignment of the propyl group for the 1-propenyl group in the paper chromatography work. Table 1 also lists the relative alliinase enzyme activities for several of the alliums, and the ratios of volatile thiosulfinates directly analyzed by cold trapping (Ferary, 1997).

HPLC analysis showed that garlic contains contains an alliin : methiin : isoalliin (50 : 2 : 5) mixture. This should be converted by alliinase into the three sulfenic acids: $CH_2{=}CHCH_2SOH$, CH_3SOH and (*E*)-$CH_3CH{=}CHSOH$ (the latter in the absense of the LFS enzyme should not rearrange to the LF). As discussed previously, these three sulfenic acids should combine to afford a total of nine different thiosulfinates, RS(O)SR′ (where R and R′ represent the allyl, methyl and 1-propenyl groups). It is worthwhile discussing practical details of LC-MS analysis of garlic and a related plant, ramp (*A. tricoccum*). While solvents such as diethyl ether are typically used to extract homogenized alliums, the technique of supercritical fluid extraction (SFE) with liquid carbon dioxide (CO_2) as the solvent has advantages. SFE avoids the use of toxic or inflammable solvents as well as problems encountered with emulsion formation, which complicates separation of the organic solvent from the aqueous plant homogenate. Garlic and ramp homogenates were separately extracted at 240 atmospheres pressure under SFE conditions with 40 grams of liquid CO_2 per gram of plant, and the resulting product dissolved in methanol for analysis by LC-MS using RP HPLC.

Figure 4.12 shows the graphical output from the LC-MS analysis. While less than 15 min is required for each analysis, the separation of the thiosulfinates is incomplete. Better separation of the garlic thiosulfinates is achieved with normal-phase HPLC (Figure 4.13), but then LC-MS cannot be used. Table 4.2 indicates the molar percentages of all of the thiosulfinates in each sample. There is excellent agreement between the data in Table 4.2, and the results of separate analyses of ether-extracted garlic and ramp by HPLC, with a UV detector calibrated using pure standards of each thiosulfinate (Calvey, 1997; Block, 1992a; Block, 1996).

Calibration is required because when a UV detector is used, the signal intensity for each compound depends on the molar UV extinction coefficients, which vary from compound to compound depending on the nature of the conjugation present in each case. Similar analyses of garlic extracts have been reported by others (Iberl, 1990a,b; Jansen, 1987; Lawson, 1991a,b).

Allicin is clearly the dominant thiosulfinate present in both plant extracts. This is not surprising since in both plants alliin is by far the major precursor, reaching concentrations of 1.4% in fresh bulbs of garlic (Koch, 1996) and 2.6% of the dry weight (Yamazaki, 2005). The ratio of the thiosulfinates in the extracts reflects the relative abundance of the precursor compounds, alliin, methiin and isoalliin, in each plant; allicin is thus statistically favored. The other minor thiosulfinates would have been missed in the isolation procedure employed by Cavallito. What is noteworthy

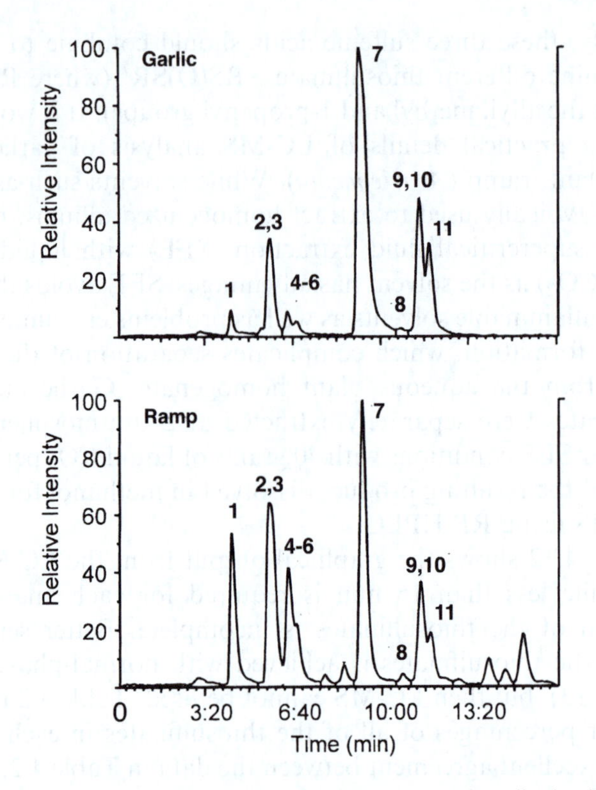

Figure 4.12 Total ion chromatographs of extracts of garlic (top) and ramp (bottom) from LC-MS analysis on a reversed-phase column. Compounds are identified in Table 4.2. (Image from Calvey, 1997).

is the significantly higher percentage of thiosulfinates containing methyl groups in ramp compared with garlic, which is consistent with the stronger taste of ramp compared to garlic.

The order of elution of the compounds in Figure 4.12, with the more polar methyl compounds eluting ahead of the less polar diallyl and allyl propyl thiosulfinates, is characteristic of RP-HPLC. Each thiosulfinate RS(O)SR′ has a different odor and taste. These thiosulfinates, as well as the polysulfides resulting from thiosulfinate decomposition upon heating or cooking, all contribute to the overall characteristic flavor of freshly cut garlic. If we are keeping track of the non-hydrogen atoms, the thiosulfinates have either two, four or six carbons, two sulfurs and one oxygen atom, with the number of carbon atoms reflecting use of one-carbon methyl groups together with three-carbon propyl or

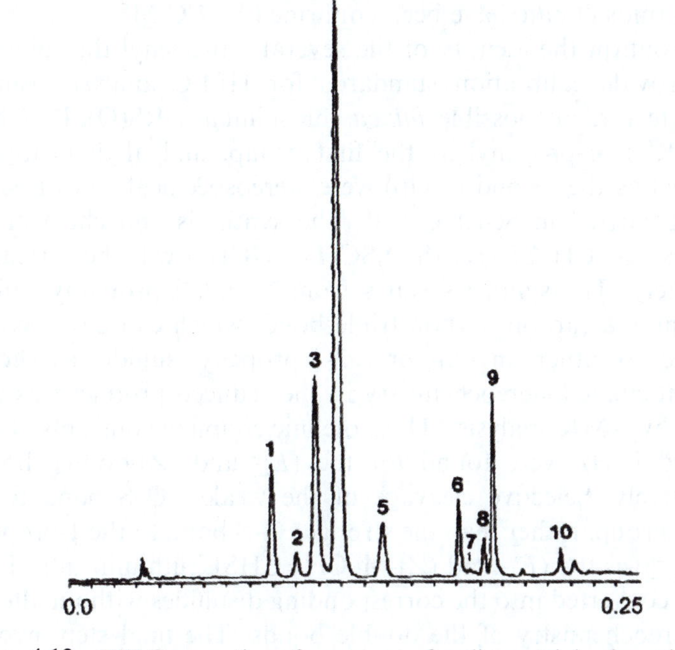

Figure 4.13 HPLC separation of an extract of garlic containing benzyl alcohol as an internal standard. Conditions: Rainen Microsorb silica gel (250 × 4.6 micrometer) column, with a 2-propanol : hexane gradient (2 : 98 to 20 : 60 during 10 min), and UV detection at 254 nm;. Peak identification: **1**, (*E*)-AllSS(O)CH=CHMe; **2**, (*Z*)-AllS(O)SCH=CHMe; **3**, (*E*)-AllS(O)SCH=CHMe; **4**, allicin; **5**, benzyl alcohol (standard); **6**, AllS(O)SMe; **7**, (*Z*)-MeS(O)SCH=CHMe; **8**, (*E*)-MeS(O)SCH=CHMe; **9**, MeS(O)SAll; **10**, MeS(O)SMe. (Image from Block, 1992).

Table 4.2 Percent peak areas for thiosulfinates in garlic and ramp extracts (Figure 4.12).

Compd No.	*Compound*	*Garlic*	*Ramp*
1	MeS(O)SMe	1.8	15.1
2,3	MeS(O)SAll, MeSS(O)All	18.0	17.2
4,5,6	MeS(O)SCH=CHMe-(*E,Z*), MeSS(O)CH=CHMe-(*E*)	7.2	25.3
7	AllS(O)SAll	53.4	26.7
8	*n*-PrS(O)SAll	trace	1.2
9,10,11	AllS(O)SCH=CHMe-(*E,Z*), AllSS(O)CH=CHMe-(*E*)	19.7	14.5

propenyl groups. In both garlic and ramp extracts, trace levels of a propyl thiosulfinate have been confirmed by LC-MS.

To confirm the identity of the several 1-propenyl thiosulfinates and provide calibration standards for HPLC analysis using a UV detector, all possible *mixed* thiosulfinates RS(O)SR′ (either R or R′ is 1-propenyl as the first group, and allyl, methyl or *n*-propyl as the second group) were stereospecifically synthesized, as summarized in Scheme 4.20. The synthesis and chemistry of isomers of $CH_3CH{=}CHS(O)SCH{=}CHCH_3$ will be discussed separately. The synthesis starts from an alkyl propargyl sulfide containing a carbon–carbon triple bond, which can be selectively reduced to either an (*E*)- or (*Z*)-1-propenyl sulfide, as shown. The anticipated stereochemistry in the reduced product was confirmed by NMR analysis,. Thus, olefinic coupling constants of 15.5 Hz and 9 Hz were found for the (*E*)- and (*Z*)-double bonds, respectively. Selective cleavage of the weaker C–S bond to the propyl group, rather than the stronger C–S bond to the 1-propenyl group, gives the (*E*)- and (*Z*)-MeCH=CHSLi lithium salts. These can be converted into the corresponding disulfides without altering the stereochemistry of the double bonds. The final step involves mono-oxidation of the unsymmetrical disulfides. Formation of unequal quantities of the pairs of isomeric thiosulfinates was expected since the sulfur attached to the electron-withdrawing 1-propenyl group should be less reactive toward electron-poor oxidizing agents compared to the sulfur attached to the electron-donating saturated alkyl group. NMR analysis of the individual mixtures, (*E*)-MeCH=CHS(O)SR/(*E*)-MeCH=CHSS(O)R and (*Z*)-MeCH=CHS(O)SR/(*Z*)-MeCH=CHSS(O)R, confirmed that isomerization about the double bonds had *not* occurred during oxidation.

Preparative HPLC or TLC could be used to separate isomeric mixtures from the above syntheses. Pure samples of (*E*)- or (*Z*)-MeCH=CHS(O)SR isolated by chromatography were configurationally stable, that is, the double bond did not isomerize. Chromatographed samples of MeCH=CHSS(O)R, with the sulfinyl group remote from the 1-propenyl group, could only be isolated as (*E*)-MeCH=CHSS(O) : (*Z*)-MeCH=CHSS(O)R (8 : 5) mixtures. Other studies confirmed that MeCH=CHSS(O)R undergoes a rapid *E*/*Z* isomerization at room temperature, presumably as shown in Scheme 4.21. As a consequence of this

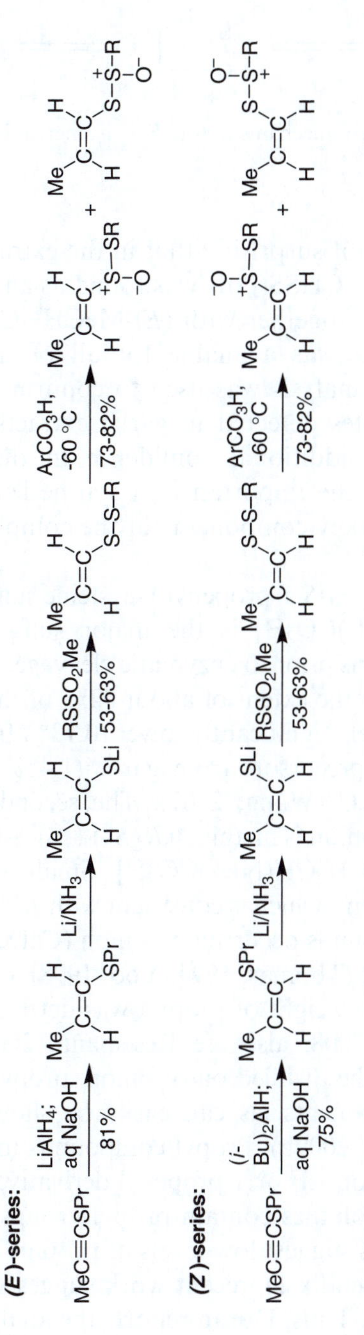

Scheme 4.20 Stereospecific syntheses of 1-propenyl thiosulfinates.

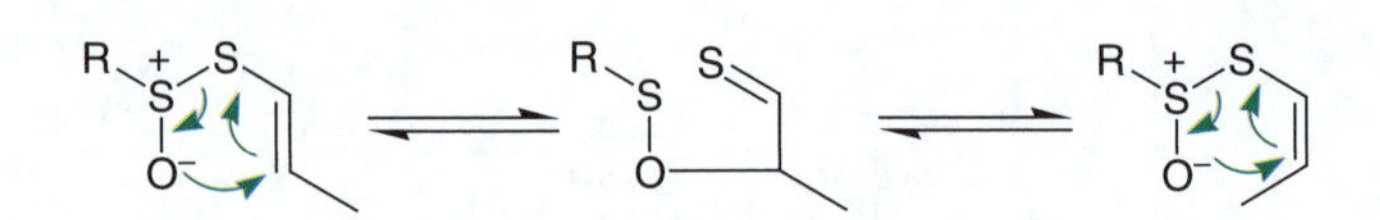

Scheme 4.21 Proposed mechanism for *E*/*Z* isomerization of RS(O)SCH=CHMe.

isomerization, it is not surprising that in the extracts of garlic and ramp, (*E*/*Z*)-MeCH=CHSS(O)R was found as an *E* : *Z* mixture of approximately 8 : 5, together with (*E*)-MeCH=CHS(O)SR. Since detailed NMR data was available for all of the thiosulfinates synthesized, NMR analysis was used to confirm the ratios of the different thiosulfinates detected in garlic extracts by HPLC and LC-MS, providing additional confidence in the results. These findings underscore the importance of synthesizing and studying the reactivity of the key components of the complex mixtures that are being analyzed.

Isoalliin [(R_CR_S)-(+)-*S*-1-propenyl-L-cysteine sulfoxide; CH_3CH=CHS(O)$CH_2CH(NH_2)CO_2H$] is the major sulfur-containing precursor found in onions prior to enzymatic cleavage. In yellow onions, isoalliin is present to the extent of about 82% of the total precursor fraction, at total levels significantly lower (0.13% fresh weight) than those of the flavor precursors from garlic (1.4% fresh weight), as determined by HPLC (Wang, 2007). The second most abundant precursor (14%) in onion is methiin [(R_CS_S)-(+)-*S*-methyl-L-cysteine sulfoxide; $CH_3S(O)CH_2CH(NH_2)CO_2H$]. Small amounts of alliin (2.5%) are also present, which is consistent with earlier reports of low levels of allyl compounds occurring in onion (Calvey, 1997). Propiin appears to be absent (Thomas, 1994; Yoo, 1998), or present at trace levels (0.00018% dry weight of propiin was detected in onion powder; Starkenmann, 2008; also see Resemann, 2004). This is very surprising, since in the distilled oil of onion, propyl polysulfides are the dominant components, as can easily be shown by ^1H NMR spectroscopy (Block, 2009b). Propyl compounds may be formed by reduction of the onion LF or 1-propenyl derivatives (Wang, 2007).

The major leek volatiles contain propyl groups (Nielsen, 2004). While earlier studies suggest low levels of propiin in the fresh plant (see Table 1 in Appendix 1), recent work suggests that this in fact may not be the case. Thus, Doran reports the analysis of the bulbs, pseudostem and leaf of fresh leek (Tadorna cultivar; all values mg g^{-1} fresh weight): for bulbs, (+)-methiin, 3.56; (–)-methiin, 0.71; (+)-ethiin, 0.50; (–)-ethiin, 0.34; (+)-propiin, 3.78; (+)-isoalliin, 0.69; total cysteine sulfoxides, 9.58; for leaves (green, photosynthetic tissue), (+)-methiin, 1.17; (–)-methiin, 0.35; (+)-ethiin, 0.53; (–)-ethiin, 0.44; (+)-propiin, 2.4; (+)-isoalliin, 1.54; total cysteine sulfoxides, 6.44; for pseudostem, (+)-methiin, 0.70; (–)-methiin, 0.45; (+)-ethiin, 0.45; (–)-ethiin, 0.45; (+)-propiin, 5.55; (+)-isoalliin, 1.12; total cysteine sulfoxides, 8.78. What is notable in this work is the high levels of propiin and ethiin and the occurrence of *both* diastereomers of methiin and ethiin, although the (+)-isomer dominates. Doran distinguishes his work from that of others, by employing a very mild extraction method and HPLC with a long retention time, to optimize peak separations (Doran, 2007).

Other *Allium* species can have different ratios of alliin, methiin and isoalliin, as well as small amounts of the minor flavor precursor propiin, and occasionally the very rare precursors ethiin, butiin, *S*-1-butenylcysteine sulfoxide (Kubec, 2000, 2002, 2009) and *S*-3-pentenylcysteine sulfoxide (Dini, 2008). Each precursor, alliin, methiin and isoalliin, contributes to the observed mixtures of thiosulfinates (RS(O)SR′), as well as the polysulfide decomposition products (RS_nR'). The unique mix of RS(O)SR′ and RS_nR' determines the overall flavor for each *Allium*-derived sample. Ultimately, the specific flavor of individual alliums is due to their characteristic ratios and amounts of ACSOs. Several types of *Allium* species, among 55 species surveyed (Fritsch, 2006; Keusgen, 2002; Storsberg, 2003), can be distinguished based both on the dominant sulfur-containing flavor precursor and from their taste. The precursor methiin dominates in chives and Chinese chives among the edible alliums. Many of the ornamental alliums also contain methiin as the dominant ACSO. The precusor isoalliin dominates in the "onion-type" alliums, which includes onion, chive and top onion, Japanese branching onion, shallot and leek. There is a great difference in the isoalliin levels between mild onions and the extremely pungent dehydrator onion. Alliin dominates in "garlic-type" alliums, which includes garlic, elephant garlic, wild leek (*A. obliquum*) and sand leek (*A. scorodoprasum*). Regular garlic contains more alliin than the milder elephant garlic, which is a member of the leek family. Alliin and isoalliin rarely co-dominate, doing so only in the Chinese chives (*A. tuberosum*). Alliin and methiin are equally high in

A. victorialis. A triple mix of equal amounts of methiin, alliin and isoalliin is present in ramson (*A. ursinum*).

Methiin dominates in the majority of *Allium* species that are rarely used by man, because methyl derivatives are associated with an unpleasant, "hard" smell. The rare butiin and *S*-1-butenylcysteine sulfoxide (Kubec, 2009), both of which occur in *A. siculum* (also known as *Nectaroscordum siculum*, Sicilian Honey Garlic or Mediterranean Bells), give the cut plant a strong, natural gas-like odor. The plant is reported to be used as a seasoning in Bulgaria and is said to be deer resistant (Block, 2009a). *S*-3-Pentenylcysteine sulfoxide has been recently identified in *A. cepa* var. *tropeana* (red onion) seeds (Dini, 2008). Relatively low levels of methiin ($<20\%$ of total precursors) are found in many *Allium* species used as vegetables or medicinal plants (*A. ampeloprasum*, *A. sativum*, *A. cepa*, *A. galanthum*, *A. proliferum* and *A. tuberosum*). Very high levels of cysteine sulfoxides are associated with a very hot taste (thiosulfinates), presumably protecting these species against herbivory. The occurrence of different ratios for the four flavor precursor compounds, alliin, methiin, isoalliin and propiin, is very helpful in identifying the different *Allium* species. The use of chemical markers in classifying plants is known as chemotaxonomy. Variations have been found in the amounts and ratios of the different flavor precursors from *Allium* bulbs of the same type (accession) grown under different environmental conditions, for example, Israel *vs.* The Netherlands (Kamenetsky, 2005), and based on the plants parts examined (leaves, stems, bulbs).

4.5.3 Gas Chromatographic Analysis of Distilled *Allium* Oils; Artifact Problems

Gas chromatography (GC) has long been used for the analysis of volatiles, such as those in distilled *Allium* oils. GC entails introducing a small sample into a heated inlet port, which is then carried by a flow of helium gas through a glass column coated on the inside with a stationary phase. The compounds move through the column at different rates, depending on boiling points and interaction with the stationary phase. Individual peaks can be identified by their mass spectrum using GC-MS. If compounds are relatively stable in the gas phase, preparative gas chromatography can be used to purify compounds, as was the case with the onion LF and its homologs (Brodnitz, 1971a). The strengths of the GC technique include simplicity of operation, outstanding separating ability (allowing separation of very similar compounds) and ease of compound identification by interfacing with mass spectrometers.

4.5.3.1 Distilled Onion Oil. Onion oil is typically obtained, in an average yield of 0.015%, by distillation of crushed fresh onions, which have been allowed to stand for a few hours before distillation. One pound of onion oil has the flavoring strength of two tons of fresh onions (Shaath, 1998). Early GC-MS analyses of commercial onion oil identified the major components (in decreasing abundance) as: PrSSPr, PrSSSPr, MeSSPr, MeSSSPr, (*E,Z*)-MeCH=CHSSPr, (*E,Z*)-MeCH=CHSSMe, 3,4-dimethylthiophene, MeSSSMe and MeSSMe, along with traces of (*E,Z*)-MeCH=CHSSSPr, (*E,Z*)-MeCH=CHSSSMe, AllSSMe and AllSSPr (Brodnitz, 1969; Carson, 1961a). These values agree with the results of recent analyses, identifying *n*-propyl polysulfides as the most abundant compounds present. Brodnitz suggests that these various divalent sulfur compounds originate from decomposition of thiosulfinates containing the respective alk(en)yl group; he did not discuss the origin of 3,4-dimethylthiophene, which will be considered below. His paper predates, by many years, HPLC studies confirming the presence of 1-propenyl thiosulfinates, along with very low concentrations of the corresponding allyl compounds in fresh onion extracts. Supercritical CO_2 extracts of onions had far lower levels of *n*-propyl sulfur compounds than found in onion oil by the same GC-MS methods (Sinha, 1992).

The precise composition of onion oil, as determined by GC-MS, is important, because pure onion oil is an expensive flavoring agent that is subject to adulteration. A non-sulfur containing compound, 2-*n*-hexyl-5-methyl-3(2*H*)furanone, which is abundant in distilled onion oil and which can also be identified by IR spectroscopy, is useful in authenticating onion oil suspected of being adulterated (Losing, 1999). Later GC-MS work on onion oil led to the identification of as many as 84 compounds, including those of type RSCH(Et)SSR′, structurally related to cepaenes (see Section 4.10; Farkas, 1992; Shaath, 1998; Kuo, 1990, 1992a).

4.5.3.2 Artifact Formation. In order to rapidly volatilize the injected samples, the GC inlet port is typically heated to temperatures as high as 250 °C. Unfortunately, many *Allium* compounds decompose at these high temperatures. For example, when Brodnitz subjected a freshly prepared garlic extract to GC-MS analysis, two major unknown peaks at *m/z* 144 ($C_6H_8S_2$) appeared. Injection of a synthetic sample of allicin gave the same two peaks, which were assumed to result from dehydration of allicin (Brodnitz, 1971b). However, when synthetic allicin was allowed to completely decompose at 20 °C for 20 hours, the two peaks at *m/z* 144 were not found, the only products being diallyl sulfide, disulfide and trisulfide, along with sulfur dioxide. Importantly, Brodnitz concludes that the compounds at *m/z* 144 "are products of diallyl thiosulfinate [allicin] decomposition during glc [gc] and are not components of garlic;" that is, they are *artifacts* (Block, 1993, 1994a).

A second example of problems associated with breakdown of thermally sensitive compounds under GC conditions, comes from studies of Auger and colleagues on identification of the odorant given off by crushed leek (*A. porrum*), which the leek moth, *Acrolepiopsis assectella*, finds most attractive (Auger, 1989). With a very short 2.5 m column (instead of the more usual 25 m) and a cooler injection port (70 °C instead of 250 °C), the major volatile (>90%) from crushed leek was shown to be PrS(O)SPr, which the leek moth found more attractive than dipropyl disulfide (PrSSPr), the minor volatile found by GC analysis. When injected into a 25 m column, PrS(O)SPr decomposed to PrSH, PrSSPr and PrSSSPr, among other compounds. Auger notes that PrS(O)Pr is stable in the vapor state and can persist in the environment (Auger, 1990).

Auger concludes, "the majority of sulfur volatiles identified by GC-MS in *Allium* spp. are thus artifacts produced during the isolation of the sample and during chromatography" (Auger, 1989). The author has reached a similar conclusion, that thiosulfinates originally present would decompose in a GC under high temperature conditions giving polysulfides as artifacts (Block, 1992b). To avoid artifact problems, it is best to test the GC properties of known standards, and compare the analytical results with those obtained by a separation method that does not involve heating, such as HPLC. Indeed, use of room temperature HPLC-APCI-MS conditions for direct analysis of *Allium* odors confirmed the presence of thiosulfinates and the absence of di- and polysulfides and their rearrangement products (Ferary, 1996a,b). If a compound is unstable under one set of GC conditions, it is possible to achieve better results by cooling the injection port or even using cryogenic (low temperature) injection conditions; shortening the length of the column and increasing the flow rate so that the sample spends less time in the column; or using a GC column with a thicker (4 μm) stationary phase (Arnault, 2000).

Examination of the decomposition products of allicin under a variety of conditions revealed that the compounds at *m*/*z* 144, originally seen by Brodnitz and suggested to be dehydration products of allicin, are in fact the minor and major Diels–Alder dimers of thioacrolein, 2-vinyl-4*H*-1,2-dithiin (**57a**) and 3-vinyl-4*H*-1,3-dithiin (**57b**), respectively (Scheme 4.22; Block, 1984, 1986). Brodnitz incorrectly identified the minor compound at *m*/*z* 144 as 3-vinyl-6*H*-1,2-dithiin. Thioacrolein is thought to be formed by a process

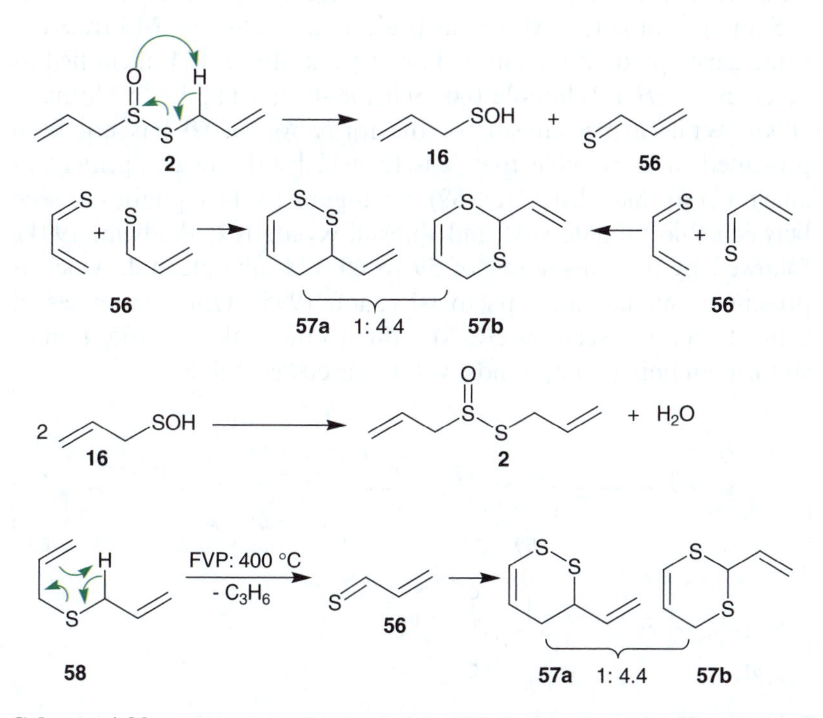

Scheme 4.22 Formation of thioacrolein (**56**) from allicin and from FVP of diallyl sulfide (**58**), and dimerization giving vinyl dithiins (**57a,b**).

analogous to that affording thioformaldehyde, further facilitated by weak allylic SC–H bonds (BDE of 85.8 kcal mol^{-1}; Okada, 2005, 2006; Scheme 4.8). Thioacrolein can also be generated by FVP at 400 °C of diallyl sulfide (**58**; Bock, 1982), using the pyrolysis system shown in Figure 4.5 attached to the trap shown in Figure 4.7. When frozen in liquid nitrogen, thioacrolein is sapphire blue; on warming **57a** and **57b** are formed in a ratio of 1 : 4.4, which is the same ratio of the mixture of these two compounds isolated from room temperature decomposition of allicin. Dithiins **57a** and **57b** are found in some types of garlic health supplements. Biological testing of **57a** and **57b** shows that they have interesting types of biological activity, discussed in Chapter 5. The decomposition of allicin (**2**) to 2-propenesulfenic acid (**16**) and thioacrolein (**56**), shown in Scheme 4.22, explains why allicin is more stable in hydrogen-bond-donating solvents such as water (Vaidya, 2008): hydrogen bonding ties up electron pairs on oxygen, retarding their hydrogen-abstracting ability.

Unsaturated heterocyclic thioacrolein dimers **57a,b**, formed by unimolecular decomposition of allicin, were first observed in the GC-MS analysis of garlic extracts as peaks at *m*/*z* 144. GC-MS traces of some garlic preparations also show a peak at *m*/*z* 104, identified as heterocycle 3*H*-1,2-dithiole (**60**; Scheme 4.23; Kim, 1995; Dittman, 2000). While a mechanism for the formation of **60** has not been presented, it is possible that it is formed by the rearrangement of allicin (**2**) to thiosulfoxylate (**59**), analogous to the equilibrium seen between thiosulfinate (**61**) and thiosulfoxylate (**62**; Baldwin, 1971), followed by decomposition of **59** to **60** and allyl alcohol, which is present in similar amounts to **60** (Kim, 1995). Other examples of artifact products seen by GC-MS, due to thermal decomposition of sulfur-containing compounds, will be discussed below.

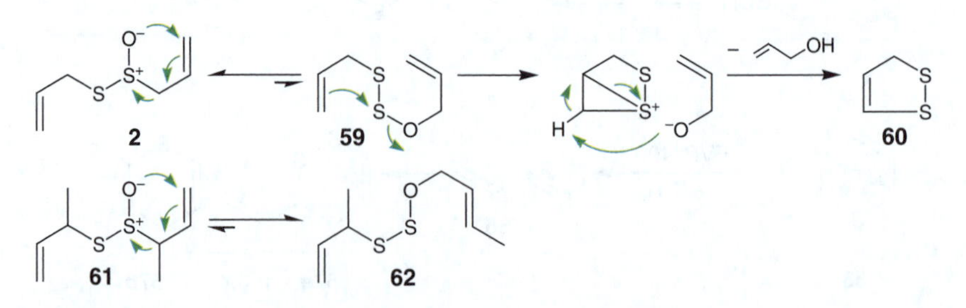

Scheme 4.23 Proposed mechanism for formation of 3*H*-1,2-dithiole (**60**) from allicin (**2**).

4.5.4 Ambient Mass Spectrometric Study of Alliums

Mass spectra are typically obtained by introducing samples into the vacuum environment of a mass spectrometer or by using specially prepared samples. A new family of techniques, ambient mass spectrometry, permits direct sampling of molecules in their native environment without sample preparation or separation by creating ions outside the mass spectrometer (Cooks, 2006). These techniques include desorption electrospray ionization (DESI; Takats, 2004), direct analysis in real time (DART; Cody, 2005), plasma-assisted desorption/ionization (PADI; Ratcliffe, 2007), and extractive electrospray ionization (EESI; Chen, 2007). The DESI and DART techniques are illustrated schematically in Figure 4.14. All of these techniques have been used to examine *Allium* chemistry.

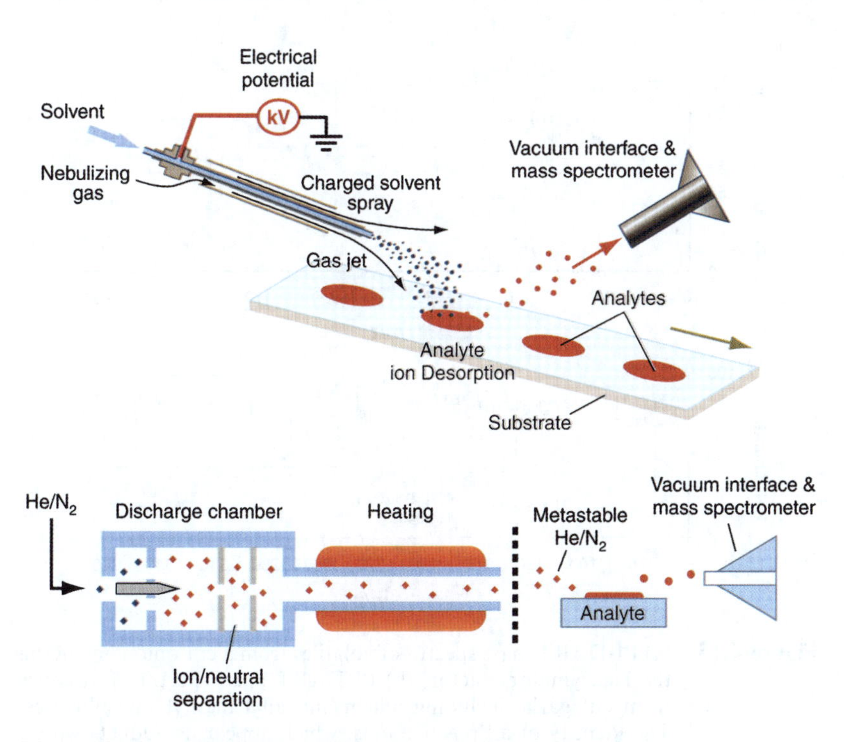

Figure 4.14 Schematic of DESI (upper) and DART (lower) analyses. (Image from Cooks, 2006).

4.5.4.1 Direct Analysis in Real Time (DART): A Mass Spectrometric Method for Direct Observation of 2-Propenesulfenic Acid, Propanethial S-Oxide, Allicin and Other Reactive Allium Sulfur Compounds. The mass spectrometric technique DART (Direct Analysis in Real Time) can be used to analyze samples at atmospheric pressure and ground potential by simply placing them between a DART ion source and a mass spectrometer (Figure 4.14; Cody, 2005). When an onion was abraded with a glass capillary, under positive ion (PI) DART conditions, a strong signal was seen for the onion lachrymatory factor (LF) as its adducts with a proton (*m*/*z* 91, $[C_3H_6SO+H]^+$) and an ammonium ion (*m*/*z* 108, $[C_3H_6SO+NH_4]^+$; ammonia is present from decomposition of aminoacrylic acid) along with the species involving two molecules of the LF plus a proton (*m*/*z* 181, $[2(C_3H_6SO)+H]^+$; Figure 4.15a). In addition small peaks were seen corresponding to protonated zwiebelanes (**84a,b**, see Section 4.11.1 below; *m*/*z* 163, $[C_6H_{10}S_2O+H]^+$), protonated thiosulfinates

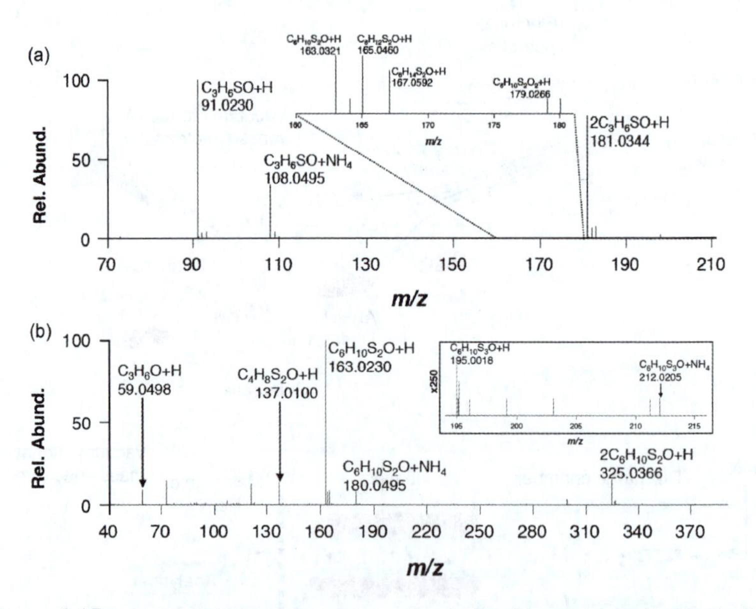

Figure 4.15 (a) PI-DART mass spectra of volatiles from a cut onion, including the lachrymatory factor; (b) PI-DART mass spectra of volatiles from cut garlic including allicin and allyl methyl thiosulfinates. The identity of all positive ions, which appear as adducts with a proton or ammonium ion, is confirmed by high resolution mass measurements. (Image from Block, 2009a).

MeCH=CHS(O)SPr-*n*, MeCH=CHSS(O)Pr-*n* (*m*/*z* 165, $[C_6H_{12}S_2O+H]^+$) and *n*-PrS(O)SPr-*n* (*m*/*z* 167, $[C_6H_{14}S_2O+H]^+$) and protonated bis-sulfine, O=S=CHCH(Me)CH(Me)CH=S=O (**92**, see Scheme 4.38 below; *m*/*z* 179, $[C_6H_{10}S_2O_2+H]^+$). When a garlic clove was sampled, PI-DART showed strong signals for allicin as its adducts with a proton (*m*/*z* 163, $[C_6H_{10}S_2O+H]^+$; Figure 4.15b) and with NH_4^+, a protonated dimeric species (*m*/*z* 325, $[2(C_6H_{10}S_2O)+H]^+$), weaker signals for protonated allyl methyl thiosulfinate (~5% abundance; *m*/*z* 137, $[C_4H_8S_2O+H]^+$) and protonated allyl alcohol (*m*/*z* 59), and very weak signals for protonated diallyl trisulfide *S*-oxide (**72**, *m*/*z* 195; Scheme 4.26 below) and its NH_4^+ adduct (*m*/*z* 212). All of the above peaks were characterized by high resolution mass measurements (work with R. B. Cody and A. J. Dane, JEOL USA; Block, 2009a). As expected, no signals for polysulfides were seen, since they are not primary products formed in when garlic or onions are cut.

Using negative ion (NI) DART methods, in addition to pyruvate, $CH_3C(O)CO_2^-$, major amounts of SO_2 were detected in onion volatiles (Figure 4.16a). This is consistent with Virtanen's proposal (Scheme 4.5) for decomposition of the onion lachrymatory factor through loss of "H_2SO". Hydrolysis of the LF (Scheme 4.24) nicely accounts for formation of H_2S, SO_2 and trithiolanes **63** (Kuo, 1992a,b), all found in onion extracts and distillates. Analysis of garlic volatiles under NI-DART conditions (Figure 4.16b) shows significant signals corresponding to SO_2 and pyruvate as well as the anions of 2-propenesulfenic acid, $CH_2=CHCH_2SO^-$, 2-propenesulfinic acid, $CH_2=CHCH_2SO_2^-$, and a species of formula $CH_2=CHCH_2SO_3^-$. Under these same conditions, an authentic sample of allicin only shows signals for 2-propenesulfinic acid and $CH_2=CHCH_2SO_3^-$. The mass spectral signal for sulfenate is very short-lived, disappearing within less than one second. This is the first direct observation of formation of a sulfenic acid upon cutting an *Allium* plant as well as the first observation of 2-propenesulfinic acid from garlic. The SO_2 presumably arises from decomposition of 2-propenesulfinic acid (**67**; Scheme 4.26 below) to propene, which is detected in trace amounts (Block, 2009a).

When freshly cut garlic is examined under PI-DART conditions at 24 °C, the summed ratio of peaks associated with allyl methyl thiosulfinates at *m*/*z* 136 to those associated with allicin and isomeric thiosulfinates at *m*/*z* 162, is 5 : 100 (Figure 4.15b). This ratio

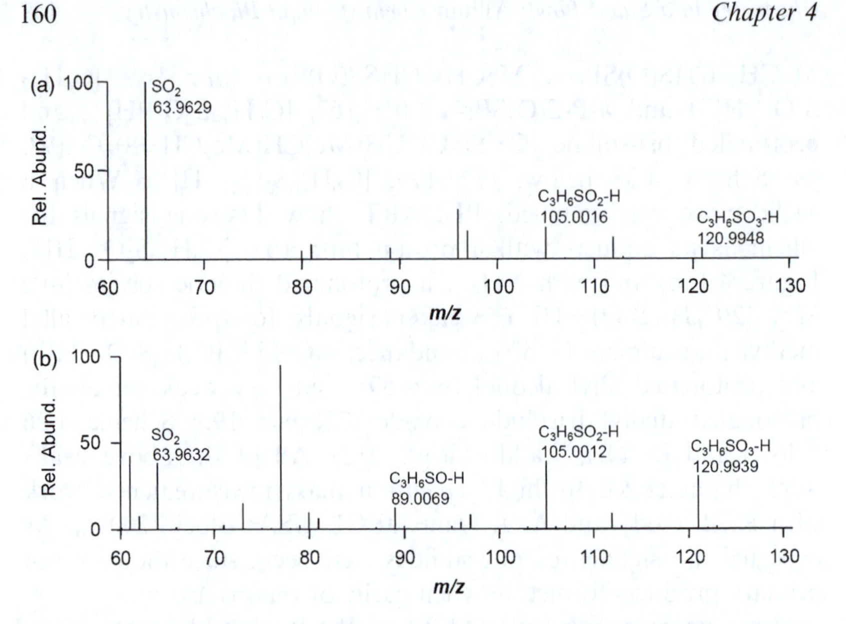

Figure 4.16 (a) NI-DART mass spectra of volatiles from cut onions. (b) NI-DART mass spectra of volatiles from cut garlic showing the 2-propenesulfenic acid anion. The identity of all anions, which are formed by deprotonation of the parent acids, was confirmed by high resolution mass measurements. (Image from Block, 2009a).

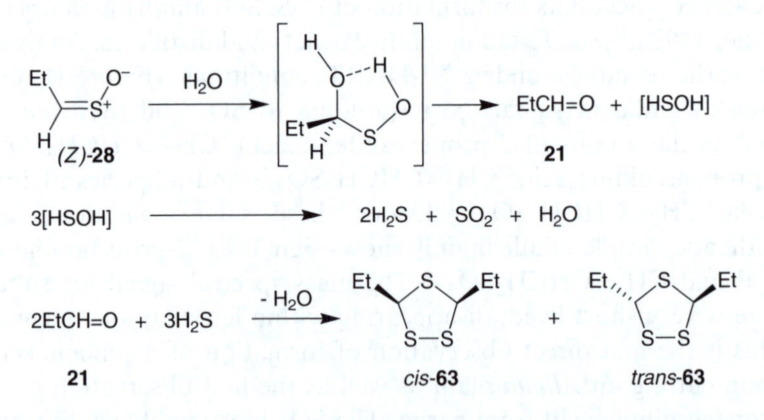

Scheme 4.24 Proposed mechanism for hydrolysis of the onion LF and formation of isomeric 3,5-diethyl-1,2,4-trithiolanes (*cis*- and *trans*-**63**).

increases to 6.8 : 100 at 40 °C. The 24 °C data compares well with data on ratios of methiin, alliin and isoalliin. The slightly higher relative methyl thiosulfinate levels seen with DART, may reflect the higher volatility of these compounds compared to allicin. The increase in this ratio with temperature is consistent with expectations that, while methiin is cleaved more slowly than alliin (Schmitt, 2005), a greater percentage of methiin reacts during the sampling time at 40 °C compared to 24 °C. Thus, from Lawson's work, the time to achieve a maximum yield of 50% for isomeric methyl thiosulfinates at m/z 136, decreases from 0.6 to 0.2 min as the temperature increases from 23 °C to 37 °C, while formation of allicin is almost instantaneous at both temperatures (Lawson, 1992).

The slower rate of enzymatic cleavage of methiin compared to alliin merits further comment. If methiin is cleaved ten times more slowly than alliin, then as methiin is being cleaved there should be no remaining alliin to produce the mixed allyl methyl thiosulfinates by the process shown in Equation (2). To rationalize the formation of the mixed allyl methyl thiosulfinates, it is suggested that methanesulfenic acid rapidly exchanges with allicin to produce MeS(O)SAll, see Equation (3), and that the resultant AllSOH combines with MeSOH as in Equation (2). Indeed it has been reported by Lawson that: (1) after allicin is maximally formed in 30 seconds it gradually decreases until the mixed allyl methyl thiosulfinates are formed at 5 minutes; and (2) for each mole of allicin lost, 1.9 moles of the mixed allyl methyl thiosulfinates are formed. This proposal requires that MeS(O)SAll be formed in excess over AllS(O)SMe, which is indeed the case for garlic (~2 : 1 ratio; Block, 1974b, 1992b; Lawson, 1992; Shen, 2000, 2002). The above proposal is also consistent with the finding that when powdered garlic is added to excess aqueous MeS(O)SMe, the major product is AllS(O)SMe (Equation (4); Block, 1992b).

$$\mathrm{AllSOH + MeSOH \rightarrow MeS(O)SAll + AllS(O)SMe} \quad (2)$$

$$\mathrm{AllS(O)SAll + MeSOH \rightarrow MeS(O)SAll + AllSOH} \quad (3)$$

$$\mathrm{MeS(O)SMe + AllS(O)CH_2CH(NH_2)CO_2H + alliinase} \\ \rightarrow \mathrm{AllS(O)SMe\ (major)} \quad (4)$$

4.5.4.2 Desorption Electrospray Ionization (DESI). In this technique a fine spray of charged droplets hits the surface of the sample, from which it picks up molecules and ionizes them, and delivers them into the mass spectrometer. In a novel use of this method, allicin from freshly crushed garlic was introduced into the mass spectrometer together with a cysteine spray solution.

This led to the detection of signals for the protonated molecular ion of allicin (m/z 163), the adduct of ammonia and allicin (m/z 180) and, significantly, a signal for an allicin–cysteine complex (m/z $284 = 162 + 121 + 1$), which is thought to precede formation of *S*-allylmercaptocysteine (7; Scheme 3.1; Zhou, 2008).

4.5.4.3 Extractive Electrospray Ionization (EESI). This technique was used to study the breath of volunteers who consumed garlic 30 minutes prior to breath analysis and then blew through a gas transfer line into the ESI source, which was maintained at 80 °C. Protonated allicin (m/z 163) and even γ-glutamyl *S*-methylcysteine (m/z 265) could be directly detected in the breath along with numerous other signals. However the spectrum showed no signals corresponding to nonpolar organosulfur compounds. To remedy this situation a silver nitrate water solution was used to generate the electrospray. This method proved to be highly selective for sulfur compounds, including both the nonpolar allyl methyl sulfide and a compound claimed to be methyl vinyl sulfide, as well as the polar allicin and γ-glutamyl *S*-methylcysteine, whose intensities, as ^{107}Ag and ^{109}Ag adducts, were enhanced ten-fold (Chen, 2007). The authors provide no proof that the compound at m/z 74, alleged to be methyl vinyl sulfide, is not in fact the isomeric 2-propenethiol, which has previously been shown to be present in garlic breath, and is thus a more likely choice for this peak.

4.5.4.4 Plasma-Assisted Desorption/Ionization Mass Spectrometry (PADI). This technique uses a nonthermal ("cold"), radio frequency-driven atmospheric plasma which is directed onto the analyte surface without charged particle extraction and without heating the sample. While a strong peak is seen for the onion LF, in contrast to the above DART studies, only a very weak signal $[M+H]^+$ peak is seen for allicin from garlic at m/z 163 (Radcliffe, 2009).

4.5.5 X-ray Absorption Spectroscopic Imaging of Onion Cells

How can living *Allium* tissue be examined, for example to obtain *in situ* information on the location within cells of different forms of sulfur, such as the LF precursor isoalliin? Traditional methods of analysis based on wet chemistry techniques would disrupt the cells leading to rapid reaction between alliinase and isoalliin, altering the chemical forms in which sulfur is present. Nondestructive analysis of sulfur is difficult since, as an element, sulfur lacks well-established spectroscopic probes and is therefore often called a spectroscopically "silent" element. Useful information on the speciation of sulfur in alliums and other plants can be obtained using the nondestructive technique of sulfur K-edge X-ray absorption spectroscopy (XAS; Sneeden, 2004) and X-ray fluorescence imaging (XFI). The technique of XAS involves tuning photon energy in a range where bound electrons can be excited (0.1–100 keV photon energy), and typically employs a synchrotron radiation facility having intense and tunable X-ray beams. When this technique is used with sulfur-containing systems, the near-edge portion of the X-ray absorption spectrum is dominated by dipole-allowed transitions to unoccupied levels possessing significant *p*-orbital character. XAS is very sensitive to electronic structure and is a particularly useful probe of sulfur since the spectra of various chemical forms are generally quite distinct. Both qualitative and quantitative analysis of different organosulfur functionalities (thiols, sulfides, sulfoxides, sulfones, *etc.*; Figure 4.17) can be achieved. By conducting these studies using a microscope, that is using spectromicroscopy with a microfused X-ray beam, it is possible to directly visualize the distribution of different chemical forms of sulfur within plant tissue, such as onion cells.

At the Stanford Synchrotron Radiation Laboratory, the K near-edge X-ray spectra of whole and homogenized *A. cepa*, *A. sativum* and *A. tuberosum* samples were examined. For the whole tissues of all three species, the spectra clearly show the presence of the sulfoxide precursor (*ca.* 2473.5 eV), the more reduced sulfur forms including thiols, disulfides and sulfides (2469–2472 eV), and small quantities of sulfate (2479.6 eV). X-ray absorption spectroscopy of onion sections showed increased levels of lachrymatory factor (LF) and thiosulfinate (RS(O)SR′), and decreased levels of sulfoxide (LF precursor; *ca.* 2473.5 eV) following cell breakage (Figure 4.17; Pickering 2009). In whole cells, X-ray fluorescence spectroscopic imaging showed elevated levels of sulfoxides in the cytosol, and elevated levels of reduced sulfur in the central transport vessels and bundle sheath cells.

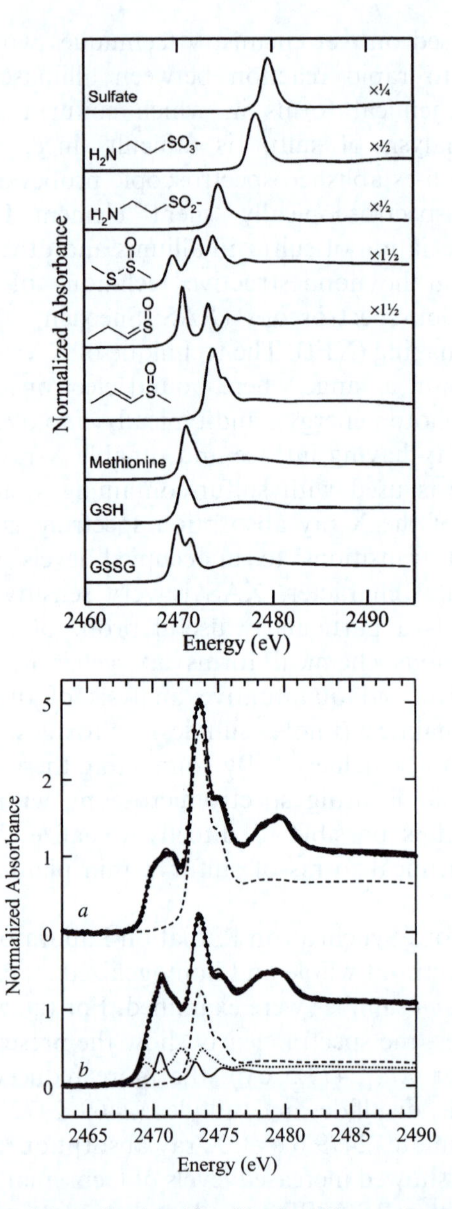

Figure 4.17 Top: Normalized sulfur K-edge X-ray absorption spectra of solutions of sulfur species relevant to onion. Bottom: Sulfur K-edge X-ray absorption spectra of onion tissue (a) before, and (b) after, rubbing to induce cell breakage. Experimental data are shown as points (· · · ·) and the linear combination fits as solid lines (—). In both (a) and (b) the sulfoxide components of the linear combination fits are shown as broken lines (--), and in (b) thiosulfinates, a dotted line (·········), and LF ((*Z*)-propanethial *S*-oxide) as a solid line (—). (Image from Pickering, 2009).

4.5.6 Other Separation and Analysis Methods: Supercritical Fluid Chromatography (SFC), Capillary Electrophoresis and Cysteine Sulfoxide-Specific Biosensors

Chromatography using supercritical fluid carbon dioxide (SFC) coupled to a mass spectrometer was used to characterize allicin from freshly cut garlic. Low oven (50 °C) and restrictor tip temperatures (115 °C) were needed in order to obtain a chemical ionization (Cl) mass spectrum of allicin, showing the protonated molecular ion at m/z 163 as the major ion, just as in Figure 4.15b (Calvey, 1994). Capillary electrophoresis (CE) separates species based on their size-to-charge ratio in the interior of a small capillary. The capillary is filled with an electrolyte where electrically charged analytes move in a conductive liquid medium under the influence of an electric field. The CE technique offers excellent resolution and selectivity, allowing for separation of analytes with very little physical difference. CE has been used to separate FMOC derivatives of alliin, isoalliin, methiin and propiin from different alliums (Kubec, 2008). Derivatives are prepared through reaction of fluorenylmethyl chloroformate with homogenates prepared by extracting the fresh alliums with 90% aqueous methanol, acidified with 10 mM HCl (to denature alliinase) and containing *S*-isobutylcysteine sulfoxide as internal standard. The analytical data on garlic, onion, shallot, leek and chive is included in Table 1 in Appendix 1 along with earlier CE data obtained by Horie (2006).

Keusgen and coworkers have developed a biosensor specific for cysteine sulfoxides, using a technique based on immobilized alliinase combined with an ammonia-gas electrode. The enzyme was either placed in a small cartridge or immobilized in direct contact with the electrode surface, giving detection limits of 0.37 and 5.9 μM, respectively (Keusgen, 2003).

4.6 PRECURSORS OF THE PRECURSORS, ALLIIN, ISOALLIIN AND METHIIN: BIOSYNTHETIC ORIGINS OF *ALLIUM* SULFUR COMPOUNDS

We have focused on alliin, isoalliin and methiin as the most significant precursors of the *Allium* flavor compounds. However, most of the γ-glutamyl derivatives of these compounds, with or without oxygen on sulfur (Figure 4.18), have also been isolated from

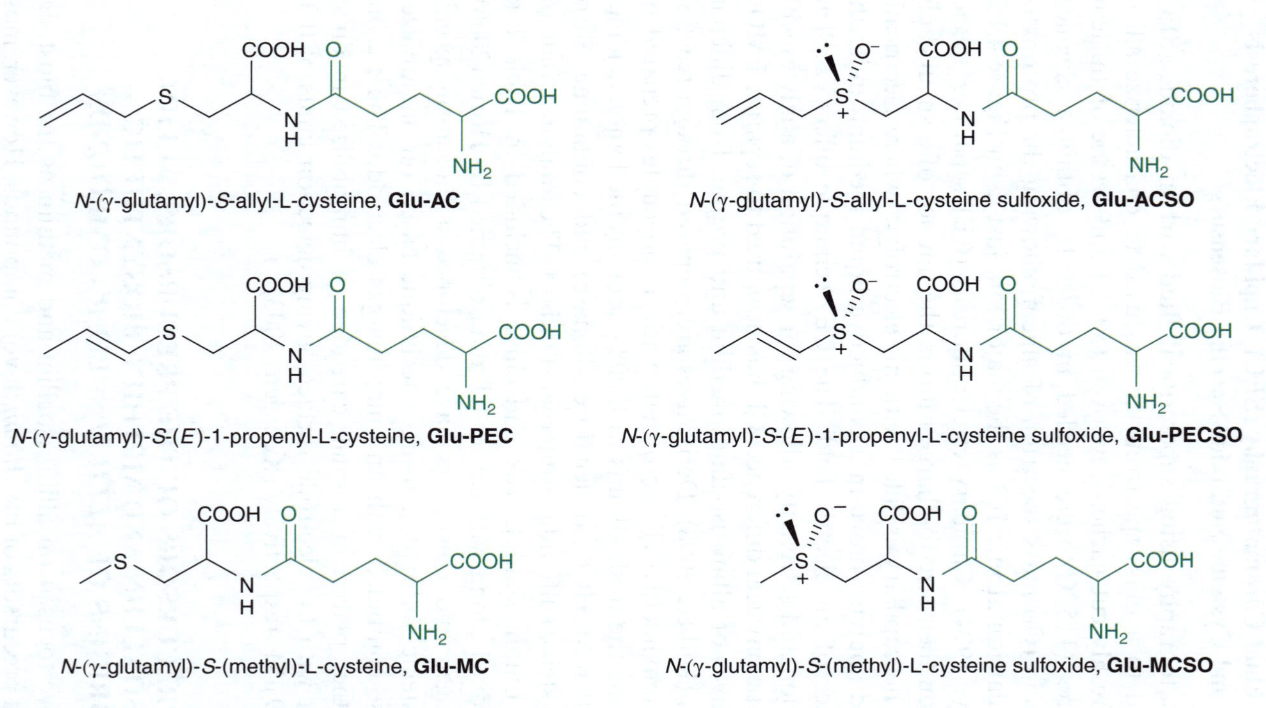

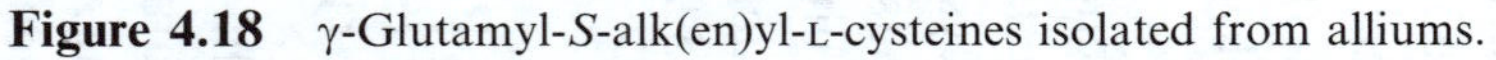

Figure 4.18 γ-Glutamyl-*S*-alk(en)yl-L-cysteines isolated from alliums.

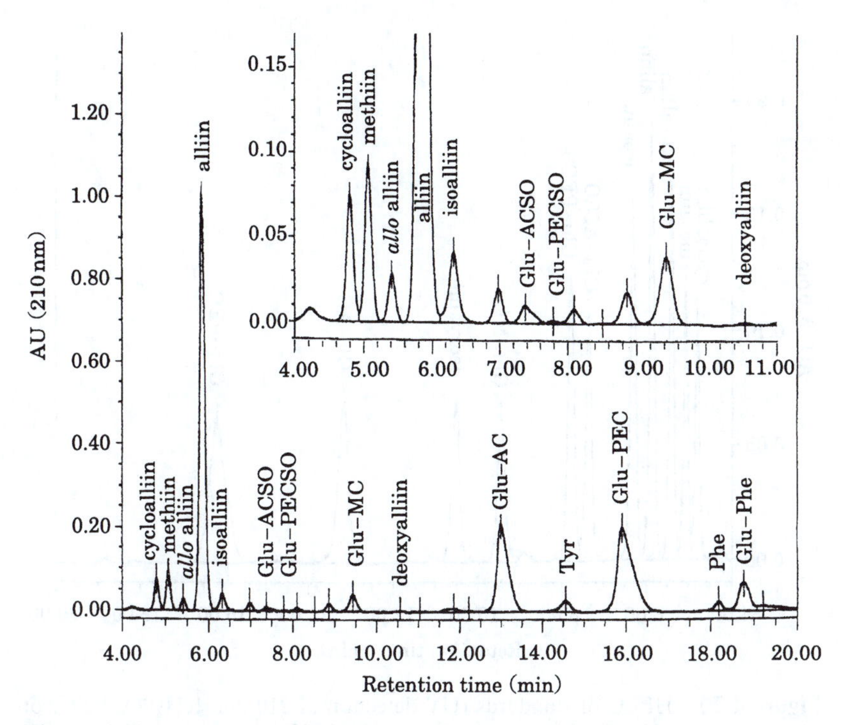

Figure 4.19 HPLC of alliinase-deactivated garlic extract. (Image from Yamazaki, 2005).

alliums and characterized. This was achieved in 1960 to 1962, by Virtanen (Virtanen, 1960, 1961, 1962) and Kyoto University chemists (Suzuki, 1961) using paper and ion-exchange column chromatography, among other techniques. Using HPLC with a cation-exchange column, excellent separation of standards has been achieved (Figure 4.19). All of the compounds shown in Figure 4.18 have been found in extracts of alliinase-deactivated fresh garlic (Figure 4.20), except *N*-(γ-glutamyl)-*S*-(methyl)-L-cysteine sulfoxide, which had been previously found in extracts of other alliums (Yamazaki, 2005). The average %dry weight values for the various components were: alliin, 2.62; (–)-*S*-(2-propenyl)-L-cysteine sulfoxide ("*allo* alliin"), 0.06; *S*-(2-propenyl)-L-cysteine (deoxyalliin), 0.02; methiin, 0.36; isoalliin, 0.13; cycloalliin, 0.16; Glu-MC, 0.43; Glu-AC, 2.34; Glu-PEC, 2.13; Glu-ACSO, 0.16; Glu-PECSO, trace (see Figures 4.11 and 4.18 for abbreviations, and Schemes 4.1 and 4.3 for structures of (–)-*S*-(2-propenyl)-L-cysteine sulfoxide (**12**) and cycloalliin (**19**), respectively).

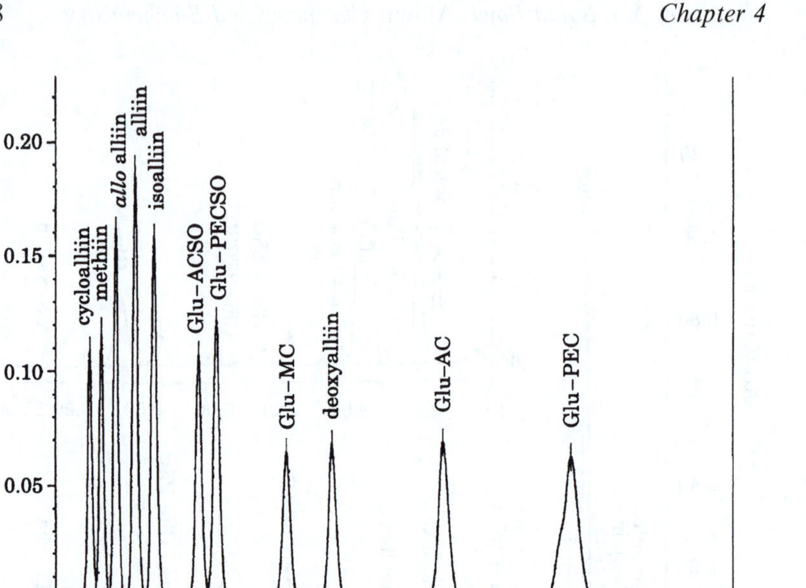

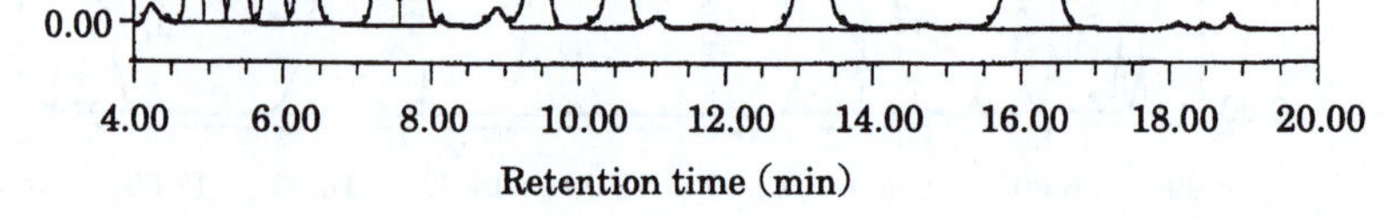

Figure 4.20 HPLC of standards (UV detection at 210 nm; KH_2PO_4 buffer of pH 2.5). (Image from Yamazaki, 2005).

Glu-AC and Glu-PEC are almost as abundant in fresh garlic bulbs as alliin. Similar results were obtained by normal-phase HPLC, with considerable variation in the composition of the compounds due to origin, variety, cultivation and storage conditions (Ichikawa, 2006). When fresh garlic is well crushed, γ-glutamyl-*S*-2-propenylcysteine is released and allicin and other thiosulfinates are rapidly formed. Garlic supplement tablets, based on dehydrated, powdered garlic, behave similarly when pulverized and mixed with water. The Kyolic supplement affords *S*-2-propenylcysteine and γ-glutamyl-*S*-2-propenylcysteine but no allicin, as summarized in Table 4.3 (Lawson, 2005, 1991c). By comparison, onions release γ-glutamyl-*S*-(*E*)-1-propenylcysteine sulfoxide in addition to the LF, and small amounts of other sulfur compounds (Lawson, 1991c).

Researchers have established the mechanism for the formation of alliin, isoalliin, methiin and propiin (*i.e.*, their *biosynthesis*), from the ultimate inorganic source of sulfur for the plant, namely sulfate (SO_4^{-2}), by injecting the plant with sulfur-35 labeled sulfate

Table 4.3 Major sulfur compounds from crushed garlic, garlic supplements pulverized with water (Lawson, 2005) and crushed onions (Lawson, 1991c).

Compound	*mg g^{-1} fresh weight or mg g^{-1} of tablet*
Crushed fresh garlic	
Allicin	3.1±0.11; range: 2.3–6.6
methyl 2-propenyl thiosulfinate[a]	1.2±0.05; range: 0.4–2.1
γ-glutamyl-*S*-2-propenylcysteine	5.0±0.30; range: 0.9–6.8
γ-glutamyl-*S*-(*E,Z*)-1-propenylcysteine	(E)-isomer: 3.6±0.15; (Z)-isomer: 0.06
S-2-propenylcysteine	0.06
Garlicin (garlic supplement)	
allicin[b]	5.2±0.3
allyl methyl thiosulfinate	1.2±0.1
γ-glutamyl-*S*-2-propenylcysteine	3.3±1.2
Kyolic 100 (garlic supplement)	
S-2-propenylcysteine	0.60±0.11
γ-glutamyl-*S*-2-propenylcysteine	0.82±1.0
Fresh boiling onion [yellow onion]	
γ-glutamyl-*S*-(*E*)-1-propenylcysteine	0.27±0.10
γ-glutamyl-*S*-(*E*)-1-propenylcysteine sulfoxide	4.0±1.9
Fresh yellow onion	
γ-glutamyl-*S*-(*E*)-1-propenylcysteine	0.08±0.03
γ-glutamyl-*S*-(*E*)-1-propenylcysteine sulfoxide	0.53±0.24

[a] ~0.4 mg g^{-1} fresh weight (1 mg g^{-1} dry weight) of allyl (*E*)-1-propenyl thiosulfinates also present. [b] >95% dissolution allicin release.

($^{35}SO_4{}^{-2}$; Ettala, 1962; Suzuki, 1961). As a result of such an isotope-labeling experiment, a total of 18 different ^{35}S-containing compounds were isolated. While sulfate is mostly found in the soil (a "pedospheric" source), H_2S and SO_2 serve, albeit to a lesser extent, as atmospheric sources of sulfur, taken up *via* the leaves. Alliums, like other plants, are able to "fix" the sulfur in sulfate, converting it into the amino acid cysteine, which is then transformed into a series of sulfur storage molecules which incorporate glutamic acid as a γ-glutamyl group (Jones, 2007). It is known that the concentration of alliin in garlic increases with levels of soil sulfur fertilization up to a point (above 200 kg ha^{-1} of sulfur fertilization there is no further increase; Arnault, 2003). Sulfate fixation is guided by specific sulfate transporter proteins.

Reduction of sulfate takes place mainly in the shoot of the plant through the following steps: (1) soil sulfate taken up by the root is transported to the shoot; (2) shoot sulfate is activated to adenosine

5′-phosphosulfate (APS); (3) APS is enzymatically reduced by APS reductase to sulfite, SO_3^{-2}; (4) the enzyme sulfite reductase reduces sulfite to sulfide, S^{-2}; and (5) sulfide is incorporated into cysteine by an enzyme-directed reaction with *O*-acetylserine ($CH_3C(O)OCH_2CH(CO_2H)NH_2$; Leustek, 1999; Durenkamp, 2004, 2007). What happens next depends on the stage of development of the plant, temperature, and other environmental factors. In the leaves, cysteine, in the form of γ-glutamyl cysteine or glutathione (a tripeptide containing cysteine, glycine and glutamate), is converted into its *S*-alkyl or *S*-alk(en)yl derivatives by way of compounds such as γ-glutamyl-S-(β-carboxypropyl) cysteinyl-glycine (**64**; Scheme 4.25). As the plant develops, these compounds are transported to the bulb or cloves, where they mainly function as sulfur and nitrogen storage compounds. The mechanism for decarboxylation of **64** to the *S*-1-propenyl compounds has been established by isotope-labeling studies (Parry, 1989, 1991).

There is uncertainty about several stages of these processes, including the origin of the allyl group and the oxidation step giving the sulfoxides (Jones, 2004, 2007; Hughes, 2005). 2-Propenethiol (**65**; allyl thiol), whose biosynthesis is unknown, is thought to react with *O*-acetylserine giving *S*-allylcysteine, which is then converted to its stored γ-glutamyl derivative. Indeed, it has been shown that feeding onion root cultures 2-propenethiol, propanethiol or ethanethiol, along with serine, affords substantial amounts of allyl, propyl or ethyl alk(en)ylcysteine sulfoxides, respectively, not normally found in onions (Prince, 1997). When the plant approaches maturation, or the bulb experiences lower temperatures, two sequential processes occur: (1) the various sulfur storage compounds are converted into alliin, isoalliin and methiin by processes requiring stereocontrolled oxidation *via* oxidase enzymes giving a chiral sulfoxide group; and (2) a hydrolysis of the peptide bonds by γ-glutamyl transpeptidase enzymes occurs.

Many different precursors to alliin, methiin and isoalliin have been isolated, characterized and synthesized, and some of these compounds display an interesting biological activity, as will be discussed separately. An amino acid glycoside of alliin has been isolated from the leaves of garlic plants (Mütsch-Eckner, 1993). Methiin has also been identified in all *Brassica oleracea* L. vegetables (cauliflower, cabbage, broccoli; Jones, 2004). Compounds related to alliin and methiin and their enzymatic degradation

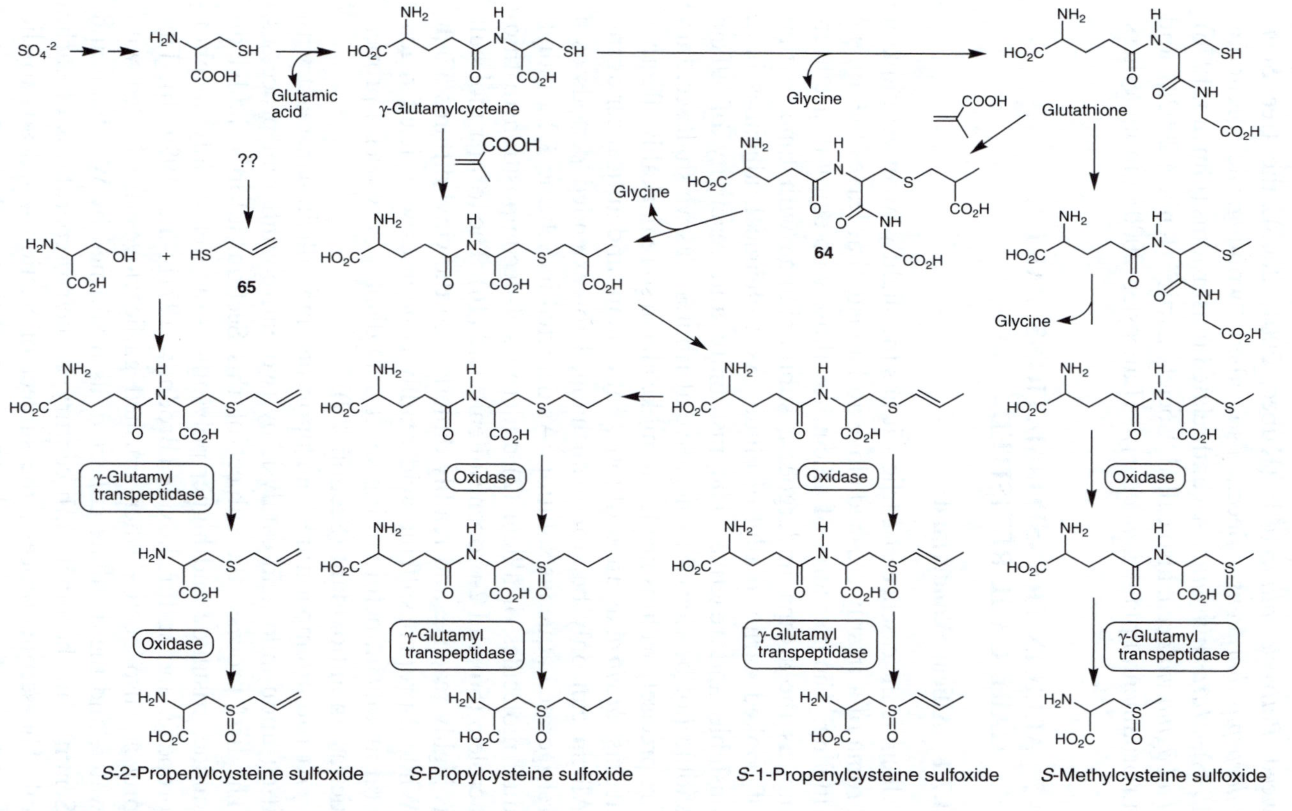

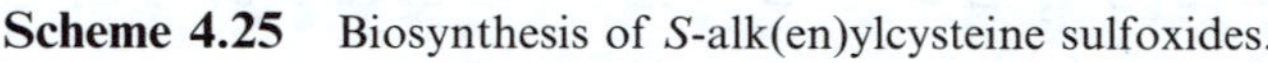

Scheme 4.25 Biosynthesis of *S*-alk(en)ylcysteine sulfoxides.

products have been isolated from other species, *e.g.*, the tropical weed *Petiveria alliacea* L (Kubec, 2001, 2005), the tree *Scorodocarpus borneensis* Becc., also called "wood garlic" (Kubota, 1998), *Marasmius alliaceus* and related mushrooms (Gmelin, 1976), *Tulbaghia violacea*, also called "society garlic" (Kubec, 2002a), and an ornamental plant *Leucocoryne* (Lancaster, 2000), among others.

4.7 ALLICIN TRANSFORMATIONS, PART 1: GARLIC OIL REVISITED

4.7.1 Allicin Wonderland

Allicin is the essence of garlic – it has the slightly sweet yet piquant aroma of a freshly crushed clove and a taste that leaves the tongue tingling. The pleasures of allicin, for those who enjoy it, are fleeting, as the aroma undergoes a subtle change with time to that of cooked garlic: diallyl disulfide. As a chemical, allicin is both unstable and reactive. The instability and reactivity of allicin should not be viewed as negative attributes – as we shall see, they are precisely what is needed to make allicin so remarkably effective, at least *in vitro*, as an antibiotic, antioxidant and anticancer agent! Allicin can only be stored unchanged for extended periods as a refrigerated aqueous solution. As depicted in Scheme 4.22, individual molecules of allicin (**2**) undergo self-decomposition to thioacrolein (**56**) and 2-propenesulfenic acid (**16**). The former molecule is highly reactive and readily dimerizes to the vinyl dithiins **57a**,**b**. While 2-propenesulfenic acid readily condenses with itself regenerating allicin, before doing so, it may display very useful properties as an antioxidant (Section 4.9).

At room temperature, and upon heating, allicin can react with itself upon mild acid catalysis by two entirely different processes. The first process, the subject of this Section, affords diallyl disulfide, trisulfide and higher polysulfides, as well as allyl alcohol, propene, and sulfur dioxide (Brodnitz, 1971b; Yu, 1989a,b). This process also occurs when crushed garlic cloves are boiled with water, and the distillate collected, as performed by Wertheim and Semmler in the nineteenth century in their preparations of garlic oil. The second process, the subject of Section 4.8, leads to the formation of a remarkable nine-carbon, three-sulfur, one-oxygen compound known as ajoene. All of the above reactions are

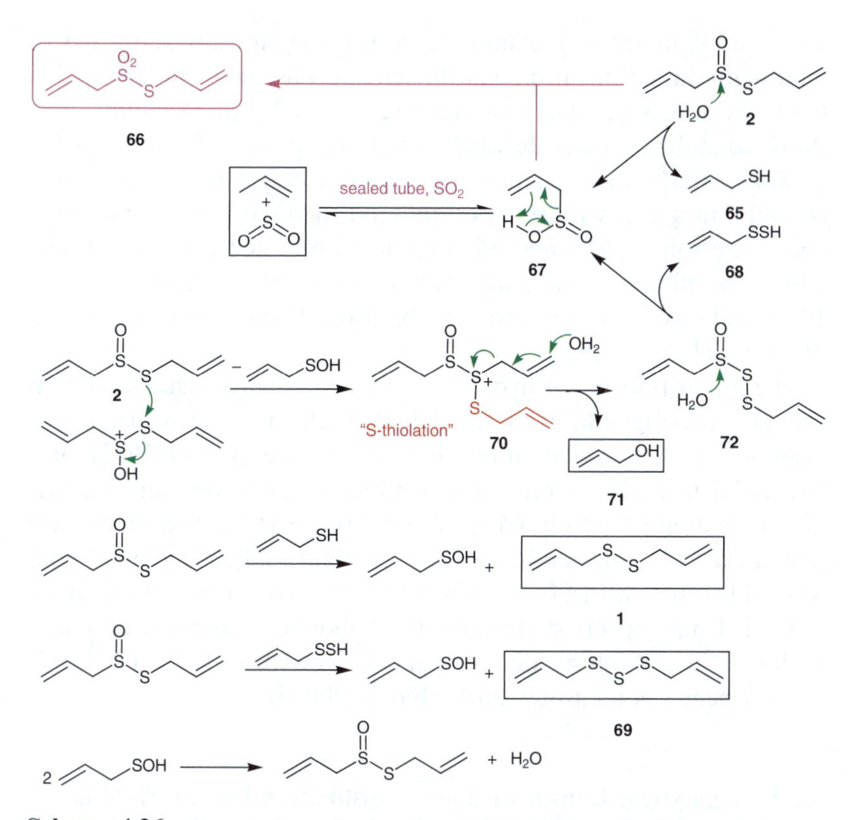

Scheme 4.26 Proposed mechanism for decomposition of allicin (**2**) at room temperature in water.

facilitated by the weakening of the central S–S bond of allicin, caused by attachment of a single oxygen that has the ability to drive chemical processes by bonding to a hydrogen, either internally (Scheme 4.22) or externally (Scheme 4.26). The fact that a relatively simple molecule such as allicin can undergo so many unusual transformations is indeed wonderous!

4.7.2 Garlic Oil Formation by Hydrolysis of Allicin

Scheme 4.26 accounts for the products formed upon decomposing allicin (**2**) in water. Amongst the products formed are propene and sulfur dioxide. In some studies, trace amounts of 2-propenyl 2-propenethiosulfonate (**66**; also known as pseudoallicin; Belous, 1950), have also been isolated (Hu, 2002). While decomposition of neat methyl methanethiosulfinate (**4**) gives significant quantities of methyl methanethiosulfonate (**40**; Scheme 4.8), **66** is only formed in trace amounts since the presumed precursor, 2-propenesulfinic acid (**67**) is unstable, rapidly decomposing to propene and sulfur dioxide as shown ($\Delta H^{\ddagger} = 11\,\text{kcal mol}^{-1}$; Hiscock, 1995). However, if allicin is sealed in a vial with excess liquid sulfur dioxide to increase the concentration of **67**, then **66** becomes the major product (Block, 1992). Pseudoallicin (**66**) was first synthesized by other means in 1950, and has been found to have biological activity similar to that of allicin (Belous, 1950).

The major product of hydrolysis of allicin, diallyl trisulfide (**69**), is thought to originate from a series of chemical transformations beginning with *S*-thiolation of allicin giving intermediate **70**. Hydrolysis of **70** gives allyl alcohol (**71**), together with a second intermediate **72**, which itself is hydrolyzed to **67** and 2-propene-1-sulfenothioic acid (**68**), a class of compounds known as perthiols. 2-Propenethiol (**65**) is formed together with **67** from direct hydrolysis of allicin. Through use of DART mass spectrometry (discussed above), it has proven possible to directly detect intermediates **67** and **72** as well as allyl alcohol, SO_2 and propene upon cutting garlic (Block, 2009a).

4.7.3 Analytical Considerations; Coordination Ion Spray-Mass Spectrometry (CIS-MS)

Scheme 4.26 simplifies the products actually found in garlic oil and the mechanisms leading to their formation. A remarkable array of straight-chain and cyclic sulfur-containing compounds are formed during the preparation of garlic oil and even just by the prolonged heating of diallyl disulfide. Of course, crushing garlic forms thiosulfinates containing methyl and 1-propenyl groups, in addition to the allyl groups. Therefore, the polysulfides found in the distilled oil of garlic contain methyl and 1-propenyl, as well as allyl groups (Equations (5) and (6); the subscript "*n*" indicates a range of values which can be as high as nine; 1-propenyl compounds are not shown). The very useful properties of garlic oil as a biologically active material are discussed in Chapters 5 and 6.

$$\begin{aligned} &CH_2{=}CHCH_2S(O)SCH_2CH{=}CH_2(\text{boil in water}) \\ &\quad\rightarrow CH_2{=}CHCH_2S_nCH_2CH{=}CH_2 \end{aligned} \tag{5}$$

$$CH_3S(O)SCH_2CH{=}CH_2 \text{ and } CH_2{=}CHCH_2S(O)SCH_3 \text{ (boil in water)} \rightarrow CH_3S_nCH_2CH{=}CH_2 \quad (6)$$

Gas chromatography, with its excellent resolving power and ability to be coupled to a mass spectrometer, has been a popular tool in the characterization of garlic oil. The earliest GC studies indicated the presence of diallyl and dimethyl sulfides, disulfides and trisulfides, as well as mixed allyl methyl polysulfides, in garlic oil and headspace (Oaks, 1964; Schultz, 1965). A recent GC study of garlic oil is informative. Steam distillation of freshly peeled and crushed Egyptian garlic cloves gave a distilled oil (2% yield), which was immediately analyzed following distillation, by GC (30 m× 0.32 mm column with 0.25 μm film thickness), indicating 2.5% All_2S, 3.2% All_2S_2, 2.1% $AllS_3Me$, 29.7% All_2S_3, 4.4% All_2S_4, along with smaller amounts of AllSMe, n-Pr_2S, $MeS_2CH{=}CHMe$, Me_2S_3, $AllS_2Pr$-n, $MeS_3CH{=}CHMe$, $AllS_3CH{=}CHMe$, $AllS_4Me$ and $AllS_4CH{=}CHMe$, among other compounds (All indicates $CH_2{=}CHCH_2$). The very high percentage of diallyl trisulfide in this sample is notable, as is the much higher yield of the distilled oil compared with the much lower yield (0.09%) obtained by Semmler in 1891. Elemental sulfur was also detected, but is not thought to be a genuine constituent of garlic oil. In fact it is generated by pyrolysis in the injection port, since the relative amount of sulfur increased with increasing temperatures of the injection port from 1.3% at 100 °C, to 12% at 200 °C (Jirovetz, 1992).

Earlier in this chapter, we presented some of the seminal studies by Arturi Virtanen and coworkers in Finland on the chemistry of *Allium* sulfur compounds. Observations published in 1970 by Virtanen's coworker, Bengt Granroth, on garlic oil are worth noting here. Granroth found that pure, colorless diallyl disulfide underwent decomposition at 100 °C "yielding a very complex pattern of products" by gas chromatography. Furthermore, while diallyl disulfide itself showed no significant antibiotic effect, higher diallyl polysulfides formed during decomposition of diallyl disulfide showed "very high" activity, *e.g.*, 4 parts per million of diallyl tetrasulfide inhibited the activity of the bacterium, *Staphylococcus aureus*. Granroth cautions: "considering the readily occurring thermal degradation of alkyl disulfides, particularly All_2S_2, many reported analyses of garlic substances may have been in part studies on artifacts." Granroth goes on to accurately predict that bis(1-propenyl) disulfide will prove to be "a still more unstable compound than the allylic isomer" (Granroth, 1970). The author was intrigued by Granroth's suggestions and undertook a detailed study of the thermal decomposition of both diallyl disulfide as well as the isomeric bis(1-propenyl) disulfide. The fascinating results of these investigations will be presented below, following a discussion of the key features of sulfur bonding and of the specialized analytical methods needed for the study of polysulfide (or polysulfanes, as they are alternatively named).

Garlic oil illustrates a key property of the element sulfur, namely the ability to *catenate*, *e.g.*, repeatedly join to itself with covalent bonds forming chains or rings (*catena* is Latin for "chain"). Elemental sulfur itself exists in nature as an eight-membered ring, that is as cyclooctasulfur or S_8 (Figure 4.21). Polysulfides with more than four sulfur atoms in a chain are thermally unstable. For this reason, the higher polysulfides in garlic oil are best analyzed using room temperature chromatographic techniques such as HPLC, rather than GC. GC requires heating, leading to decomposition, as suggested by Granroth. HPLC of garlic oil on a non-polar ("C-18") column using methanol–water, shows families of polysulfides (disulfides, trisulfides, tetrasulfides, *etc.*) as a series of evenly spaced peaks of increasing numbers of sulfurs (and increasing lipophilicity/hydrophobicity). Specific compounds found in garlic oil include All_2S–All_2S_6, $AllSMe$–$AllS_6Me$ and Me_2S_2–Me_2S_6, with pentasulfides and hexasulfides comprising 4.2 and 0.8% of the total sulfides, respectively. Trace amounts of heptasulfides and octasulfides were also present (Miething, 1988; Lawson, 1991; O'Gara, 2000). Many of these compounds were not previously seen by GC methods.

While the identity and amount of each component in a mixture can be determined by HPLC if the detector is properly calibrated, peak overlap and uncertainty concerning peak identity complicate interpretation. LC-MS using electrospray ionization mass

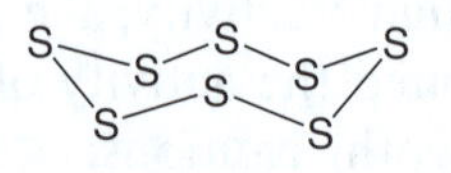

Figure 4.21 Elemental sulfur, S_8, in its "crown" conformation, showing the S–S bonding.

spectrometry (ESI-MS), as in the allicin and the onion LF DART studies (Figure 4.14), does not work well with nonpolar polysulfides, which have very low proton affinities. However, since silver ions strongly coordinate to diallyl polysulfides (Oliinik, 1997, 1998; Goreshnik, 1997, 1999; Salivon, 2006, 2007), post-separation infusion of silver salts, combined with electrospray mass spectrometry, greatly enhances sensitivity. This method, called coordination ion spray mass spectrometry (CIS-MS) or extractive electrospray ionization mass spectrometry (EESI-MS), when used to study human garlic breath (as described in Section 4.5.4.3), significantly increased the intensity of sulfur peaks (Chen, 2007). Diallyl polysulfides with up to six contiguous sulfur atoms from garlic powder were analyzed by this method, which showed the following peaks (given as All_2S_n, m/z $[M]^+$, m/z $[M+Ag]^+$ for Ag^{107}/Ag^{109} isotopes): All_2S_2, 146, 253/255; All_2S_3, 178, 285/287; All_2S_4, 210, 317/319; All_2S_5, 242, 349/351; All_2S_6, 275, 381/383 (Laisheng, 2006). While it was assumed, for purposes of quantitation, that all diallyl polysulfides had "similar molar absorption coefficients," this assumption is incorrect since coordination of diallyl polysulfides to silver ions varies with the total numbers of sulfur atoms in a chain (Block, 2009b). For this reason, the claim that diallyl tetrasulfide was the major polysulfide in a sample of garlic oil analyzed by Ag^+-CIS-MS (Laisheng, 2006) should be viewed with skepticism.

Peak separation can be improved and run times shortened by using ultra performance liquid chromatography (UPLC) instead of HPLC. This technique uses much smaller particle sizes for the chromatographic support than HPLC, and much higher flow rates and pressures, together with a shorter column, to achieve much faster separations than HPLC, along with narrower, taller peaks. Typically, separation at a flow rate three times higher than that used for HPLC with a column one third of the length, results in an analysis completed in one-ninth of the time, while maintaining peak resolution. Furthermore, the UPLC operating conditions are ideal for use with ion spray mass spectrometers capable of very rapid data collection. Further improvements can be achieved by selective ion monitoring techniques, whereby certain ions, or families of ions, are chosen for display. This latter technique is especially useful when different compounds co-elute (*e.g.*, pass through the column at the same rate).

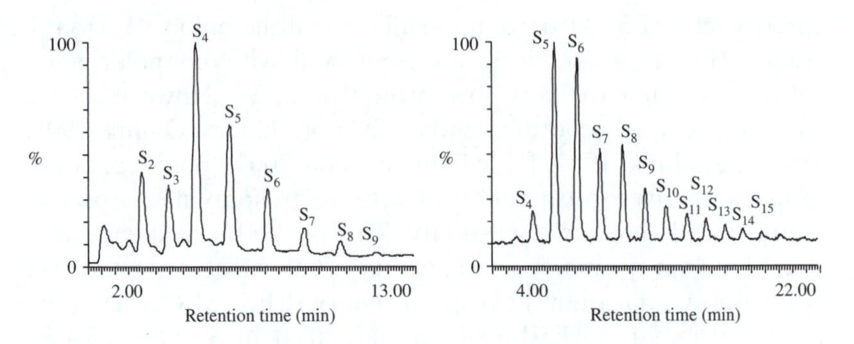

Figure 4.22 UPLC-(Ag^+)CIS-MS of diallyl polysulfides in garlic oil before (left), and after (right), liquid sulfur treatment. (Block, 2009b).

Application of the UPLC-(Ag^+)CIS-MS technique to a sample of garlic oil, with collaborator Robert Sheridan, is illustrated in Figure 4.22 (Block, 2009b). In this work, immediately following chromatographic separation, a solution of silver tetrafluoroborate ($AgBF_4$) is introduced into the liquid sample. The ions that are then observed are of the type $[M + Ag]^+$. We found $AgBF_4$ to be preferable to $AgNO_3$, since under some circumstances the nitrate ion can lead to oxidation processes. Furthermore, $AgBF_4$ is a stronger Lewis acid than $AgNO_3$. Selective ion techniques were used to generate the Figure 4.22 traces, which only show polysulfides containing two allyl groups, and not other types of compounds also present in this mixture. Enhancement of the mixture by liquid sulfur treatment will be discussed below.

4.7.4 Analysis of Garlic Oil by Nuclear Magnetic Resonance Spectroscopy

The unusual preponderance of diallyl tetrasulfide in garlic oil found by LC-(Ag^+)CIS-MS in our work, as well as by Laisheng (2006), is at odds with analyses of garlic oil by HPLC and GC. In an effort to obtain quantitative information on the composition of garlic oil by another method not requiring heating, we examined the proton NMR spectrum of the same sample of garlic oil used in the MS studies (Hile, 2004). For a complex mixture of naturally derived products, garlic oil has a relatively simple ^1H NMR spectrum (Figure 4.23). It consists of: a nicely separated series (**a–e**) of doublets (each with coupling constant $J = 7.3$ Hz) from 3.1 to 3.7 ppm for the

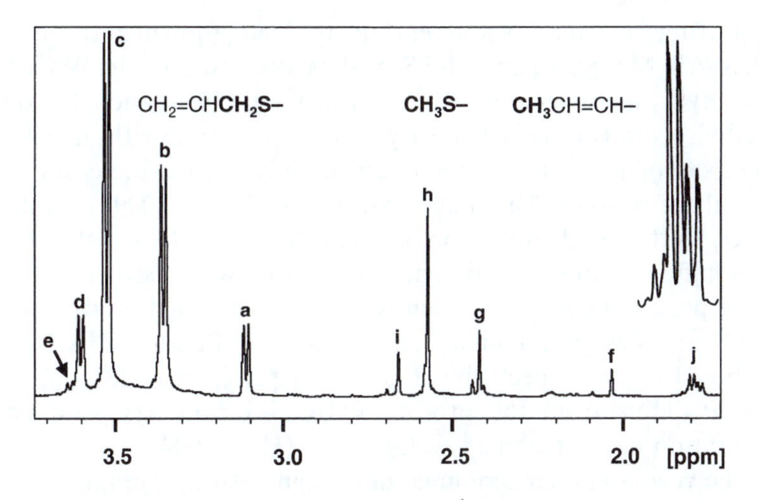

Figure 4.23 The 1.8–3.7 ppm region of the ^{1}H NMR spectrum (500 MHz; $CDCl_3$) of a representative commercial sample of distilled oil of garlic. Peaks **a–e** correspond to the allylic CH_2S protons (doublets) of All_2S, All_2S_2, All_2S_3, All_2S_4 and All_2S_5, respectively; peaks **f–i** correspond to the CH_3S protons (singlets) of MeSAll, MeS_2All, MeS_3All and MeS_4All, respectively; multiplet **j** (also shown in enlarged form; two doublets of doublets) corresponds to the CH_3 protons of the (*E*,*Z*)-$CH_3CH{=}CHS$– group. (Hile, 2004).

thioallylic protons ($CH_2{=}CHC\mathit{H}_2S$); a similarly well separated series **(f–i)** of singlets from 2.0 to 2.7 ppm for the thiomethyl protons ($C\mathit{H}_3S$); a weak but characteristic set **(j)** of doublets of doublets, centered at 1.8 ppm ($J = \sim 7$ and 1) for the methyl protons of (*E*)- and (*Z*)-1-propenyl groups (*E* : *Z* ratio of 2 : 1; $C\mathit{H}_3CH{=}CHS$); along with olefinic multiplets (not shown) at 5 to 6 ppm. There is virtually no absorption in the 0–1.8, 2.7–3.1 or 3.7–5.0 ppm regions.

It is significant that the pattern of the four singlets (peaks **f**, **g**, **h**, **i**) almost exactly parallels the pattern of the four major doublets (peaks **a**, **b**, **c**, **d**). This is consistent with the view that these sequences of peaks reflect the corresponding families of diallyl compounds: AllSAll (**a**), AllSSAll (**b**), AllSSSAll (**c**), AllSSSSAll (**d**); and methyl/allyl compounds: MeSAll (**f**), MeSSAll (**g**), MeSSSAll (**h**) and MeSSSSAll (**i**). Further scrutiny of the spectrum reveals a similar, but very tiny, peak pattern for the three major dimethyl polysulfides: MeSSMe (intermediate intensity), MeSSSMe (highest intensity) and MeSSSSMe (lowest intensity), appearing as small shoulders to the left (downfield) on the MeSSAll, MeSSSAll and MeSSSSAll peaks, respectively. While the thioallylic methylene protons (CH_2=$CH\mathit{CH_2}S_n$) and the thiomethyl protons (CH_3S_n) show greater deshielding with increasing numbers of sulfur ("n"), the effect diminishes as n increases such that there is no further change for $n \geq 6$ (Block, 2009b). In most cases, authentic samples, prepared by synthesis or isolated from polysulfide mixtures by preparative HPLC, were used to confirm these peak assignments, which were also supported by carbon-13 NMR analysis of individual compounds and of garlic oil itself. The carbon-13 NMR spectrum of garlic oil provides additional data not available from the proton NMR spectrum based on peak separation in the olefinic (C=C) region (Hile, 2004).

The relative molar amounts and weight percent composition of this representative garlic oil sample, summarized in Table 4.4, can be calculated from the integrated peak areas for compounds containing up to five sulfur atoms. This data shows diallyl trisulfide to be the major component of garlic oil, which is in good agreement with GC-MS data. The predominance of diallyl tetrasulfide indicated by the above described LC-(Ag^+)CIS-MS analysis reflects the superior coordination of silver to the tetrasulfide compared to the trisulfide, as confirmed using authentic samples (Block, 2009b). Thus, the LC-(Ag^+)CIS-MS method cannot be used for purposes of quantitation

Table 4.4 500 MHz ^{1}H NMR Data for Garlic Oil (Hile, 2004).

$\delta^1 H$	*multiplicity (J), group*	*mole (wt) %*	*compound*
3.67	d (7.3), CH_2S	trace	All_2S_6
3.63	d (7.3), CH_2S	1 (1)	All_2S_5
3.60	d (7.3), CH_2S	6 (8)	All_2S_4
3.52	d (7.3), CH_2S	33 (37)	All_2S_3
3.36	d (7.3), CH_2S	26 (23)	All_2S_2
3.11	d (7.3), CH_2S	7 (5)	All_2S
2.70	s, CH_3	trace	MeS_5All
2.67	s, CH_3	trace	Me_2S_4
2.66	s, CH_3	2 (3)	MeS_4All
2.59	s, CH_3	1 (1)	Me_2S_3
2.58	s, CH_3	11 (10)	MeS_3All
2.44	s, CH_3	0.4 (0.2)	Me_2S_2
2.42	s, CH_3	5 (4)	MeS_2All
2.04	s, CH_3	2 (1)	MeSAll
1.80	dd (6.5, 0.9), CH_3	4 (5)	(*E*)-MeCH=CHS_nAll
1.77	dd (6.9, 1,1), CH_3	2 (2)	(*Z*)-MeCH=CHS_nAll

without prior calibration using authentic samples. An advantage of the NMR method is its simplicity, the fact that it is an absolute method directly reflecting the molar amounts of each component, and the utility of the method to spot adulteration, which would appear as additional peaks not typically seen in authentic, pure samples of garlic oil. A limitation of the NMR method is that it cannot be used to differentiate polysulfides containing more than six contiguous sulfur atoms, due to the coalescence of chemical shifts above this number of sulfur atoms. Fortunately, HPLC can be used to "see" polysulfides containing virtually any number of sulfur atoms. Application of the NMR method to onion oil shows that a typical sample contains a mixture of 37% dipropyl disulfide [$(CH_3CH_2C\mathit{H}_2S)_2$; $C\mathit{H}_2S$ δ 2.66 ppm], 24% trisulfide ($C\mathit{H}_2S$ δ 2.84 ppm), 17% tetrasulfide ($C\mathit{H}_2S$ δ 2.92 ppm), 14% pentasulfide ($C\mathit{H}_2S$ δ 2.96 ppm) and 7% hexasulfide ($C\mathit{H}_2S$ δ 2.97 ppm), along with minor amounts of methyl polysulfides (Block, 2009b).

4.7.5 Synthesis of Symmetrical and Unsymmetrical Trisulfides and Heavier Polysulfides

Synthesis of individual polysulfanes, particularly those with *different* alk(en)yl groups, requires care since the allylic C=C bond represents a center of reactivity. A useful synthesis of the garlic oil component, allyl methyl trisulfide, with a 60% yield is shown in Equations (7) and (8). An amine catalyst such as *N*-methylmorpholine is required to scavenge HCl (Mott, 1984). Unsymmerical disulfides can be synthesized by thiol-induced fragmentation of sulfenyl thiocarbonates (Brois, 1970). Symmetrical polysulfanes with three to five sulfur atoms, as shown in Equation (9), can be prepared through reactions developed by David Harpp and coworkers (Hou, 2001, 2000; Rys, 2000; Jacob, 2008).

$$CH_3OC(O)SSCl + CH_3SH \rightarrow CH_3OC(O)SSSCH_3 \tag{7}$$

$$\begin{aligned} &CH_3OC(O)SSSCH_3 + CH_2{=}CHCH_2SH + R_3N \\ &\qquad \rightarrow CH_2{=}CHCH_2SSSCH_3 + R_3NHCl \end{aligned} \tag{8}$$

$$\begin{aligned} &2CH_2CH{=}CH_2SH + S_2Cl_2 + \text{pyridine/ether/} -78\,^{\circ}C \\ &\qquad \rightarrow CH_2{=}CHCH_2SSSSCH_2CH{=}CH_2 \end{aligned} \tag{9}$$

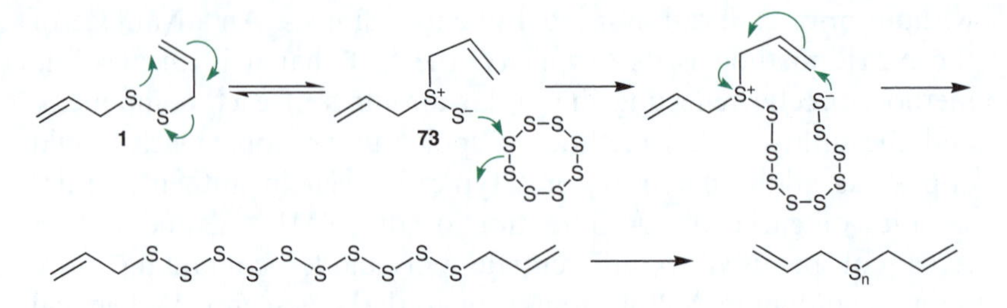

Scheme 4.27 Reaction of diallyl disulfide (**1**) with liquid sulfur giving a family of diallyl polysulfides.

In a large number of biological studies, described in Chapters 5 and 6, the biological activity of diallyl polysulfides was found to increase with increasing numbers of sulfur, at least up to four sulfurs, *e.g.*, the typical order of activity is tetrasulfides > trisulfides > disulfides. Diallyl polysulfides with more than four sulfur atoms in a chain, while they occur naturally in garlic oil, are little studied because they are only available in very small quantities by tedious isolation procedures. By exploiting a novel property of diallyl disulfide (**1**), whereby on heating it is in equilibrium with diallyl thiosulfoxide (**73**; Scheme 4.27; Barnard, 1969; Höfle, 1971; Baechler, 1973) and allowing this polar "valence isomer" to briefly react with liquid sulfur (shown as cyclooctasulfur ring) at its melting temperature of 115–120 °C, it has been possible to directly prepare mixtures enriched in the higher polysulfides, such as diallyl decasulfide, both from garlic oil or synthetic diallyl disulfide (Figure 4.22; Block, 2009b). Prior to the reaction, the disulfide and liquid sulfur exist as two separate phases. However, within three minutes at 115–120 °C, the solution suddenly becomes homogeneous as the reaction occurs. Under these reaction conditions, the diallyl decasulfide is further transformed into polysulfides with higher and lower numbers of linked sulfur atoms.

Under these conditions, a family of diallyl polysulfides containing as many as 22 sulfur atoms in a continuous chain can be prepared, as indicated by HPLC analysis (Figure 4.24). This procedure makes available, for analysis and biological testing purposes, samples of diallyl polysulfides with five to nine sulfurs which are found naturally, as well as the previously unknown diallyl polysulfides with ten or more sulfur atoms, not yet detected in nature.

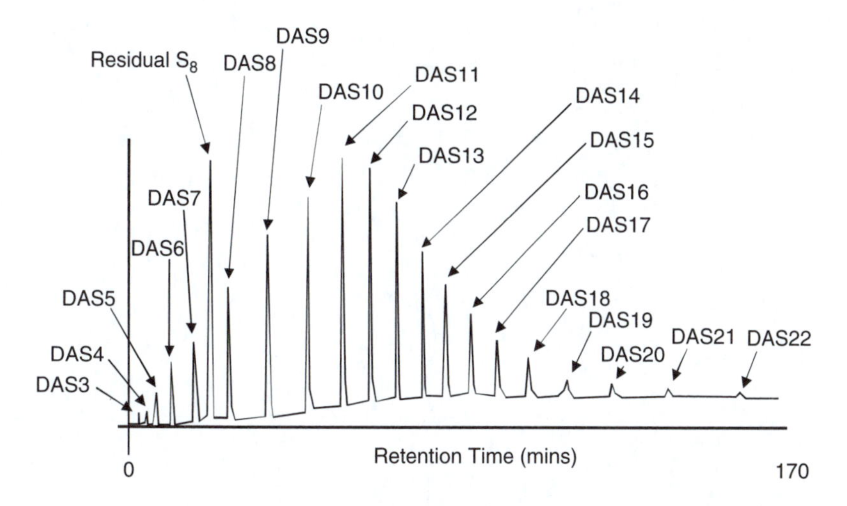

Figure 4.24 Analytical C-18 HPLC trace of liquid S_8 + diallyl disulfide reaction product. Individual diallyl polysulfides are abbreviated "DAS*n*" where "*n*" represents the number of contiguous sulfurs. More stable polysulfides can be isolated using preparative HPLC. (Block, 2009b).

Several aspects of biologically relevant diallyl polysulfide chemistry can be summarized here, prior to discussing mechanisms related to these compounds While diallyl polysulfides are formed by decomposition of allicin, as noted above, they can also be enzymatically formed by action of alliinase on a mixture of alliin (**10**) and cystine (**75**; Keusgen, 2008; Lancaster, 2000). Alliinase presumably converts cystine to disulfane (HSSH), which then reacts with 2-propenesulfenic acid (**16**) and/or allicin (**2**), formed from alliin to produce diallyl tetrasulfide (**74**), and other polysulfides (Scheme 4.28).

Diallyl disulfide can neutralize the toxic effects of cyanide. Diallyl thiosulfoxide (**73**) is thought to play a key role in this process, by functioning as a sulfane sulfur donor to the cyanide ion, thereby converting cyanide to the less toxic thiocyanate (see Chapter 5.11; Iciek, 2005). The chelating ability of diallyl polysulfides, which in the case of silver and copper involves both the sulfur atoms as well as the C=C double bonds, may explain the ability of diallyl tetrasulfide to protect cells against the toxic effects of cadmium (Pari, 2007). The varying biological activities of diallyl polysulfides also reflects their lipophilicity and ability to penetrate

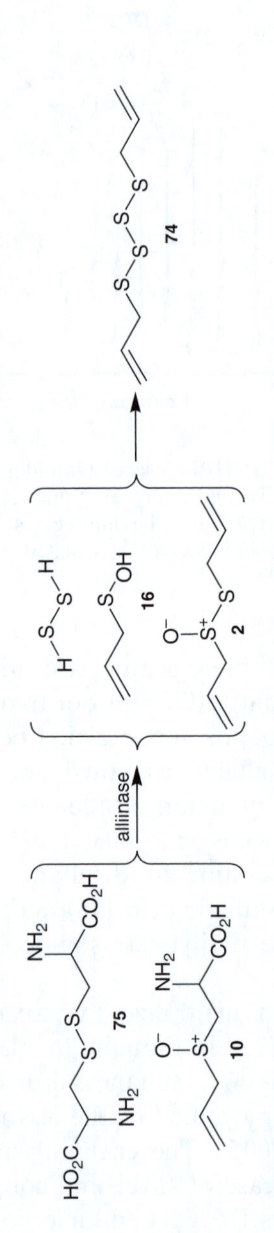

Scheme 4.28 Direct formation of diallyl tetrasulfide **(74)** from cystine **(75)**, alliin **(10)** and alliinase.

biological membranes, as well as their relative S–S bond strengths. The latter is related to the ability of these bonds to react with biological thiol groups, as will be considered below.

4.7.6 Mechanistic Considerations

As discussed in Section 4.7.3, when diallyl disulfide (**1**) is heated, it decomposes giving a complex mixture (Granroth, 1970). With graduate student, Raji Iyer, we studied the effect of heat on pure samples of **1** and sought to identify the products of this intriguing reaction. Through use of GC-MS we discovered a series of new *nine-carbon* polysulfides. It would appear that these are formed from 1.5 molecules of **1** (Figure 4.25; Block, 1988). These same compounds were also found in trace amounts in garlic oil. The concentrations of these compounds could be increased by prolonged heating of **1** or garlic oil at 80 °C. Analysis of a garlic oil sample by UPLC-(Ag^+)CIS-MS with selected ion monitoring confirmed the presence of families of compounds, just as in the case

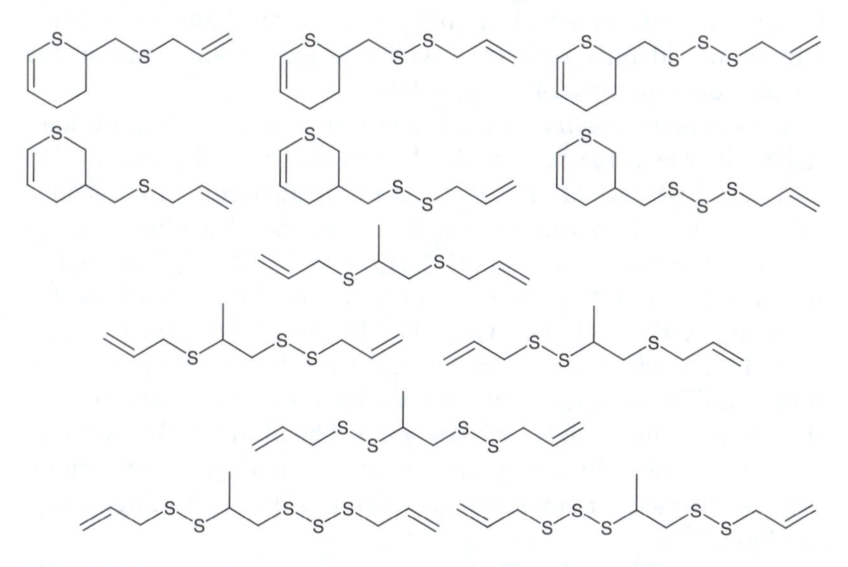

Figure 4.25 Families of nine-carbon sulfur compounds ($C_9H_{14}S_n$, top series; $C_9H_{16}S_n$, bottom series) present in garlic oil and formed upon heating diallyl disulfide. Related families of compounds ($C_7H_{14}S_n$ and $C_5H_{12}S_n$) corresponding to the bottom series are found in which one or both of the allyl groups are replaced by methyl groups.

of the diallyl polysulfides, and also showed the presence of related compounds with five and seven carbon atoms (Figure 4.26; Block, 2009b).

A detailed mechanism was published, accounting for the formation of all of the compounds (Schemes 4.29 and 4.30; Block, 1988). In **1**, the C–S bond (46 kcal mol^{-1}) is 16 kcal mol^{-1} weaker than the S–S bond (62 kcal mol^{-1}; Block, 1988; Gholami, 2004). Cleavage of the C–S bond of diallyl disulfide (**1**) is followed by a sequence of reactions initiated by addition of the allyldithio radical (AllSS•) to **1** and ultimately leading to the compounds shown in Figure 4.25. To explain the origin of these compounds, the author proposed a mechanism which represented a rare exception to the manner in which free radicals normally add to unsymmetrical carbon–carbon double bonds. Thus radicals typically add to the *less-substituted* carbon of an unsymmetrical C=C bond, so as to generate the more stable radical intermediate by a process termed *anti-Markovnikov addition*, *e.g.*, X• + CH_2=CHR → X–CH_2–CHR•. However when the allyldithio radical adds to **1**, anti-Markovnikov addition would simply be followed by β-cleavage of the intermediate, regenerating the starting radical and **1** in what is termed an "identity reaction," which would go unnoticed in the absence of some type of isotopic labeling.

Markovnikov addition would generate a less stable, primary radical in which the reactive RCH_2• radical is well positioned to abstract a hydrogen atom from the position adjacent to the disulfide sulfur. This process would trigger loss of thioacrolein (**56**) and form a new, sulfur-centered radical CH_2=$CHCH_2SSCH_2CH(CH_3)S$•, which can ultimately form various compounds containing the grouping CH_2=$CHCH_2S_nCH_2CH(CH_3)S$–. Thioacrolein reacts with the abundant molecules of diallyl disulfide and trisulfide, to form families of cyclic compounds by Diels–Alder addition to the allylic double bonds (Scheme 4.30). The above provides a very satisfying explanation for the formation of a remarkable number of unusual organosulfur compounds upon heating diallyl disulfide.

Higher diallyl polysulfides can also reversibly isomerize to thiosulfoxides, and these can transfer sulfur to other molecules. In this way, polysulfides of varying numbers of sulfur atoms can form and decompose. Higher polysulfides, such as diallyl tetrasulfide, in pure form undergo disproportionation on exposure to heat,

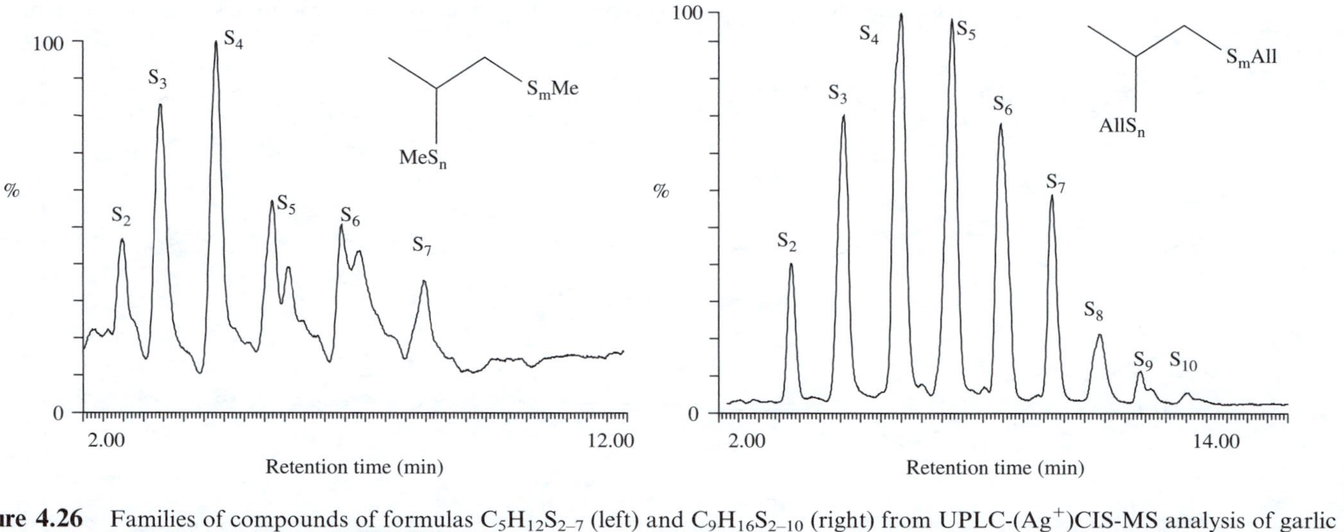

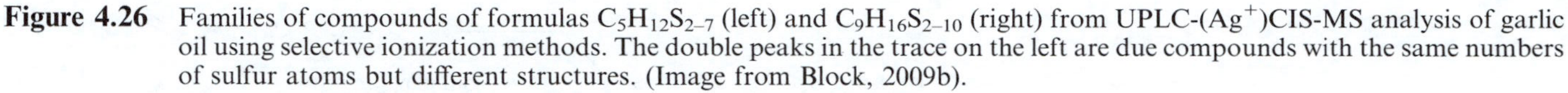

Figure 4.26 Families of compounds of formulas $C_5H_{12}S_{2-7}$ (left) and $C_9H_{16}S_{2-10}$ (right) from UPLC-(Ag^+)CIS-MS analysis of garlic oil using selective ionization methods. The double peaks in the trace on the left are due compounds with the same numbers of sulfur atoms but different structures. (Image from Block, 2009b).

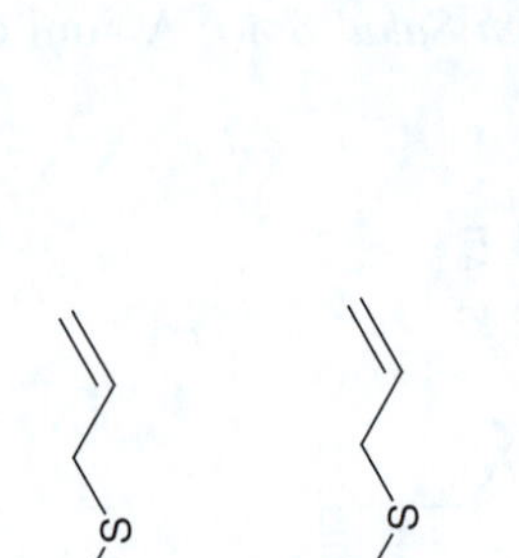

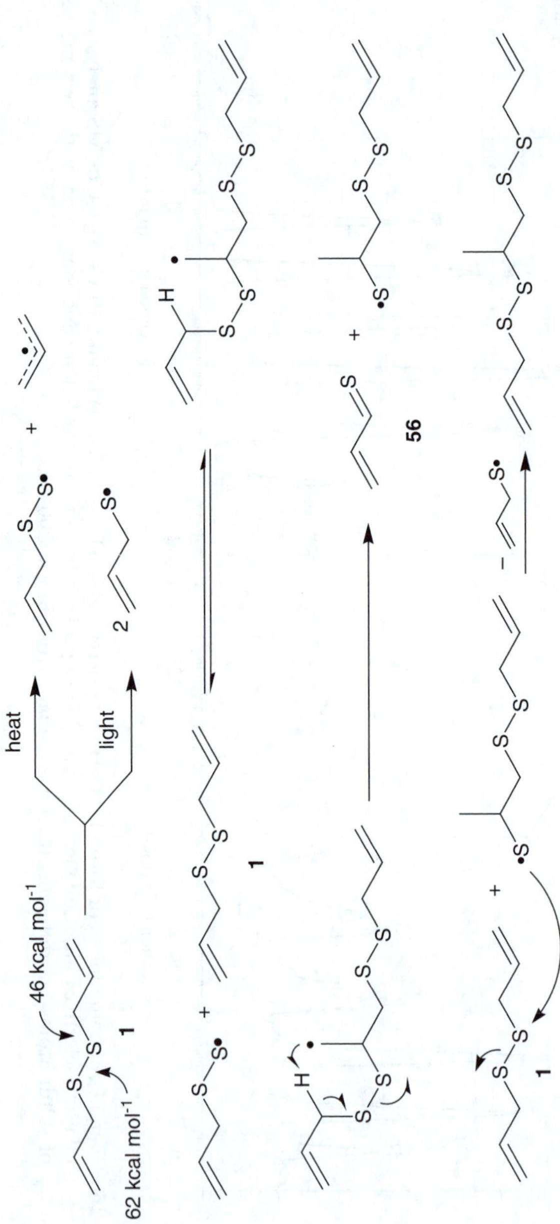

Scheme 4.29 Mechanism for decomposition of diallyl disulfide (**1**) giving a $C_9H_{16}S_4$ compound.

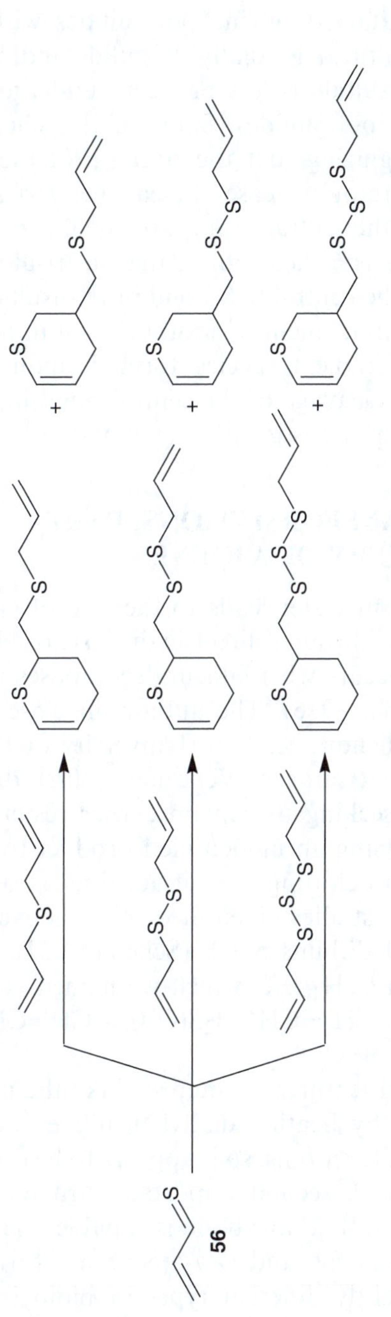

Scheme 4.30 Reaction of thioacrolein (**56**) with diallyl sulfide, disulfide and trisulfide to form cyclic nine-carbon components of garlic oil ($C_9H_{14}S_{2\text{–}4}$).

light and various catalysts, giving polysulfides with greater and fewer numbers of sulfur, *e.g.*, diallyl trisulfide and pentasulfide in this case. Diallyl trisulfide is less prone to undergo this reaction compared to higher polysulfides. Some of this chemistry can be understood by recognizing that the strength of the sulfur–sulfur (S–S) bond, and therefore its ease of cleavage, varies according to the number of the other sulfur atoms attached, *e.g.*, the S–S bond strength in disulfides is 62 kcal mol^{-1}, that in trisulfides is 45 kcal mol^{-1}, and that of the central S–S bond in tetrasulfides (SS–SS) is 35 kcal mol^{-1}. The attachment of additional sulfur atoms to a S–S bond stabilizes the radical species formed upon cleavage, for example RSS• from cleavage of the central bond in a tetrasulfide. In this way, bond rupture is greatly facilitated.

4.8 ALLICIN TRANSFORMATIONS, PART 2: THE DISCOVERY OF AJOENE

While boiling garlic in water leads to the "oil of garlic" mixture described previously, through direct hydrolysis of allicin, another important process occurs when allicin decomposes in mixtures of solvents (Block, 1984, 1986). The author, in close collaboration with colleagues Mahendra K. Jain (University of Delaware) and Rafael Apitz-Castro (Caracas, Venezuela), first discovered this process in 1984 in seeking to explain earlier results of Jain and Apitz-Castro concerning an unidentified product from garlic that prevented blood from clotting, *e.g.*, it acted as an antithrombotic agent. Laboratory studies identified the active compounds to be the two vinyl dithiins **57a,b** (Scheme 4.22), along with a molecule of formula $C_9H_{14}S_3O$ which we named "ajoene" (*ajo* is Spanish for garlic; $CH_2{=}CHCH_2S(O)CH_2CH{=}CHSSCH_2CH{=}CH_2$; **76**).

Ajoene has several features of interest. Like the major $C_9H_{14}S_3$ compounds formed by heating diallyl disulfide, ajoene has *nine* carbons and *three* sulfur atoms, so it appears to be formed from 1.5 molecules of allicin. A second important structural features of ajoene is that the central double bond is capable of existing in *E* or *Z* forms. Both (*E*)-ajoene and (*Z*)-ajoene are formed. Interestingly, they have slightly different types of biological activity. A third significant feature of ajoene is that, like alliin, it has a sulfinyl, S(O), group which can exist in two stereoisomeric forms. There are two stereoisomers for (*E*)-ajoene, (*R*)-(*E*)-ajoene and (*S*)-(*E*)-ajoene, as well as two stereoisomers for (*Z*)-ajoene, (*R*)-(*Z*)-ajoene and (*S*)-(*Z*)-ajoene. While the *R* and *S* forms are much more difficult to separate from each other than the *E*/*Z* isomers, they are also anticipated to differ from each other in biological activity. There is currently considerable interest in ajoene due to its anticancer and antifungal activity. The best yields of (*E*)- and (*Z*)-ajoenes from natural sources are (*E*)-ajoene, $172\,\mu g\,g^{-1}$ of garlic, and (*Z*)-ajoene, $476\,\mu g\,g^{-1}$ of garlic, using freshly prepared Japanese garlic with rice oil heated at 80 °C; (*Z*)-ajoene is less stable toward heat than (*E*)-ajoene (Naznin, 2008). Ajoene also forms from allicin in water (Fujisawa, 2008).

Formation of (*E*/*Z*)-ajoene from allicin can be represented in a deceptively simple manner [Equation (10)]. This equation bears a striking resemblance to that of a curious reaction discovered by the author in 1974, involving conversion of methyl methanethiosulfinate [**4**; Equation (11)] to 2,3,5-trithiahexane 5-oxide (**79**; Block, 1974). Apart from identical stoichiometries (three molecules of thiosulfinate give two molecules of the product and one molecule of water), the reactions are also similar in that the sulfinyl groups in the starting material are each bonded to sulfur as well as carbon, while in the products they are bonded only to carbon. Both reactions can be followed by infrared spectroscopy: the characteristic thiosulfinate (SS(O)C) sulfur–oxygen band of **4** and allicin at $1078\,cm^{-1}$ shifts to $1044\,cm^{-1}$, the sulfoxide band of **79** and ajoene.

$$3CH_2{=}CHCH_2S(O)SCH_2CH{=}CH_2 \rightarrow 2CH_2{=}CHCH_2S(O)CH_2CH{=}CHSSCH_2CH{=}CH_2 + H_2O \quad (10)$$

$$3CH_3S(O)SCH_3 \rightarrow 2CH_3S(O)CH_2SSCH_3(\mathbf{79}) + H_2O \quad (11)$$

It is informative to consider the mechanism proposed in Scheme 4.31 for the reaction in Equation (11), before considering how ajoene might be formed from allicin (Scheme 4.32). By analogy to extensive studies by John Kice on diaryl thiosulfinates (Kice, 1980), two molecules of **4** are thought to react in the presence of an acid catalyst giving a thiosulfonium ion intermediate (**77**) by *S*-thiolation along with simultaneous loss of one molecule of methanesulfenic acid (CH_3SOH; **36**). Ion (**77**) then eliminates a molecule of **36**, similar to

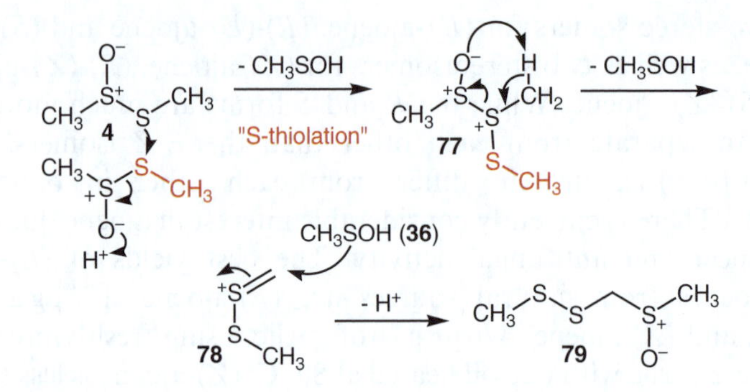

Scheme 4.31 Formation of 2,3,5-trithiahexane 5-oxide (**79**) from methyl methanethiosulfinate (**4**) *via* *S*-thiolation to **77**, followed by addition of methanesufenic acid (**36**) to ion **78**.

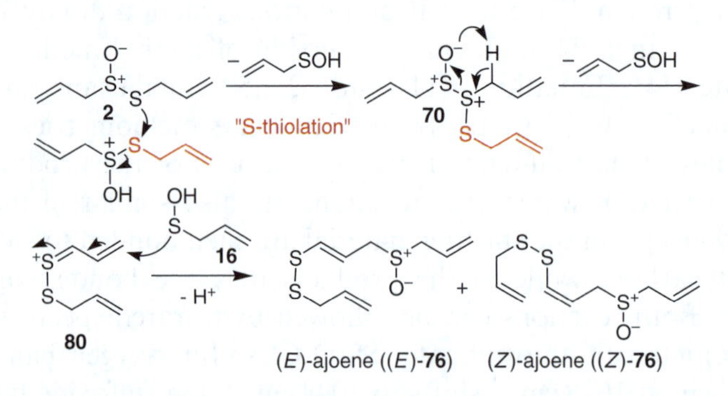

Scheme 4.32 Formation of (*E*/*Z*)-ajoene (**76**) from allicin (**2**) *via* ions **70** and **80**.

the process shown in Scheme 4.8, but leading to a very reactive α-disulfide carbocation (**78**; Block, 1974c) instead of thioformaldehyde. In an unusual twist, methanesulfenic acid adds to the carbon of carbocation **78** giving product **79** and regenerating the catalytic proton. As can be seen, the final product is formed by a 1,2-rearrangement of the thiosulfonium ion: the methylsulfinyl group migrates from sulfur to the adjacent carbon atom. Similar processes are most likely involved in the conversion of allicin to ajoene, namely *S*-thiolation giving thiosulfonium ion (**70**) followed by loss of 2-propenesulfenic acid (**16**), giving ion **80**. Because this ion contains a conjugating double bond, the positive charge on sulfur is transmitted to the terminal carbon allowing "γ-attack" or "vinylogous addition" of **16** on **80** giving (*E*/*Z*)-ajoene (**76**).

Following the discovery of ajoene from garlic, extracts of homogenized bulbs of *A. ursinum* were found to contain the *seven*- and *five*-carbon ajoene homologs: "(*E*/*Z*)-methylajoene" ($CH_3S(O)CH_2CH{=}CHSSCH_2CH{=}CH_2$) and "(*E*/*Z*)-dimethylajoene" ($CH_3S(O)CH_2CH{=}CHSSCH_3$), respectively. These two compounds, also found in low concentrations in garlic extracts, act as *in vitro* inhibitors of cholesterol synthesis and inhibitors of 5-lipoxygenase and cyclooxygenase enzymes (Sendl, 1991, 1992a,b). Formation of the methylajoenes can be explained by reactions of 2-propenyl methanethiosulfinate ($CH_3S(O)SCH_2CH{=}CH_2$), abundant in *A. ursinum*, which are analogous to those proposed for allicin.

4.9 ANTIOXIDANT AND PRO-OXIDANT ACTIVITY OF *ALLIUM* COMPOUNDS

Reactive oxygen species (ROS), including hydrogen peroxide (H_2O_2), superoxide radical ($O_2{\cdot}^-$), hydroxyl radical ($HO{\cdot}$) and peroxynitrite ($ONOO^-$) are responsible for cellular damage to nucleic acids, proteins and lipids, and are implicated as a cause of many diseases, including cancer, chronic inflammation and cardiovascular diseases, among others. Antioxidants are molecules capable of slowing or preventing the oxidation of other molecules caused by ROS. Oxidation reactions can produce free radicals, which in turn can initiate cell-damaging chain reactions. Antioxidants terminate these chain reactions by removing free radical intermediates, as well as inhibiting other oxidation reactions by being oxidized themselves. Garlic-derived compounds such as allicin, and their reaction products with glutathione and other thiols, are often described as antioxidants (Rabinkov, 1998, 2000).

The effectiveness of allicin in trapping peroxyl radicals is thought to be due to the facile formation of 2-propenesulfenic acid (Vaidya, 2008). Indeed, 2-methyl-2-propanesulfenic acid was found to react with the peroxyl radical with a rate constant $>10^7\ M^{-1}\ s^{-1}$, *e.g.*, diffusion-controlled, making sulfenic acids the most potent of all peroxyl-radical-trapping agents (Koelewijn, 1972; Vaidya, 2008). The antioxidant ability of allicin is diminished when hydrogen-bond-donor solvents, such as hexafluoroisopropanol, retard decomposition of allicin to the sulfenic acid. The effectiveness of sulfenic acids as radical traps is attributed to the weak O–H bond.

Thus, the calculated bond-dissociation enthalpy (BDE) of methanesulfenic acid ($CH_3SO–H$) is 68.4 kcal mol^{-1}, compared to the BDE of 86.2 kcal mol^{-1} for methyl hydroperoxide ($CH_3OO–H$). The O–H BDEs for sulfenic acids are among the lowest known, an effect attributed to the ability of sulfur to stabilize the sulfinyl radical resulting from O–H cleavage, RSO•, by almost equal delocalization on oxygen and sulfur. By comparison, only 30% of the unpaired spin density in the ROO• radical is delocalized on the nonterminal oxygen (Vaidya, 2008). In contrast to allicin, diallyl disulfide is an ineffective autoxidation inhibitor (Amorati, 2008), since it cannot form a sulfenic acid.

There is a second, somewhat controversial mechanism whereby *Allium* compounds can function as antioxidants, namely by metal-coordination. Cellular hydrogen peroxide, derived from oxygen metabolism, can be reduced by metal ions such as Fe(II) or Cu(I) in Fenton type reactions to generate hydroxyl radicals. Complexation of Cu by sulfur antioxidants may afford protection against cellular damage by inhibiting Fenton-like chemistry, *e.g.*, the spin forbidden reduction of molecular oxygen or peroxide, to superoxide or hydroxyl radical (Battin, 2008). While there is evidence that diallyl polysulfides, sulfoxides and thiosulfinates can all bind to Cu ions, it has been reported that the antioxidant ability of garlic compounds is "unrelated to Cu^{2+} chelation" (Pedraza-Chaverri, 2004). Flavonols, such as quercetin, represent a second category of compounds that are particularly abundant in onions (Section 3.11). It has been shown that under conditions that would otherwise cause significant DNA strand cleavage, polyphenols, such as quercetin, protected the DNA by binding to iron and preventing the generation or release of the hydroxyl radical. Quercetin showed an IC_{50} of 10.7 μM in DNA damage inhibition (Perron, 2008). Organoselenium compounds in garlic and other alliums are also effective antioxidants (see Section 4.14). A novel, active heterocyclic antioxidant, 2,4-dihydroxy-2,5-dimethylthiophene-3-one (thiacremonone), has also recently been isolated from heated garlic preparations (Hwang, 2007).

The situation with regard to the role of *Allium* sulfur compounds and metals in redox processes is somewhat more complex than that presented above, since it has been found that such sulfur compounds can function not only as antioxidants, but also as *pro-oxidants.*

Pro-oxidants are chemicals that induce oxidative stress, either through creating ROS or inhibiting antioxidant systems. The oxidative stress produced by these chemicals can damage cancer cells and thus provide the basis for chemotherapy. Some *Allium* organosulfur compounds can act as either antioxidants, or pro-oxidants, depending on the specific set of conditions, *e.g*., whether or not oxygen or transition metals are present. Diallyl trisulfide-induced ROS generation, thought to be involved in the anticancer activity of this compound, is mediated by an increase in labile iron from degradation of ferritin (Antosiewicz, 2006). Diallyl trisulfide can be converted by glutathione (GSH) to a perthiol, *e.g*., 2-propene-1-sulfenothioic acid ($CH_2{=}CHCH_2SSH$; **69**; Scheme 4.26), which should be more acidic than the corresponding thiol (Everett, 1994), and should therefore be ionized to a greater extent to the corresponding anion, $CH_2{=}CHCH_2SS^-$, see Equations (12) and (13). The latter should readily undergo one-electron oxidation by Fe^{+3} to a stable, delocalized perthiyl radical, $CH_2{=}CHCH_2SS\cdot$ (see above discussion of delocalization in the RSO• radical). The perthiyl radical could then react with glutathione giving a polysulfide radical anion, $[CH_2{=}CHCH_2SSG]^{\cdot -}$, which in turn could reduce molecular oxygen to a reactive oxygen species (ROS), see Equations (14)–(16). In summary, diallyl trisulfide, upon reduction to a perthiol, can generate ROS (H_2O_2, $HO\cdot$, $O_2^{\cdot -}$) under physiological conditions through reaction with molecular oxygen in the presence of iron, or other transition metal ions. The redox chemistry of polysulfanes and perthiols has been reviewed in detail (Iciek, 2009; Jacob, 2008b; Filomeni, 2008; Antosiewicz, 2008; Munday, 2003; Sahu, 2002; Everett, 1995).

$$GSH + CH_2{=}CHCH_2SSSCH_2CH{=}CH_2 \rightarrow GSSR + CH_2{=}CHCH_2SSH \quad (12)$$

$$CH_2{=}CHCH_2SSH \rightarrow CH_2{=}CHCH_2SS^- + H^+ \quad (13)$$

$$CH_2{=}CHCH_2SS^- + Fe^{3+} \rightarrow CH_2{=}CHCH_2SS\cdot + Fe^{2+} \quad (14)$$

$$CH_2{=}CHCH_2SS\cdot + GSH \rightarrow [CH_2{=}CHCH_2SSSG]^{\cdot -} \quad (15)$$

$$[CH_2{=}CHCH_2SSSG]^{\cdot -} + O_2 \rightarrow CH_2{=}CHCH_2SSSG + O_2^{\cdot -} \quad (16)$$

4.10 THE MUNICH/NAGOYA CONNECTION: CEPAENES, NINE-CARBON, THREE-SULFUR, ONE-OXYGEN MOLECULES FROM ONION

The previous section describes how the careful analysis of chopped garlic preparations, in 1985, led to the structural elucidation of ajoene (**76**), the *nine* carbon- and *three* sulfur-atom compound of molecular formula $C_9H_{14}S_3O$, as well as related compounds containing *seven* and *five* carbons. In 1988, the analysis of fresh onion juice led, in a similar way, to the discovery of related substances, of formula $C_9H_{16}S_3O$ and $C_9H_{18}S_3O$, containing *nine* carbons, *three* sulfurs and one oxygen, and $C_5H_{12}S_3O$, containing *five* carbons, *three* sulfurs and one oxygen. These molecules were named "cepaenes," from the botanical name for onion, as shorthand for the more cumbersome formal name for this class, 1-[alk(en)-ylsulfinyl]propyl alk(en)yl disulfides. As so often happens in science, two different research teams virtually simultaneously published the discovery of cepaenes, both in the medical journal, *The Lancet* – Kawakishi and Morimitsu in Nagoya, Japan in August 1988 ($C_5H_{12}S_3O$) and Bayer, Wagner *et al.* in Munich, Germany in October 1988 ($C_9H_{16}S_3O$). The Nagoya group tested different fractions obtained by chromatographically separating the homogenate of 15 kg of onions, using an assay that measured human blood platelet aggregation inhibition. The substance of formula $C_5H_{12}S_3O$ exhibited a median inhibitory concentration of 67.6 μM, only slightly larger than the value of 41.2 μM for aspirin, a known aggregation inhibitor. This compound was identified as MeS(O)CHEtSSMe (methyl 1-(methylsulfinyl)propyl disulfide; **81**), which is structurally similar to α-sulfinyl disulfide (**79**; Scheme 4.31), but with an ethyl group instead of hydrogens on the α-carbon. The Nagoya group suggested that **81** is formed by "carbophilic" addition of methanesulfenic acid (**36**) to the onion LF, propanethial *S*-oxide (**28**; Scheme 4.33; Kawakishi, 1988).

The Munich group similarly identified the cepaenes, $C_9H_{16}S_3O$ and $C_9H_{18}S_3O$, through their inhibitory activity toward sheep seminal microsomal cyclooxygenase and 5-lipoxygenases (5-LO), and determined the structures to be isomers (diastereomers) of MeCH=CHS(O)CHEtSSCH=CHMe (**82**; 1-propenyl 1-(1-propenylsulfinyl)propyl disulfide) and MeCH=CHS(O)CHEtSSPr (propyl 1-(1-propenylsulfinyl)propyl disulfide; Bayer, 1988a,

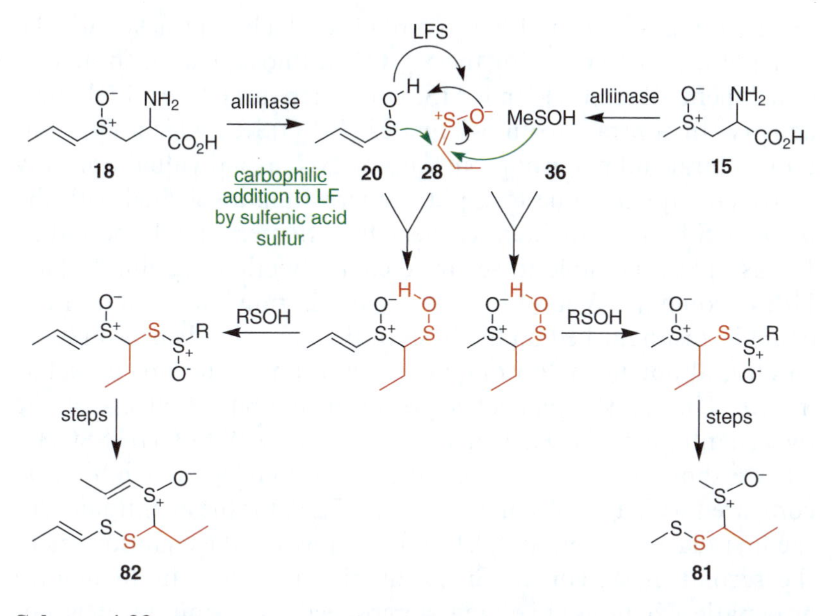

Scheme 4.33 Proposed mechanism for the formation of cepaenes **82** and **81** from products from the "carbophilic attack" of sulfenic acids **20** and **36**, respectively, on propanethial *S*-oxide (**28**).

1988b, 1989, Dorsch, 1990, Wagner, 1990). The cyclooxygenase and 5-LO are key enzymes associated with the formation of leukotrienes, prostaglandins, thromboxanes and other regulatory compounds which play an important role in asthma. Additional work by the Nagoya and Munich groups led to the identification of a family of anti-inflammatory cepaenes from onion juice of general structure: RS(O)CHEtSSR′, including the additional examples, MeS(O)CHEtSSPr and MeS(O)CHEtSSCH=CHMe (Morimitsu, 1990, 1991, 1992; Bayer, 1988b, 1989; Dorsch, 1990).

Since the sulfinyl (S^+–O^-) group and the carbon attached to the ethyl (Et) group are both chiral centers, each of stereochemistry *R* or *S*, four stereoisomers are possible, *RR*, *SS*, *RS* and *SR*. In addition, when double bonds are present, (*E*)- and (*Z*)-isomers are possible, increasing the number of individual stereoisomers for the formula MeCH=CHS(O)CHEtSSPr to 8, and for MeCH=CHS(O)CHEtSSCH=CHMe to 16. Individual pairs of enantiomers (mirror-image stereoisomers; *e.g.*, (*E*)-*RR*/(*E*)-*SS*) have been isolated using preparative HPLC; characterized by NMR,

IR (showing S^+–O^- at 1040–1055 cm^{-1}, which contrasts with the value of *ca.* 1088 cm^{-1}, for the S^+–O^- in thiosulfinates, shifted by attachment of sulfur) and MS; and then tested for biological activity. In contrast to thiosulfinates (RS(O)SR), which also contain a chiral sulfinyl group but cannot exist as individual optically active enantiomers due to rapid isomerization associated with the labile S–S bond, individual cepaene enantiomers should be stable. It was indeed possible to separate enantiomers using chiral-phase HPLC columns. When this was achieved, individual enantiomers could be distinguished on the basis of their circular dichroism (CD) spectra, although the absolute configurations could not be determined. The human platelet aggregation inhibitory effects of the two enantiomers for each diastereomer of MeS(O)CHEtSSCH=CHMe showed IC_{50} values (concentrations giving 50% inhibition compared with a blank) of 49.4 and 23.7 μM for the enantiomers of the first diastereomer, and 9.1 and 22.9 μM for the enantiomers of the second diastereomer. Since inhibition is thought to involve enzymatic reactions involving stereospecific recognition, it is not surprising that the activities of the enantiomers and diastereomers should differ (Morimitsu, 1991).

Simple syntheses of cepaenes were developed in Albany (Block, 1992a, 1997; Calvey, 1998). In collaboration with Elizabeth Calvey at the U.S. Food and Drug Adminstration, methods for the direct analysis of cepaenes and related compounds were developed using coupled LC-MS and supercritical fluid chromatography-mass spectrometry (Calvey, 1997). Auger and colleagues in France have used related analytical methods to characterize *Allium* extracts (Ferary, 1996; Mondy, 2002).

4.11 FROM MUNICH TO ALBANY: THE SOLUTION TO A CHEMICAL PUZZLE – ZWIEBELANES AND A BIS-SULFINE

4.11.1 Discovery of Zwiebelanes

In 1963, Virtanen commented on the difference between the enzymatic behavior of alliin from garlic and isoalliin from onion. In the case of garlic, alliinase cleaves alliin giving 2-propenesulfenic (**16**; CH_2=$CHCH_2SOH$; C_3H_6SO), which self-condenses with loss of water giving allicin (**2**; $C_6H_{10}S_2O$). With onion, alliinase cleaves isoalliin giving 1-propenesulfenic acid (**20**; CH_3CH_2=CHSOH; also C_3H_6SO), which Virtanen identified as the onion lachrymatory factor (LF), but which we now know is its immediate precursor. Virtanen expressed surprise that "no allicin-like 'double-molecule' is formed" from sulfenic acid **20** (Spåre, 1963). As we shall see, this puzzle was finally solved many years later.

In the late 1980s, the author initiated collaboration with Professor Hildebert Wagner of the Institute of Pharmaceutical Biology of the University of Munich and his doctoral student Thomas Bayer. Our mutual interests concerned a pair of molecules of formula $C_6H_{10}S_2O$, which Bayer in Munich had isolated from extracts of freshly crushed onions, while searching for cepaenes and other antiasthmatic agents, as described in Section 4.10. The Albany team isolated the identical compounds from oxidation of bis(1-propenyl) disulfide (**83**; CH_3CH=CHSSCH=$CHCH_3$) while attempting to prepare Virtanen's elusive "allicin-like 'double-molecule,'" CH_3CH=CHS(O)SCH=$CHCH_3$ (1-propenyl 1-propenethiosulfinate; **27**). Isolation of the $C_6H_{10}S_2O$ isomers was an exciting discovery since their empirical formulas were identical to both that of allicin and **27**. Like allicin, the new compounds showed the presence of a sulfinyl group in their IR spectra. However, while both the proton and carbon NMR spectra of allicin show the expected olefinic bands (^{1}H NMR: multiplets from δ 6.07 to 5.37 ppm; ^{13}C NMR: δ 135.98, 128.03, 127.85, 121.97 ppm), the new onion compounds show only saturated carbon atoms, with ^{1}H and ^{13}C NMR signals upfield from δ 4.3 and δ 80 ppm, respectively.

With the collaboration of NMR spectroscopist Andras Neszmelyi and Thomas Bayer's move to Albany as a postdoctoral fellow, the new compounds were characterized as isomers of 2,3-dimethyl-5,6-dithiabicyclo[2.1.1]hexane 5-oxides **84a** and **84b**, and were given the trivial names *cis*- and *trans*-zwiebelanes ("Zwiebel" is German for onion), respectively (Figure 4.27). The structural assignments were made by detailed spectroscopic studies, by chemical reactions and by X-ray crystal structures of derivatives (Block, 1996a).

cis-Zwiebelane can be distinguished from *trans*-zwiebelanes since the former has a plane of symmetry (containing the oxygen and sulfur atoms) with only three unique ^{13}C signals while the latter has six unique ^{13}C signals. Furthermore *trans*-zwiebelane is chiral

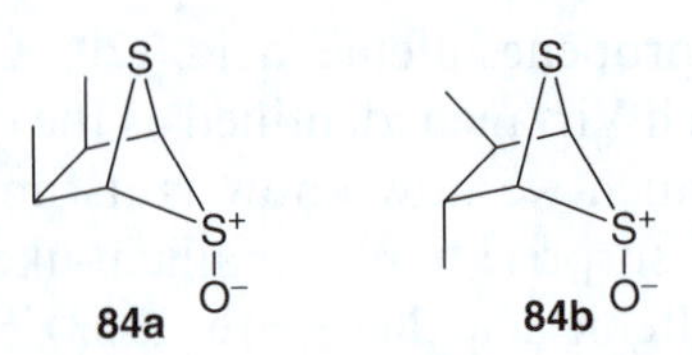

Figure 4.27 *cis*- and *trans*-zwiebelanes, **84a** and **84b**, respectively.

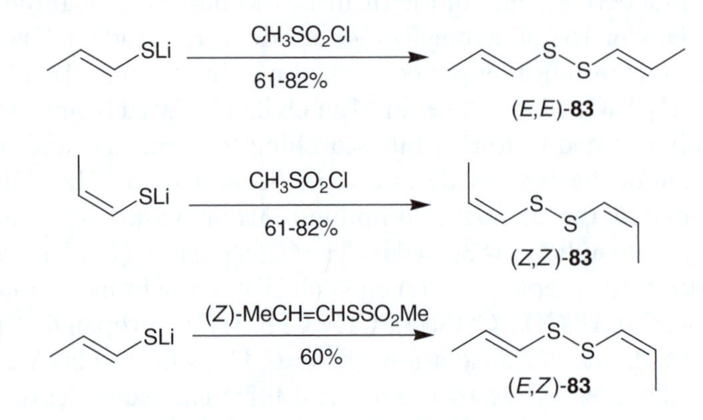

Scheme 4.34 Selective syntheses of individual isomers of bis(1-propenyl) disulfide [(*E,E*)-, (*Z,Z*)-and (*E,Z*)-**83**].

whereas *cis*-zwiebelane is not. Chromatography of natural *trans*-zwiebelane (from extracts of cut onion) on a chiral GC column, which resolves the enantiomers, shows it to be completely racemic. As described below, this observation is consistent with the formation of the *trans*-zwiebelane from racemic thiosulfinate, in turn formed by the condensation of achiral 1-propenesulfenic acid without enzymatic involvement. These compounds represented a novel, previously unknown natural product ring system (Bayer, 1989). Zwiebelanes are stable enough to be easily seen and separated by GC. Recently a third zwiebelane has been detected in onion extracts, although its complete structure has not yet been determined (Mondy, 2002). To fully describe the structures of the two zwiebelane isomers, a series of stereochemical descriptors are needed to define the relative positions of the oxygen and methyl groups. Thus, **84a** is (1α, 2α, 3α, 4α, 5β)- and **84b** is ($\pm$)-(1α, 2α, 3β, 4α, 5β)-2,3-dimethyl-5,6-dithiabicyclo[2.1.1]hexane 5-oxide.

Individual pure isomers of bis(1-propenyl) disulfide, (*E,E*)-**83**, (*Z,Z*)-**83** and (*E,Z*)-**83**, were synthesized as shown in Scheme 4.34.

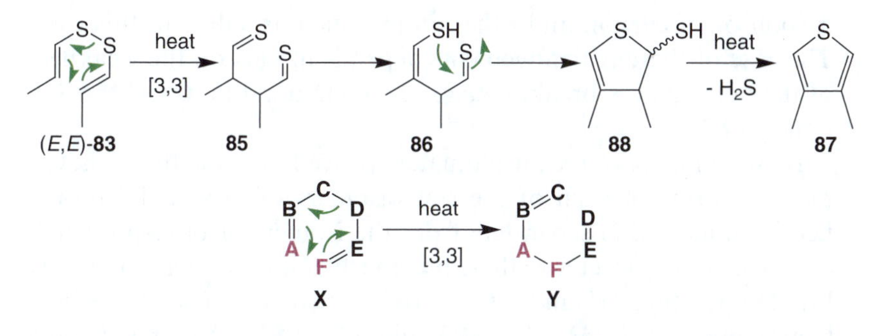

Scheme 4.35 Rearrangement of bis(1-propenyl) disulfide (**83**) on heating to **87** and **88**. Generalized [3.3]-sigmatropic rearrangement.

Before considering the oxidation of isomers of **83** to isomers of **27** and the subsequent rearrangement of the latter compounds, it is helpful to first discuss related rearrangement processes involving conversion of **83** to 3,4-dimethylthiophene (**87**), a well known minor component of distilled onion oil. When individual isomers (*E,E*)-, (*Z,Z*)- and (*E,Z*)-**83** were heated in benzene at 85 °C, a 1 : 1 mixture of *cis*- : *trans*-2-mercapto-3,4-dimethyl-2,3-dihydrothiophene (**88a** : **88b**), along with **87** was isolated (Scheme 4.35). Prolonged heating of the mixture gave **87** as the major product. It is likely that **83** undergoes [3,3]-sigmatropic rearrangement giving bis(thial) (**85**). The occurrence of this process is consistent with the fact that if the stereoisomers of **83** [(*E,E*)-, (*E,Z*)- and (*Z,Z*)-**83**], are separately heated, no interconversion occurs, precluding a free-radical mechanism.

Rearrangement of **85** to **86**, followed by ring closure, would explain the formation of **88**, which can easily lose H_2S giving the very stable 3,4-dimethylthiophene (**87**; Block 1990b, 1996a). Thus **87** represents another example of an artifact, found in *Allium* distilled oils, formed by decomposition of a thermally unstable compound (in this case **83**) in the hot inlet port of a gas chromatograph. [3,3]-Sigmatropic rearrangements (illustrated in general form by X → Y) are a well-known type of concerted reaction included in the Woodward–Hoffman rules, of a chain of six atoms with double bonds at each end in which the double bonds come together forming a new single bond, and the bond between the third and fourth atoms breaks, creating a set of new double bonds (Woodward, 1970). The six-atom chain can consist just of carbon or of carbon and other atoms, such as sulfur in this case. The ease of these rearrangements depends in part on the weakness of the bond that is breaking, *e.g.*, the C–D bond in **X** and the S–S bond in **83**.

By working at –60 °C, it ultimately proved possible to synthesize racemic forms of each of the stereoisomers of propenyl 1-propenethiosulfinate (**27**) separately from the isomers of bis(1-propenyl) disulfide, and characterize these previously unknown compounds by low temperature ^{1}H and ^{13}C NMR spectroscopy (Block, 1996a). Compounds (*E*,*E*)-**83**, (*Z*,*Z*)-**83** and (*E*,*Z*)-**83** were individually treated at –60 °C, with chilled solutions of the oxidizing agent *meta*-chloroperbenzoic acid (MCPBA) in the presence of dry sodium carbonate (Scheme 4.36). After 5 min, the reaction mixtures were analyzed by NMR at –60 °C. From both the ^{1}H and ^{13}C NMR data, we concluded that the oxidation of the individual isomers gives isomers of thiosulfinate MeCH=CHS(O)SCH=CHMe (**27**) with retention of double bond stereochemistry in each case. Unsymmetrical (*E*,*Z*)-**83** gave a 2 : 1 mixture of (*E*,*Z*)-**27** : (*Z*,*E*)-**27**.

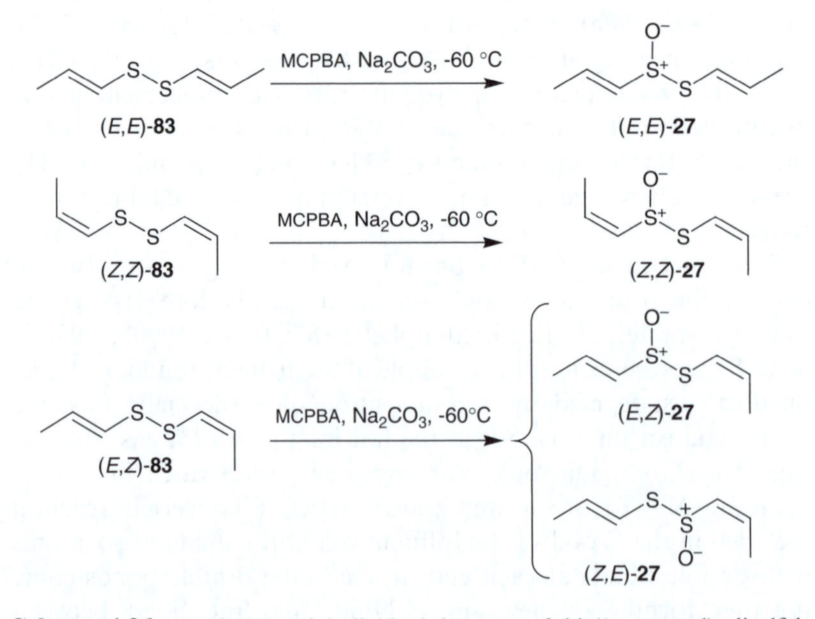

Scheme 4.36 Oxidation of individual isomers of bis(1-propenyl) disulfide [(*E*,*E*)-, (*Z*,*Z*)- and (*E*,*Z*)-**83**] giving individual isomers (*E*,*E*)-, (*Z*,*Z*)-, (*E*,*Z*)- and (*Z*,*E*)-**27**.

The relative rates of oxidation were (*E*,*E*)-**83** > (*E*,*Z*)-**83** > (*Z*,*Z*)-**83**, as might be expected based on the hindrance of a (*Z*)-double bond toward the approaching peracid. The several isomers of **27** were quite unstable, disappearing with half-lives of about 17 min at −15 °C. To examine the products formed on warming isomers **27**, individual isomers of **83** were oxidized at −40 °C with peracetic acid, followed by warming to 8 °C with rapid workup to avoid decomposition of unstable products. In this manner it was found that (*E*,*Z*)-**27** afforded *cis*-zwiebelane (**84a**) while (*Z*,*Z*)-**27** gave *trans*-zwiebelane (**84b**). An isomeric set of different bicyclic compounds was simultaneously formed from each isomer of **27**. For example, oxidation of (*Z*,*Z*)-**27** gave minor amounts of isomers **90a** and **90b** (Scheme 4.37). These two isomers, like *trans*-zwiebelane, each have *trans*-methyl groups, but unlike the zwiebelane each have a carbon–oxygen bond.

Just as in Scheme 4.35 we saw that (*E*,*E*)-bis(1-propenyl) disulfide [(*E*,*E*)-**83**] undergoes a [3.3]-sigmatropic rearrangement on heating to give bis-thial (**85**), so too (*Z*,*Z*)-**27** should undergo an analogous [3,3]-sigmatropic rearrangement. While Scheme 4.35 depicts the six-atom chain as flat, stereochemical features of the rearrangement are better understood if the six-atom chain is drawn to resemble a cyclohexane chair conformation with one of the

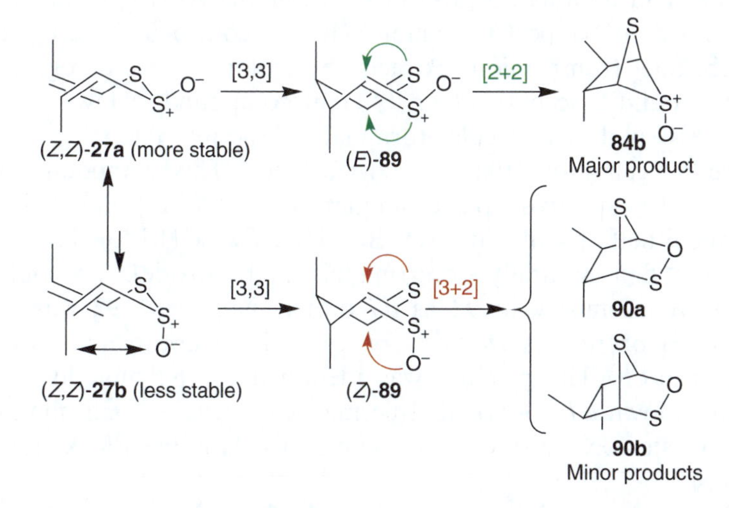

Scheme 4.37 Cyclization of (*Z*,*Z*)–**27** to *trans*-zwiebelane (**84b**) and compounds **90a** and **90b**.

bonds absent, a so-called pseudo-chair conformation. From Scheme 4.37, it can be seen that in the pseudo-chair structure, the oxygen on sulfur can occupy two different positions, pseudo-equatorial, (*Z*,*Z*)-**27a**, or pseudo-axial, (*Z*,*Z*)-**27b**. These forms differ in energy due to repulsive interactions between the oxygen and the nearby methyl groups (double-headed arrow). In the highly reactive [3,3]-rearranged thial/thial *S*-oxide intermediate [(*E*)-**89**], bonding can occur between the opposite sulfur and carbon atoms (see arrows) *via* a so-called intramolecular [2+2]-cycloaddition, leading to *trans*-zwiebelane (**84b**). Alternatively, in (*Z*)-**89**, the pseudo-axial oxygen participates in a process known as an intramolecular [3+2]-cycloaddition (see arrows) giving rise to **90a** and **90b**. The ultimate fate of the highly reactive compounds of type **90a** and **90b** in onion extracts is not known. Similar processes occur with each of the other isomers of **83** with the distribution between zwiebelanes and compounds of type **90**, depending on the position of the equilibrium between the two pseudo-chair conformers. The rearrangement of isomers of **83** is more complex than indicated here; the full paper should be consulted for complete details (Block, 1996a).

4.11.2 Discovery of a Bis-sulfine

Nature had another surprise in store for the Albany team as we continued to "unpeel the onion." In the course of isolating *cis*-zwiebelane from onion extracts by column chromatography, Bayer found a second rather different compound, a low-melting, colorless solid of molecular formula, $C_6H_{10}S_2O_2$, *e.g.*, having one more oxygen atom than the zwiebelanes. This compound was identified by spectroscopic techniques as (*Z*,*Z*)-*d*,*l*-2,3-dimethyl-1,4-butanedithial 1,4-dioxide (O=S=CHCH(Me)CH(Me)CH=S=O; **92**), the first naturally occurring bis(thial *S*-oxide). To confirm the stereochemistry of **92** as *d,l* rather than *meso*, **92** was subjected to ozonolysis followed by oxidative workup and Fischer esterification. The product was identified as the dimethyl ester of the known *d,l*-2,3-dimethylsuccinic acid, thereby confirming the relative stereochemistry of the methyl groups in **92** (Block, 1990a, 1996a).

Based on the close relationship of **92** to the proposed intermediate in the formation of zwiebelanes, we suspected that **92** was formed by a [3,3]-sigmatropic rearrangement of bis-sulfoxide (**91**; Scheme 4.38). The doubly axial arrangement shown for the sulfinyl groups in **91** is an energy minimum for α-disulfoxides (Freeman, 1984; Ishii, 2006), and would be expected to lead to *Z*-geometry for the CH=S=O groups. Indeed, oxidation of (*E*,*E*)-bis(1-propenyl) disulfide, (*E*,*E*)-**83**, with two equivalents of peracid at −60 °C, afforded pure **92** with a 34% yield.

Based on the observation that synthetic dipropyl disulfoxide decomposes giving propanethial *S*-oxide (Scheme 4.39; Freeman, 1984), the reversal of this reaction involving "thiophilic addition" of 1-propenesulfenic acid (**20**) to propanethial *S*-oxide (**28**) should give an α-disulfoxide, which on reaction with a second molecule of **20** should give disulfoxide **91**. (For "carbophilic addition" to the LF, see Scheme 4.33). In the low-temperature double oxidation of (*E*,*E*)-**83** or single oxidation of (*E*,*E*)-**27**, it was not possible to

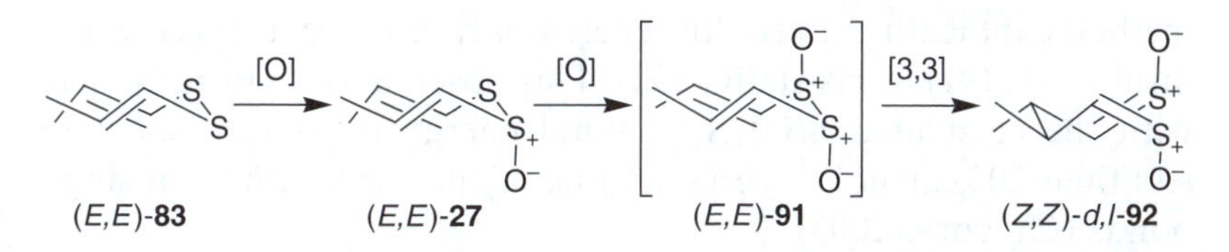

Scheme 4.38 Double oxidation of (*E*,*E*)-bis(1-propenyl)disulfide [(*E*,*E*)-**83**] to bis(thial *S*-oxide) (**92**).

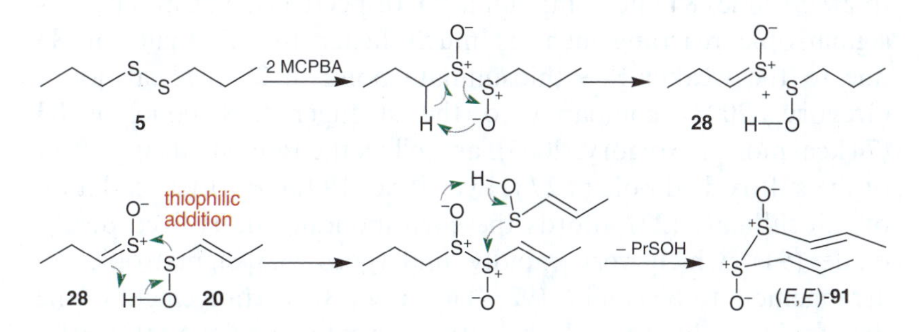

Scheme 4.39 Top: decomposition of di-*n*-propyl α-disulfoxide to the LF and propanesulfenic acid. Bottom: proposed mechanism for formation of α-disulfoxide (*E*,*E*)–**91** *via* thiophilic addition of **20** to LF (**28**).

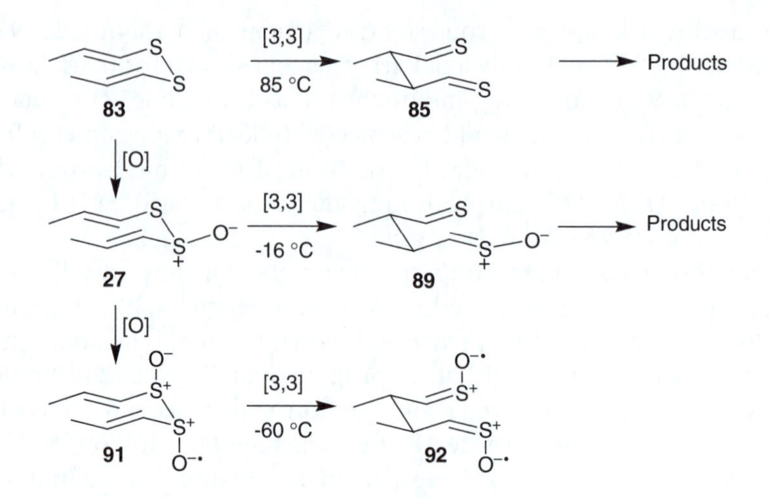

Scheme 4.40 General mechanism for [3.3] processes.

detect the intermediate disulfoxide using low temperature NMR methods, indicating a very low energy barrier for rearrangement of disulfoxide (**91**) to bis-sulfine (**92**). This observation is in agreement with the calculated S(O)–S(O) bond energy in α-disulfoxides of less than 20 kcal mol^{-1}, certainly one of the weakest known single bonds (Gregory, 2003).

To summarize (Scheme 4.40), mono-oxidation of disulfide (**83**) affords thiosulfinates (**27**) which undergo rapid, sequential [3,3]-sigmatropic rearrangement followed by both [2+2] and [3+2] intramolecular cycloaddition reactions of intermediates **89**, leading to zwiebelanes **84** and compounds **90**, respectively. The initial [3,3]-sigmatropic rearrangement is much faster for **27** than for **83** due to the weaker S–S thiosulfinate bond in **27** (47 kcal mol^{-1}; Gregory, 2003) compared to the stronger S–S bond in **83** (74 kcal mol^{-1}; Gregory, 2003), as well as the rate-enhancing effect of the sulfoxide dipole in **27** (Hwu, 1986, 1991). Further oxidation of thiosulfinate (**27**) affords the corresponding short-lived disulfoxide (**91**), which very rapidly undergoes [3,3]-sigmatropic rearrangement to bis-sulfine (**92**). The very fast rearrangement of the disulfoxide is due to both the extreme weakness of the S(O)–S(O) bond as well as the strong, rate-enhancing effect of the two sulfoxide dipoles.

4.11.3 Super Smelly Onion Compounds

Another Munich connection should be mentioned. Peter Schieberle and coworkers at the Technischen Universität München used a stable isotope dilution assay procedure to determine an organsulfur compound from onion and onion-like alliums called 3-mercapto-2-methylpentan-1-ol ($CH_3CH_2CH(SH)CH(CH_3)CH_2OH$), which has an extraordinarily low odor threshold of *ca.* $1\,ng\,L^{-1}$ (Granvogl, 2004). To put this number in some perspective, 0.08 mg of this compound in an average size (16×32 ft) swimming pool would give a detectable odor. This compound is thought to be formed *via* addition of hydrogen sulfide to 2-methyl-2-pentenal, following by reduction (Widder, 2000) The very sensitive isotopic dilution technique involved preparation of $CH_3CH_2CH(SH)CH(CH_3)CD_2OH$, which has the same GC retention time as the deuterium-free compound but a different mass, facilitating quantitation when both compounds are present.

Disproportionation of thiosulfinates to disulfides and thiosulfonates, a common reaction seen with both aliphatic and aromatic thiosulfinates (Equation (17); Barnard, 1957; Block, 1974b), is less common with *Allium* thiosulfinates due to more competitive alternative processes. However, trace amounts of thiosulfonates ($MeSO_2SMe$, $MeSO_2SPr$ and $PrSO_2SPr$) have been found in extracts of fresh onions (Boelens, 1971). Other thiosulfonates are found in supercritical CO_2 onion extracts (Sinha, 1992) and scallion and Welsh onion extracts (Nakamura, 1996; Kuo, 1992b). Since the odor threshold for these thiosulfonates (1.5–1.7 ppb) is several hundreds times lower than their concentrations, these compounds will make significant contributions to the overall flavor.

$$2RS(O)SR \rightarrow RSSR + RSO_2SR \quad (17)$$

4.12 SILENCING GENES ALTERS NATURAL PRODUCTS CHEMISTRY: THE TEARLESS ONION

Two developments in onion chemistry reveal the broader significance of the discovery of zwiebelanes and the bis-sulfine, the first involving tearless onions, discussed in this section, and the second involving the pinking of onions, discussed in Section 4.13.

The complexity of onion sulfur chemistry is due primarily to the fact that *both* the onion LF and its precursor, 1-propenesulfenic acid, are formed when onions are cut, and both of these small molecules undergo diverse reactions. Under normal circumstances, the levels of 1-propenesulfenic acid are kept very low due to very rapid conversion to the onion LF by action of the lachrymatory factor synthase (LFS) enzyme. However, New Zealand researchers have found that it is possible to "silence" the LFS gene using molecular biology methods, producing tearless onions (Eady, 2008). This discovery allows study in onion extracts of reactions associated solely with 1-propenesulfenic acid, particularly self-condensation to 1-propenyl 1-propenethiosulfinate (**27**; $CH_3CH=CHS(O)SCH=CHCH_3$), which can then undergo a cascade of non-enzymatic reactions of the type predicted based on work with the synthetic thiosulfinate (Section 4.11; Block, 1996a).

Current "tearless" onion cultivars, *e.g.*, Vidalia, are achieved by deficient uptake and partitioning of sulfur and/or growth in sulfur-deficient soils. Such onions accumulate fewer secondary sulfur metabolites in their bulbs, leading to an onion that is "sweet [with reduced] sensory and health qualities compared to more pungent high-sulfur cultivars" (Eady, 2008). Genetic engineering offers an alternative method of producing "tearless" onions without diminishing the levels of beneficial sulfur compounds. Thus, through use of the technique of RNAi silencing, the LFS gene was suppressed in six different onion cultivars resulting in onions with highly reduced levels of LFS activity in the leaves and particularly in the bulbs, where the LFS levels are highest. In these LFS-diminished plants, the levels of both isoalliin and alliinase were found to be within the normal range (4–13 $mg\,g^{-1}$ dry weight for isoalliin; Eady, 2008). Analysis of leaf and bulb preparations from LFS-diminished plants by GC-MS showed that LF levels were reduced by as much as 30-fold.

The chemical consequences of the reduction in LFS levels are striking: (1) by GC analysis there is a dramatic increase in the level of zwiebelanes, as well as 1-propenyl *n*-propyl disulfide, bis-1-propenyl disulfide and 2-mercapto-3,4-dimethyl-2,3-dihydrothiophene; (2) the LFS-diminished plants give a pink color when treated with glycine and formaldehyde; and (3) there is a very significant decrease in the level of dipropyl disulfide. With regard to the above observations, zwiebelanes are formed from the self-condensation product of 1-propenesulfenic acid, 1-propenyl 1-propenethiosulfinate, which also gives a pink color with glycine and formaldehyde (Section 4.13). Increased concentrations of condensation products of 1-propenesulfenic acid with methanesulfenic acid would indirectly be reflected in increased formation of disulfides containing the 1-propenyl group seen on GC analysis. 2-Mercapto-3,4-dimethyl-2,3-dihydrothiophene is a major gas chromatographic decomposition product of bis-1-propenyl disulfide (Scheme 4.35). Although it is not well understood, *n*-propyl polysulfides are probably formed by reduction of the onion LF, whose concentration is reduced in LFS-deficient onions. Finally, sensory evaluation of the modified onions revealed an aroma that was less pungent and sweeter than that given off by the non-transgenic counterparts, with no tearing or stinging of the eyes (Eady, 2008).

4.13 GARLIC GREENING, ONION PINKING AND A NOVEL RED PYRROLE PIGMENT FROM "DRUMSTICK" ALLIUMS

The tendency of preparations of garlic to turn green or blue-green and those of onion and leek to turn pink, has been recognized and discussed in the scientific literature for 50 years (Joslyn, 1958; Shannon, 1967a,b). It has alarmed cooks who sometimes found their dishes combining garlic and onion purées turning turquoise blue or their sourdough garlic bread becoming laced with blue-green specks. Until recently the cause remained enigmatic. Some cooks with a background in chemistry were concerned that their garlic or onions were contaminated with toxic salts of metals such as copper or cadmium (Kubec, 2007). Once specific details emerged, it became clear just how extraordinarily complex the process is. Considerable effort has been expended trying to understand this phenomenon since formation of these pigments both diminishes the flavor of the preparations and, more significantly, may substantially reduce their economic value. This is particularly true in the case of ready-to-use prepeeled and crushed garlic products.

An intricate sequence of chemical steps takes place (summarized in Scheme 4.41), which involves many of the sulfur compounds discussed in Section 4.12 (Imai, 2006a, 2006b; Block, 2007; Li,

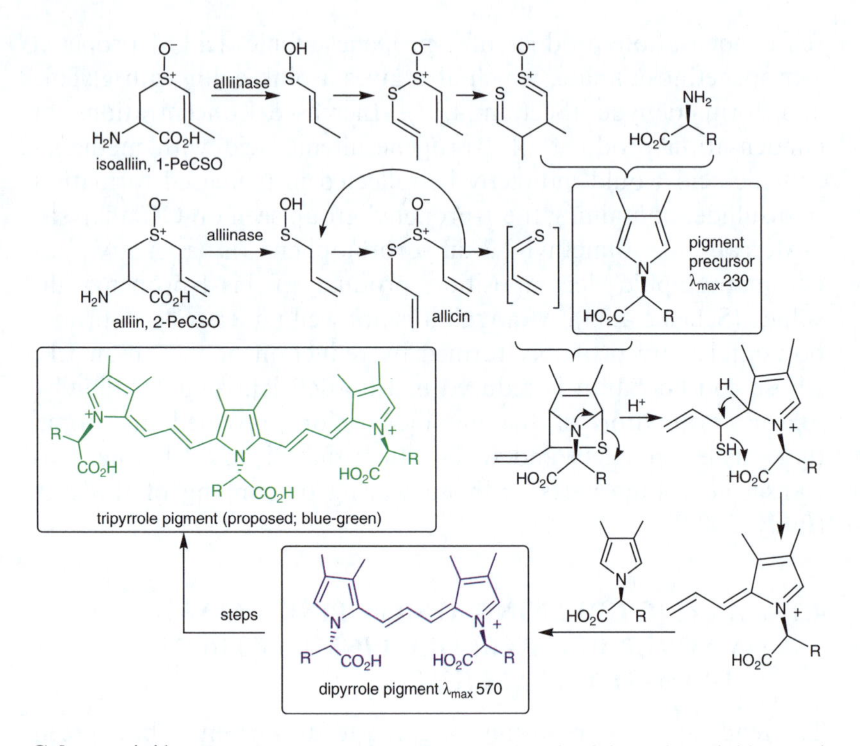

Scheme 4.41 Chemical steps thought to be involved in onion-pinking and garlic greening.

2008; Wang, 2008; Cho, 2009): (1) catalytic action of γ-glutamyl transpeptidase (EC 2.3.2.2) cleaves the γ-glutamyl group from γ-glutamyl alk(en)yl cysteine sulfoxides [a process favored at colder storage temperatures, which also favors the formation of isoalliin (*trans*-(+)-*S*-(1-propenyl)-L-cysteine sulfoxide; 1-PeCSO); Lukes, 1986; Lawson, 1991c]; (2) catalytic action of alliinase on isoalliin forms *S*-1-propenyl 1-propenethiosulfinate (**27**; $CH_3CH{=}CHS(O)SCH{=}CHCH_3$); (3) the latter rearranges to thial-thial *S*-oxide $S{=}CHCH(CH_3)CH(CH_3)CH{=}S{=}O$ (**89**) or a bis-thial *S*-oxide (**92**); (4) these latter two thiocarbonyl compounds condense with several common amino acids to form pyrroles, nitrogen-containing heterocyclic pigment precursors; (5) allicin is formed by the catalytic action of alliinase on alliin (*S*-allyl-L-cysteine sulfoxide; 2-PeCSO); (6) Diels–Alder reaction of the thioacrolein thiocarbonyl group with the pyrrole forms an electrophilic pyrrole species (proposed by the author); and (7) the pigment precursors are linked

Figure 4.28 Color variation by reaction of mixed garlic and onion thiosulfinates with individual amino acids, identified by number and amino acid abbreviation: 1) Cys, 2) Phe, 3) Gly, 4) Met, 5) Arg, 6) Val, 7) Ile, 8) Pro, 9) Lys, 10) Ser, 11) Trp, 12) h-Pro, 13) Ala, 14) Asp, 15) His, 16) Thr, 17) Leu, 18) Asn, 19) Gln, 20) Cyt, 21) Glu, 22) Tyr, 23) unspecified. (Image from Cho, 2009).

with the electrophilic pyrrole species forming deeply colored multi-pyrrole compounds.

The two-pyrrole compound is red (consistent with the 528 nm absorption found in pink onion purees; Joslyn, 1958), the three-pyrrole compound is blue (favored by garlic and onion mixtures) and the four-pyrrole molecule is green (much like the structurally-related compound, chlorophyll). Garlic discoloration has been shown to be due to the formation of as many as *eight* different blue and green pigments derived from different amino acids (Figure 4.28). The blue pigments show maximum UV absorption at 580 nm, while the green pigments are formed from a mixture of yellow pigments (λ_{max} at 440 nm) and blue pigments (Cho, 2009).

While chlorophyll also contains four pyrrole rings, they occur in a "super-ring" rather than in a chain. *Allium* pyrrole pigments, present in very low concentrations, are perfectly safe to eat. The greening of garlic can be prevented by adding 1% of the common amino acid cysteine to the garlic paste as it is being prepared

(Acquilar, 2007). The garlic-greening phenomenon is exploited by cooks in northern China who make intensely green "Laba" pickled garlic by aging garlic heads for several months and then immersing the cloves in vinegar for a week. Laba pickled garlic is served with dumplings to celebrate the Chinese New Year (Bai, 2006; cited by Kubec, 2007).

"Drumstick onions," ornamental alliums of the subgenus *melanocrommyum* such as *A. giganteum*, *A. macleanii*, *A. rosenbachianum*, and *A. rosenorum,* upon heating or wounding of plant material produce a deep red (λ_{max} at 519), biologically active tricyclic 3,3′-dithio-2,2′-dipyrrole pigment (**98**; ~0.01% of fresh weight; Figure 4.29). The formation of **98** can be explained by postulating alliinase-induced cleavage of the unusual L-(+)-*S*-(3-pyrrolyl)cysteine sulfoxide precursor (**93**) to sulfenic acid (**94**), which self-condenses giving thiosulfinate (**95**). The latter could then undergo [3,3]-sigmatropic rearrangement giving **96**, followed by thioenolization and intramolecular coupling of sulfenic acid/thiol (**97**), with loss of water to afford pigment **98** (Scheme 4.42; Jedelska, 2008). Precursor **93** is mainly located in the outer parts of the bulbs, with the highest concentrations occurring in cells surrounding plant vessels (which transport nutrients). It is suggested that the production of **98** after damage of the cell, has a function to protect the important transportation vessels. Plants producing

Figure 4.29 Time dependence of formation of the red dye after cutting a bulb of *A. macleanii*. (Image from Jedelska, 2008).

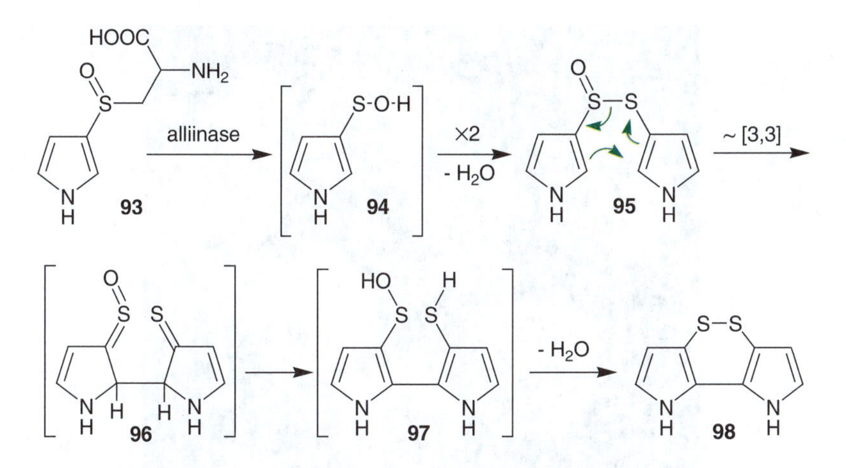

Scheme 4.42 Proposed biogenetic scheme for 3,3′-dithio-2,2′-dipyrrole (**98**).

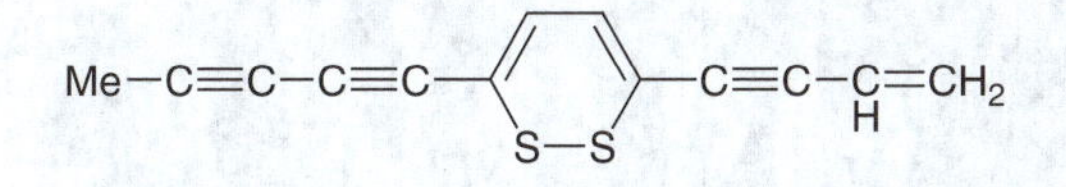

Figure 4.30 Thiarubrine B from giant ragweed (*Ambrosia trifida*).

pigment **98** are used in traditional medicine. This pigment resembles the thiarubrines, ruby-red, light-sensitive 1,2-dithiin pigments found in plants such as the black-eyed Susan, also used in traditional medicine (Figure 4.30; Block, 1994b).

A. stipitatum is another "drumstick onion," with sweetly-scented lilac-purple or white flowers, collected at higher elevations in the Tien Shan Mountains of central Asia. It was identified in 1881 by Eduard Regel and is now readily available for gardeners (Figure 4.31). The chromatography of macerated bulbs afforded, with a yield of approximately 0.01%, a series of dithio pyridine *N*-oxides (Figure 4.32), active at levels of $0.1\,\mu g\,mL^{-1}$ against *Mycobacterium tuberculosis*, including 2-(methyldithio)pyridine *N*-oxide (**99**), 2-[(methylthiomethyl)dithio]pyridine *N*-oxide (**100**) and 2,2′-dithio-bis-pyridine *N*-oxide (**101**; O'Donnell, 2009). It would be of interest to search for possible *S*-cysteine sulfoxide precursors for **99–101**, as well as related thiosulfinates in *A. stipitatum*.

Figure 4.31 *A. stipitatum*, a source of novel pyridine *N*-oxide disulfides. (Image from Van Bloem Gardens, Copyright 2008).

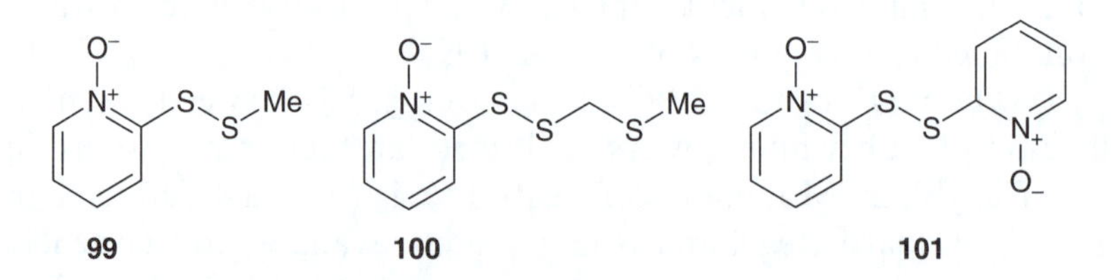

Figure 4.32 Pyridine *N*-oxide disulfides from extracts of *A. stipitatum*.

4.14 SELENIUM COMPOUNDS IN ALLIUMS

Selenium (Se), lying just below sulfur in the periodic chart of elements and sharing many properties with its "sister element," plays an important role in biology. It has long been recognized as an essential micronutrient whose dietary absence causes skeletal and

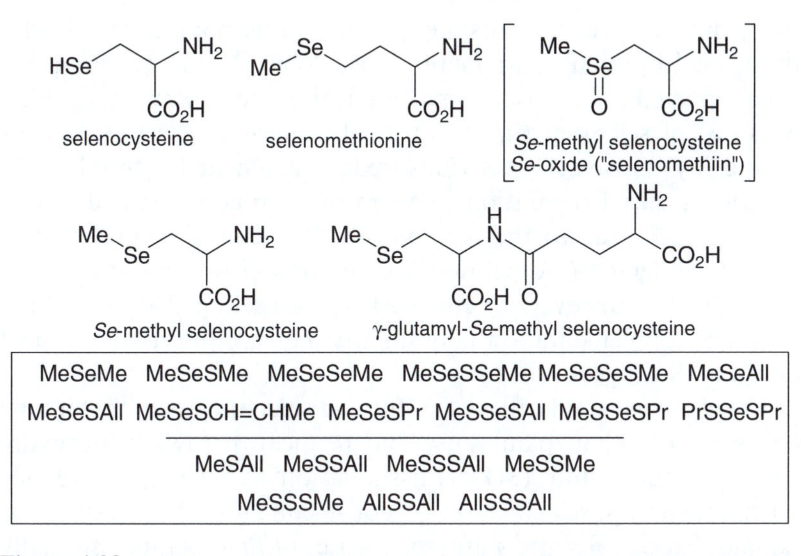

Figure 4.33 Nonvolatile (top) and volatile (top half of box) organoselenium compounds from genus *Allium* plants (All = allyl; Pr = *n*-propyl).

cardiac muscle dysfunction. Selenium is required for the correct functioning of the immune system and for cellular defense against oxidative damage, and may play a role in the prevention of cancer and premature aging. Most known selenium-dependent enzymes contain selenocysteine (Figure 4.33), a cysteine analog in which selenium replaces sulfur. Selenocysteine, called the twenty-first amino acid and essential for ribosome-directed protein synthesis, is incorporated into proteins by tRNA directed by a UGA codon (Axley, 1991; Hatfield, 2006). More than 25 selenoproteins (selenium-containing proteins) are now known (Papp, 2007).

A 1996 clinical trial suggested that selenium administered in 200 μg doses per day as selenized yeast, decreased the onset of colon, prostate and lung cancer by as much as 50% (Clark, 1996). Prospective studies on oesophageal, gastric and lung cancer were interpreted as reinforcing the conclusions of earlier trials (Rayman, 2005). However, a recent report on the Selenium and Vitamin E Cancer Prevention Trial (SELECT), which used 200 μg day^{-1} of L-selenomethionine (the major form in which selenium exists in selenized yeast) as the sole selenium source in a trial involving 35 533 men for a median of 5.5 years, concluded that "selenium or vitamin E, alone or in combination at the doses and formulations used, did not prevent prostate cancer in this population of relatively healthy men" (Lippman, 2005, 2009). While selenium has been reported to be toxic, a small clinical trial examining the effects of doses of selenium (in the form of selenized yeast) as high as 3.2 mg day^{-1} revealed no serious toxicity associated with relatively long-term supplementation. The most common complaint was garlic breath, due to the metabolism of selenium compounds to dimethyl selenide (see below for a discussion of "garlic breath;" Reid, 2004). However, recent work suggests that 200 μg per day selenium supplementation may slightly increase the risk for type 2 diabetes (Stranges, 2007).

The chemical similarities of sulfur and selenium mean that it may be plausable that plants use biochemical pathways normally employed for sulfate (SO_4^{-2}) assimilation to take up analogous selenium species, selenate (SeO_4^{-2}) or selenite (SeO_3^{-2}). It can thus be understood why the sulfur-rich genus *Allium* plants, especially garlic, contain higher levels of selenium than most vegetables (Morris, 1970). There is an interest in using selenium-enriched garlic to treat Se-deficiency diseases and inhibit tumorigenesis (Ip, 1992, 1993), and, as a consequence, in the identification and associated chemistry of selenium compounds in alliums (Arnault, 2006; Block, 1996d, 1998, 2001; Uden, 2001). It is well known that both the beneficial and toxic effects of selenium are based both on its concentration, as well as its chemical form, which can vary from one food source to another (Kápola, 2007; Fox, 2005).

In 1964, Virtanen suggested that selenium-containing analogs of γ-glutamyl *S*-alk(en)ylcysteine and *S*-alk(en)ylcysteines sulfoxides may also be present in garlic and onions (Spåre, 1974). Onions were injected with 15 ppm of Se, as selenite containing 10.4 μCi of ^{75}Se. Paper chromatographs of extracts revealed as many as 13 radioactive spots, including selenomethionine and selenocystine, and probably *Se*-methylselenocysteine selenoxide, *Se*-(β-carboxypropyl)selenocysteine, and γ-glutamyl-*Se*-propenylselenocysteine. The identity of selenocystine and selenomethionine was confirmed through comparison with the chromatographic behavior of authentic compounds. In an effort to confirm the identity of the other species, the researchers synthesized *Se*-methyl-, *Se*-propyl- and *Se*-2-propenylselenocysteine but were unsuccessful in their attempts to oxidize these compounds to the corresponding selenoxides. More recently, garlic is said to contain selenoproteins

(Wang, 1989) and a selenopolysaccharide (Yang, 1992), although detailed structures are lacking.

Intrigued by the possibility suggested by the above seminal studies by Virtanen that there might be a selenium-based flavor chemistry in *Allium* spp. parallel to that based on sulfur, *e.g.*, originating from soil selenate (SeO_4^{-2}) or selenite (SeO_3^{-2}), the author and colleagues sought to obtain detailed information on the identity, reactions, and biological activity of the organoselenium compounds naturally present in garlic and other genus *Allium* plants. The experimental problems associated with this work were daunting: the organoselenium compounds are obscured by approximately 12 000 times higher levels of organosulfur compounds, with quite similar physical properties (selenium and sulfur differ primarily by a single filled electron orbital with otherwise similar bonding). A solution to this dilemma involved the use of specialized instrumentation. Initial efforts examined volatile organoselenium compounds from cut *Allium* species by employing coupled headspace gas chromatography-atomic emission detection (HS-GC-AED). Headspace-gas chromatography (HS-GC) reduces or eliminates sample preparation, minimizing chemical changes in volatile aroma components during analysis, and is ideal for trace analysis of highly volatile compounds. Atomic plasma spectra emission provides powerful element-specific detection for gas chromatography (GC-AED). It is a highly sensitive, element-selective technique permitting simultaneous multi-element analysis (Cai, 1994a, 1994b). Since GC-AED is element-specific, the technique can flag compounds in the GC effluent containing specific elements, even if these compounds co-elute with other components.

Application of the HS-GC-AED method to the analysis of volatiles from elephant garlic (*A. ampeloprasum*) led to the identification of the series of selenium compounds shown in Figure 4.34 in the selenium channel (monitored at 196.1 nm), along with much larger quantities of the sulfur compounds shown in the sulfur channel (180.7 nm) and carbon channel (193.1 nm). The sensitivity of this method can be mathematically enhanced by taking advantage of the large mass defect of heavier elements, such as selenium, in distinguishing mass spectral peaks with identical nominal masses. This technique, requiring high resolution MS and a mass defect-based algorithm, has been used to establish the presence of MeSeSeSMe in green onions, *A. fistulosum* (Shah, 2007).

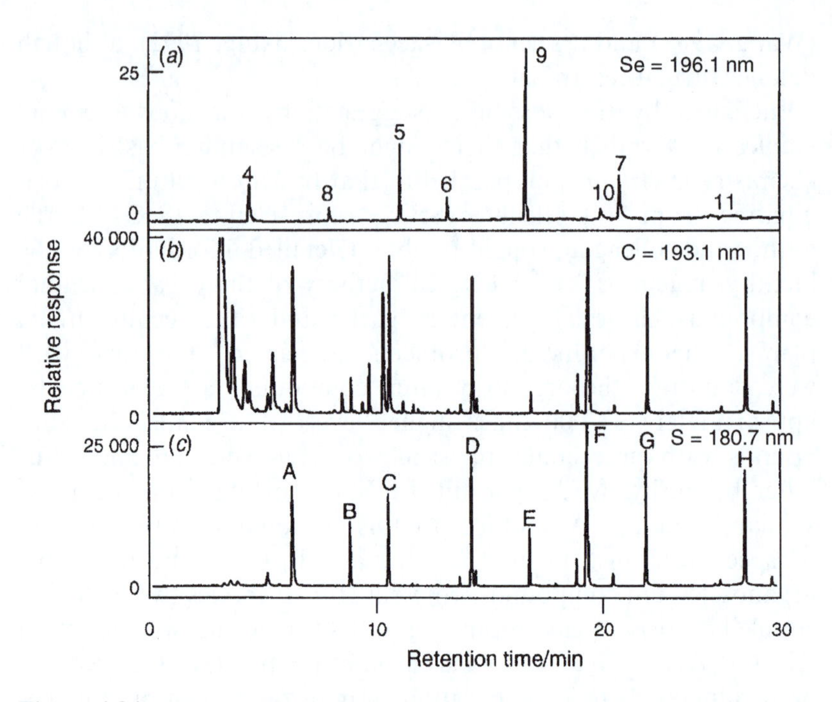

Figure 4.34 Organosulfur and organoselenium compounds in elephant garlic determined by HS/GC-AED-MSD. Organoselenium compounds (trace a): **4**, MeSeMe; **5**, MeSeSMe; **6**, MeSeSeMe; **7**, MeSSeSMe; **8**, MeSeAll; **9**, MeSeSAll; **10**, MeSeCH=CHMe; **11**, MeSSeSAll. Organosulfur compounds (trace c): **A**, MeSAll; **B**, MeSSMe; **C**, AllSAll; **D**, MeSSAll; **E**, MeSSSMe; **F**, AllSSAll; **G**, MeSSSAll; **H**, AllSSSAll. (Image from Cai, 1994a).

Addition of selenomethionine as a source of selenium to the elephant garlic homogenates increased the intensity of all of the selenium peaks so that identification by GC-MS could be accomplished. Alternatively, as will be discussed below, garlic and onions could be grown in the presence of inorganic selenium to increase the levels of the selenium compounds naturally present. Similar compounds were found in the headspace volatiles of garlic and onion homogenates (Cai, 1994a, 1994b). It is significant that, despite the fact that in garlic allyl-group-containing flavorants are more abundant than the analogous methyl compounds, of the compounds of type RSeSR, only the compounds MeSeSMe and MeSeSAll are seen, suggesting a preference for the attachment of methyl over allyl to Se in *Allium* species. While MeSeAll is detected, it could well arise out of the desulfurization of MeSeSAll upon heating.

By analogy to the well-studied organosulfur chemistry of alliums, it was postulated that the volatile selenium compounds, described above, originate from the decomposition of selenium-containing amino acids. By treating *Allium* preparations with reagents that derivatize amino acids and make them volatile, such as ethyl chloroformate, this hypothesis can be tested. Garlic was enriched with selenium, from the natural level of 0.02 ppm to 100–1355 ppm Se dry weight, by fertilizing the growing plant with selenate and/or selenite salts. Similarly, onions can be enriched to 96–140 ppm Se. A 1355 ppm Se garlic preparation treated with ethyl chloroformate and analyzed by both GC-AED and GC-MS showed the presence of *Se*-methyl selenocysteine as the major selenium compound, along with lesser levels of selenocysteine. Unenriched, or weakly-enriched (68 ppm Se), garlic showed only selenocysteine (Cai, 1995). A synthetic standard of *Se*-allyl selenocysteine was prepared. No evidence could be found, however, for the presence of this compound in enriched or unenriched garlic using the above GC-AED and GC-MS methods.

The above analytical methods all involve heating and volatilizing samples, precluding study of non-volatile components. An alternative analytical technique is that of HPLC-inductively coupled plasma mass spectrometry (HPLC-ICP-MS). Like GC-AED, this method is element specific. When garlic was analyzed by HPLC-ICP-MS it was found that γ-glutamyl-*Se*-methylselenocysteine was the major Se-containing component of unenriched or weakly-enriched (68 ppm Se) garlic (Kotrebai, 1999; Dumont, 2006). In garlic containing 1355 and 205 ppm Se, the major Se-containing component was *Se*-methylselenocysteine (Kotrebai, 1999; McSheehy, 2000). Analysis of Se-enriched ramp (*A. tricoccum*; 77, 230 and 524 ppm Se) showed *Se*-methylselenocysteine to be the major product, along with small amounts of selenomethionine, *Se*-cystathionine and γ-glutamyl-*Se*-methylselenocysteine (Whanger, 2000), while analysis of Se-enriched onions (*A. cepa*; 96 and 140 ppm Se) and green onions (*A. fistulosum*) showed γ-glutamyl-*Se*-methylselenocysteine to be the major Se-containing component (Kotrebai, 2000; Shah, 2004). Analysis of Se-enriched chives (*A. schoenoprasum*; 200–700 ppm Se) using HPLC-ICP-MS with a chiral column showed the presence of L-*Se*-methylselenocysteine and L-selenomethionine;

γ-glutamyl-*Se*-methylselenocysteine was also thought to be present (Kápolna, 2007).

What about the claim of Virtanen that *Se*-methylselenocysteine *Se*-oxide (selenomethiin) is naturally present in onions? Oxidation of synthetic *Se*-methylselenocysteine with hydrogen peroxide at pH 8 initially afforded *Se*-methylselenocysteine *Se*-oxide (bracketed in Figure 4.33). However, this compound was unstable at room temperature, resulting in a mixture of methaneseleninic acid ($MeSeO_2H$) and MeSeSeMe after five days (Block, 2001). While the peak of the so-prepared selenoxide could be seen by HPLC, there was no evidence for a peak at the corresponding position for extracts of enzyme-deactivated garlic or onion. Since little change was seen following the treatment of garlic and onion extracts with thiosulfate, a compound known to readily reduce selenoxides (including those derived from *Se*-alk(en)yl selenocysteines and selenomethionine), it was concluded that selenoxides may be present in *Allium* species at levels below the detection limits of ICP-MS. These conclusions are also supported by X-ray absorption spectroscopy (XAS) studies of the forms of selenium in normal and Se-enriched garlic (I. Pickering and E. Block, unpublished studies, Stanford Synchrotron). The same can also be said about the presence in *Allium* species of *Se*-alk(en)yl selenocysteines other than the *Se*-methyl derivative described above. The thermal instability that we found for *Se*-alk(en)yl selenocysteine selenoxides also makes their natural occurrence improbable, other than as transient intermediates in redox systems or as artifacts formed by air oxidation during analysis.

It has also become possible to directly study the distribution and local speciation (chemical form) of selenium in living plants, such as onions, using sophisticated microscopic x-ray methods (microscopic X-ray absorption near-edge structure spectroscopy, or μ-XANES, and confocal microscopic X-ray fluorescence analysis; Bulska, 2006). The use of such techniques in the analysis of growing onion roots and leaves of plants exposed to selenite and selenate, showed that most of the more toxic selenite [Se(IV)] was transformed into organoselenium compounds such as *Se*-methylselenocysteine [Se(II)], while most of the less toxic selenate [Se(VI)] remained unchanged. It is interesting to note, that no compounds of intermediate oxidation state (*e.g.*, selenoxides) were reported. The μ-XANES method is a nondestructive technique, which avoids cutting the plant, offering trace-level sensitivity, as well as microscopic lateral resolution

(Bulska, 2006). A weakness of the XANES technique is that it is not possible to distinguish different compounds of Se(II), *e.g.*, *Se*-methylselenocysteine *versus* selenomethionine.

The above analytical work directed toward identifying the organoselenium compounds in genus *Allium* plants, was specifically motivated by the discoveries that Se-enriched garlic (Se-garlic) is very effective in mammary cancer chemoprevention in the rat model; that synthetic *S*e-2-propenylselenocysteine is very potent *in vivo* in inhibiting the development of experimentally induced breast cancer; and that diallyl selenide is significantly more active as an anticancer agent than diallyl sulfide. For example, rats fed a diet with Se-enriched ramps showed a ~43% reduction in chemically induced mammary tumors (Whanger, 2000). While selenium in the form of selenomethionine (typically from selenized yeast) is often used as a natural source of selenium, a disadvantage is that selenium from this source tends to accumulate in tissues (Whanger, 1988).

It is desirable to find effective selenium sources which have high anticarcinogenic activity, but which do not lead to the accumulation of selenium in tissue, as is the case with *Se*-methyl selenocysteine from alliums. This compound is transformed through a β-lyase reaction in the liver into methylselenol (MeSeH), the assumed biologically active selenometabolite responsible for the anti-carcinogenicity and antioxidant actions of selenium. *Se*-Methylselenocysteine produces methylselenol much more efficiently than selenomethionine (Suzuki, 2007). On the basis of studies with mice, it is suggested that *Se*-methyl selenocysteine can serve as a therapeutic agent which disrupts androgen receptor signaling for prostate cancer (Lee, 2006). It is also concluded that γ-glutamyl *Se*-methylselenocysteine is an effective anticancer agent, with a mechanism of action very similar to that of *Se*-methylselenocysteine (Dong, 2001).

Organoselenium compounds of the sort present in garlic have been found to prevent oxidative DNA damage from copper-generated hydroxyl radicals at biologically relevant concentrations. The observed oxidative DNA damage inhibition required selenium-metal coordination. Since Cu^+ is a soft metal ion and selenium is a soft ligand, selenium compounds should coordinate better to Cu^+ than the harder Fe^{+2} ion. This is important since copper generates hydroxyl radicals 60 times faster than Fe^{+2} (Battin, 2006). It is noteworthy that using a similar assay, it was found that Na_2SeO_4 (sodium selenate) was unable to inhibit iron-mediated DNA damage and that Na_2SeO_3 (sodium selenite) was found to behave as both pro-oxidant and anti-oxidant, depending on the concentrations of the selenium compound and H_2O_2. Based on these observations it is preferable to use an organic rather than inorganic source of selenium as an antioxidant (Ramoutar, 2007), *e.g.*, the types of organoselenium compounds found in alliums.

Given that selenium occurs naturally in garlic, is tellurium also present? Tellurium is a rare element that lies just below selenium in the periodic chart of elements, and shares some of the properties of both selenium and sulfur. Through use of hydride-generation atomic absorption spectroscopy (HG-AAS), tellurium has been found to be present at levels of 55–65 ppb in garlic grown in Argentina (Kaplan, 2005). Despite claims to the contrary, there is no evidence for the presence of compounds of the element germanium in garlic or other alliums.

4.15 SYNOPSIS OF *ALLIUM* CHEMISTRY

The importance in *Allium* chemistry of three-carbon units attached to sulfur, such as the allyl group bearing the name of the genus, was first discovered by nineteenth century chemists. Many thousands of scientific studies later we are reminded of this earliest work by the profusion of important organosulfur molecules whose detailed chemistry we have just considered, containing three, or a multiple of three, carbon atoms: the three-carbon compounds – 1-propenesulfenic acid, propanethial *S*-oxide (the onion LF), 2-propenesulfenic acid, *n*-propanesulfenic acids, thioacrolein, the various precursor compounds with three-carbon units attached to the sulfur atom of the amino acid cysteine, and 2-propenethiol, precursor of the much-maligned metabolite of dietary garlic in garlic breath, allyl methyl sulfide; the six-carbon compounds – allicin, diallyl disulfide and polysulfides, the zwiebelanes, the bis-sulfine and the LF-dimer; and the nine-carbon compounds – ajoene and the cepaenes. Lower concentrations of sources of one-carbon methyl groups are also present, allowing formation of compounds based on various combinations of one and three carbons, *e.g.*, two $(1+1)$, four $(1+3)$, five $(1+1+3)$, and seven $(1+3+3)$.

While the alliinase enzyme very rapidly cleaves alliin to form 2-propenesulfenic acid, no enzyme is needed to convert the latter compound to allicin since self-condensation is extremely fast.

The appealing, tongue-tingling effect of allicin can quickly become an unpleasant burning sensation with prolonged exposure to raw garlic, reflecting the presumed purpose of this compound in defending garlic plants against attack through its irritant properties. The activated sulfur–sulfur bond in allicin is able to react with key thiol groups which are essential for the growth of microorganisms, as well as other thiol groups which function in animals as pain-sensing receptors. Allicin is unstable and regenerates 2-propenesulfenic acid, as well as thioacrolein, a very reactive molecule, only prevented from combining with itself to form dithiins at –200 °C (Scheme 4.22). Allicin reacts with water (Scheme 4.26) generating a family of diallyl polysulfides, whose biological activity increases with the number of linked sulfur atoms. Of these, diallyl trisulfide and tetrasulfide show promise as nematicides for agricultural applications (see Chapter 6) while diallyl trisulfide is used in China as an antibiotic (see Chapter 5).

Rapid self-condensation of 1-propenesulfenic acid from onions leads to pleasantly smelling zwiebelanes (Scheme 4.37), which presumably have little if any growth-inhibitory or repellent activity. Most of the 1-propenesulfenic acid formed by the onion alliinase enzyme is prevented from self-condensing, through conversion by the onion LF synthase enzyme into the intensely lachrymatory propanethial *S*-oxide (onion LF), certainly one of the more unusual molecules found in nature. If the onion sulfenic acid escapes the clutches of the LF synthase enzyme and encounters an LF molecule, two different types of reactions occur: so-called "carbophilic addition" leading to the cepaenes (Scheme 4.33) or "thiophilic addition" leading to a structurally unique bis-sulfine (Schemes 4.38 and 4.39). Self reaction of the LF is also possible giving an unusual four-membered ring with two adjacent sulfur atoms (Scheme 4.17). Under the correct conditions, products from the rearrangements associated with the above described chemistry can lead to the curious pinking of onions and greening of garlic (Scheme 3.40). The nine-carbon sulfur compounds, ajoene and cepaenes, derived from garlic (Scheme 4.32) and onion (Scheme 4.33), respectively, are of considerable current interest for their medical applications, discussed in Chapter 5. Details on various novel types of non sulfur-containing organic compounds found in *Allium* species can be found in recent reviews (Mskhiladze, 2008; Corzo-Martínez, 2007; O'Donnell, 2007; Lanzotti, 2006).

CHAPTER 5

Alliums in Folk and Complementary Medicine

Since things that here in order shall ensue,
Against all poysons have a secret power.
Peare, garlicke, reddish root, nuts, rape and rue –
But garlicke chiefe; for they that it devoure
May drinke, and care not who their drinke do brewe.
May walk in aires infected, every houre
Sith garlicke them have power to save from death,
Bear with it though it maketh unsavory breath,
And scorne not garlicke, like some that thinke
It only maketh men winke, and drinke – and stinke.

J. Harington (1561–1612), *The Englishman's Doctor*

The desire to take medicine is perhaps the greatest
feature which distinguishes man from animals.

William Osler (1849–1919)

5.1 EARLY HISTORY OF ALLIUMS IN FOLK MEDICINE

Early folk medicine is based on experience with herbs and natural substances and their curative powers as well as magico-religious beliefs. Belief in the protective and healing power of garlic may have been based on assumptions derived from its pungent odor: if

Garlic and Other Alliums: The Lore and the Science
By Eric Block

Published by the Royal Society of Chemistry, www.rsc.org

garlic-laden breath kept other humans at bay, it should also have the same effect on diseases and animals (Gowers, 1993; Mezzabotta, 2000). Among the early writings on uses of garlic and onion are the *Codex Ebers* (*Ebers Papyrus*), an Egyptian holy book dating from about 1550 BCE, which details more than 875 therapeutic formulas of which 22 contain garlic, and the so-called "Greek magical papyra" from the second century BCE to the fifth century CE, which contains a variety of magical spells and formulas, hymns and rituals. Among these is a spell of coercion involving ". . . salt, fat of a dead doe, and mastic, and myrtle, dark bay, barley, and crab claws, sage, rose, fruit pits and a single onion/garlic, fig meal, a dog-faced baboon's dung, and egg of a young ibis" (Betz, 1992). Onion and garlic were used by the Egyptians in embalming and mummification (Manniche, 2006; Nicholson, 2000).

The Greek writer Aristophanes in his play *The Ecclesiazusae* (390 BCE) gives a recipe for problems with the eyelids: "Pound together garlic and laserpitium juice, add to this mixture some Laconian spurge, and rub it well into the eyelids at night." One of the most comprehensive compilations of remedies associated with genus *Allium* plants is to be found in Pliny the Elder (Gaius Plinius Secundus), *The Natural History*, first published in 77 CE and subsequently translated many times (Rackham, 1971). One of the most spectacular editions was translated into Florentine Italian by Cristoforo Landino, and illuminated, bound and printed in Venice by Nicolaus Jenson around 1480 (Figure 5.1). It is one of the most treasured volumes of Oxford University's Bodleian Library. Book 19 of *The Natural History* includes a discussion of the cultivation and botanical characteristics of garlic, onion, shallot, leek, chives and several types of wild garlic, while Book 20, dealing with remedies derived from the garden plants, gives separate remedies for onion, leek and garlic. Much of what Pliny writes has been repeated by others in later herbals.

Pliny writes that "importance has recently been given to chives by the emperor Nero, who on certain fixed days of every month always ate chives preserved in oil, and nothing else, not even bread, for the sake of his voice." Pliny describes 39 different remedies involving leek. He observes that "leeks stops bleeding at the nose if the nostril is plugged with leek pounded, or mixed, with gall-nut or mint. It cures chronic cough, and affections of the chest and lungs. The leaves, applied topically, are employed for the cure of pimples,

LIBRO PRIMO DELLA NATVRALE HISTORIA DI C. PLINIO SECONDO TRADOCTA IN LINGVA FIORENTINA PER CHRISTOPHORO LANDINO FIORENTINO AL SERENISSIMO FERDINANDO RE DI NAPOLI.

PREFATIONE

ITERMINAI O GIOCONDISSIMO imperadore con epiſtola forſe di troppa licẽtia narrarti elibri della hiſtoria naturale: opera nouella alle muſe romane: nata apreſſo di me nel lultima genitura. Sia adunq; queſta prefatiõe ueriſſima di te mẽtre che gia inuecchia nel grãdiſſimo tuo padre: per che uſando el uerſo di Catullo mio compatriota tu ſoleui pure ſtimare qualche choſa le mie ciãcie. Tu conoſci queſta caſtrenſe & militare parola. Et lui chome tu ſai mutando le prime ſyllabe ſi fece alquanto piu duro che non uolea eſſere ſtimato da tuoi familiari & ſerui. Per queſto adunq; diterminai ſcriuerti: & ãchora per che le noſtre choſe appariſchino & ſieno manifeſte p queſta mia audacia maxime dolẽdoti tu che pel paſſato non lhabbi facto in una altra noſtra procace epiſtola. Et accio che tutti glhuomini ſappino quanto di pari lomperio techo uiua: Tu elquale hai triomphato & ſe ſtato cenſore & ſei uolte cõſolo & participe della tribunitia poteſta: Se ſtato prefecto del pretorio: ilche hai facto piu nobile che tutti glaltri magiſtrati: perche per piacere a tuo padre & allordine equeſtre lacceptaſti: Et tutte queſte coſe per riſpecto della republica hai facto: Et me chome nel contubernio caſtrenſe tractaſti? Et certo niẽte ha mutato inte lamplitudine & grandezza della tua fortuna: ſe non che tanto piu poſſi & uogla giouare: quãto quella e maggiore. Adũq; bẽche a tutti glaltri huomini ſia aperta la uia a impetrare ogni choſa da te uenerãdoti: Niente di meno ſolo laudacia fa che io piu familiarmente te honori. Queſta audacia adunq; imputerai a te medeſimo: & a te medeſimo nel noſtro fallo perdonerai. Io miſtroppicciai la faccia: & niente di meno neſſuno proficto ho facto: perche per unaltra uia mappariſti grande: & di lontano mi rimuoui con le faccelline del tuo ingegno. Et certo in nexuno piu ſfolgora quella: laquale piu ueramente e decta in te che in altri forza deloquentia. In te e quella facundia che alla tribunitia poteſta ſi conuiene: Con q̃ta riſonantia tuoni tu le laude paterne? Cõ quanta (non ſanza amore) dimoſtri quelle di tuo fratello? Quanto ſe excellente & ſublime nella poetica faculta? O gran fecondita danimo. Certo hai trouato inche modo poſſi imitare tuo fratello. Ma queſte choſe chi potrebbe ſanza paura conſiderare: hauendo a uenire al giudicio dellongegno tuo: maxime eſſendo quello dame prouocato? Certamente non ſono in ſimile conditione quegli che publicano alchuno libro: & quegli che ate glintitolano. Impero che ſe io lo publicaſſi & non lo intitolaſſi ate: potrei dire perche leggi tu queſte choſe o imperadore? lequali ſono ſcripte albaſſo uulgo & alla turba de glagricultori & de glartefici & a quegli che cõſumano elloro otio negli ſtudii? Perche adunq; ti fa tu giudice: concio ſia che quando io ſcriueuo queſta opera: non thaueuo poſto nella tauola doue ſono deſcripti egiudici: Et eri di tanta excellentia: che non ſtimauo che tu ti degnaſſi ſcendere ſi baſſo? Preterea quando bene non fuſſi in ſi excelſo grado: nientedimeno gli ſcriptori comunemente fuggono el giudicio de docti. Queſto fa Cicerone: elquale e di tanta eloquentia: che puo ſottomettere longegno al giuocho della fortuna: & quel

Figure 5.1 Front page of Pliny the Elder *Historia Naturalis*. (Courtesy of Oxford University).

burns, and [styes]... Other kinds of ulcers, too, are treated with leeks beaten up with honey: used with vinegar, they are extensively employed also for the bites of wild beasts, as well as of serpents and other venomous creatures. Leeks are eaten, too, in cases of poisoning by fungi, and are applied topically to wounds: they act also as an aphrodisiac, allay thirst, and dispel the effects of drunkenness; but they have the effect of weakening the sight and causing flatulency, it is said, though, at the same time, they are not injurious to the stomach, and act as an aperient. Leek greatly improves the voice, and acts as an aphrodisiac, and as a promoter of sleep."

Pliny describes 61 different remedies involving garlic (Volume 20, Chapter 23), in part writing: "Garlic has powerful properties, and is of great benefit utility against changes of water and of residence. It keeps off serpents and scorpions by its smell... It cures bites when drunk or eaten, or applied as ointment... it is an antidote against the poisonous bite of the shrew-mouse ... pounded garlic has been given in milk to asthmatics...The ancients used also to give it raw to madmen." With regard to onion, Pliny lists 27 remedies: "The cultivated onion is employed for the cure of dimness of sight, the patient being made to smell it till tears come into the eyes: it is still better even if the eyes are rubbed with the juice. It is said, too, that onions are soporific, and that they are a cure for ulcerations of the mouth, if chewed with bread. Fresh onions in vinegar, applied topically, or dried onions with wine and honey, are good for the bites of dogs, care being taken not to remove the bandage till the end of a couple of days. Applied, too, in the same way, they are good for healing excoriations. Roasted in hot ashes, many persons have applied them topically, with barley meal, for defluxions of the eyes and ulcerations of the genitals. The juice, too, is employed as an ointment for sores of the eyes, albugo, and argema. Mixed with honey, it is used as a liniment for the stings of serpents and all kinds of ulcerous sores. In combination with woman's milk, it is employed for affections of the ears; and in cases of singing in the ears and hardness of hearing, it is injected into those organs with goose-grease or honey. In cases where persons have been suddenly struck dumb, it has been administered to them to drink, mixed with water. In cases, too, of toothache, it is sometimes introduced into the mouth as a gargle for the teeth; it is an excellent remedy also for all kinds of wounds made by animals, scorpions more particularly.

In cases of alopecy and itch-scab, bruised onions are rubbed on the parts affected: they are also given boiled to persons afflicted with dysentery or lumbago. Onion peelings, burnt to ashes and mixed with vinegar, are employed topically for stings of serpents and millipedes."

Dioscorides (*ca.* 40–90 CE) was a Greek physician who served as a military surgeon under the Roman Emperor Nero. He traveled widely in southern Europe and northern Africa, observing and recording information about the medicinal value of plants, which he expanded upon in a five volume book written in Greek, but then translated into Latin as *De Materia Medica.* Considered to be the authoritative work on medicinal plants for 1500 years, *Materia Medica* has been called "the most successful botanical textbook ever written." It has been translated into many languages and is beautifully illustrated, as in the case of the extraordinary copy dating from 512 CE in the Austrian National Library in Vienna. The *Materia Medica* is important not just for the history of herbal science, but because it also gives us a knowledge of the herbs and remedies used by the Greeks, Romans, and other cultures of antiquity, and records the ancient names for some plants which would otherwise have been lost.

The work presents about 500 plants in all. Not surprisingly there is rather extensive discussion of medicinal properties of the alliums, including garlic, leek and onion. With regard to garlic, Dioscorides describes 23 medicinal uses including "cleaning the arteries," driving out broadworms (intestinal parasites), acting as a mild diuretic and as an antidote for various poisons, and "injurious change in waters" when taken internally (diarrhea, amoebic dysentery). It can also be used externally for various skin afflictions. A modern edition of Dioscorides' herbal, first translated into English by John Goodyer from 1652 to 1655, was published in 1933 (Gunther, 1933). With regard to garlic (*Allium* in Latin), Dioscorides writes that "it hath a sharp, warming biting qualitie, expelling of flatulencies, and disturbing of the belly, and drying of the stomach causing of thirst .... dulling the sight of the eyes...and drawes away the urine...Being dranck with decoction of Origanum, it doth kill lice and nitts...kept in the mouth it does assuage the paine of ye teeth...it open ye mouths of ye veins" (Gunther, 1933).

Another treasure of Oxford's Bodleian Library is the group of medical texts in the Bower Manuscript, dating from the first half of the sixth century CE (Wujastyk, 2003). In 1890, British lieutenant Hamilton Bower obtained the manuscripts found in the ruins of a Buddhist monastery near the old Silk Road trading stop of Kuqa in far western China, close to the geographic point where Kazakhstan, Kyrgyzstan and China meet (map: Figure 1.9 in Chapter 1). After they had been translated, Bower sold the manuscripts, now bearing his name, to the Bodleian Library. These birch-bark manuscripts, three of which deal with ayurvedic medicine (the ancient art of healing, practiced in India for more than 2000 years), were written in Sanskrit. It is thought that the manuscripts were copied by Buddhist monks from what is now northern Pakistan. The Bower Manuscript includes a self-contained tract on the mythical origin of garlic and its medical uses in the form of 43 verses.

An amusing section of this manuscript describes how Brahmins, normally forbidden to eat garlic, may avoid this prohibition by feeding it to a cow and then consuming the animal's milk products: "When a cow has been kept waiting for three nights with almost no grass, one should give her a preparation made of two parts grass to one part garlic stalks. A Brahmin can then partake of her milk, curds, ghee, or even buttermilk, and banish various diseases while maintaining propriety" (Wujastyk, 2003). The Bower Manuscript goes on to describe garlic as a "universal remedy" that "drives away pallid skin disease, appetite-loss, abdominal lumps, coughs, thinness, leprosy, and weak digestion. It removes wind, irregular periods, gripes, phthisis, stomach-ache, enlarged spleen, and piles. It takes away paralysis of one side, lumbago, worm disease, colic, and urinary disorders. It completely conquers lassitude, catarrh, rheumatism of the arms or back, and epilepsy… I have declared the use of garlic, as it was seen in olden times by the sages. One should practise it properly." There is also a description of a spring garlic festival when garlands of garlic are displayed in front of houses and on the gates and are worn by occupants of the house. A rite of worship is performed in the courtyard during the festival (Wujastyk, 2003).

John Parkinson, Apothecary of London and The King's Herbalist, writes in his 1656 book *Paradisi in Sole* in the section titled "The Kitchen Garden": "The juice of onions is much used to be applied to any burnings with fire, or with gunpowder, or to any scaldings with water or oil, and is most familiar for the country, whereupon such sudden occasions they have not a more fit or speedy remedy at hand. The strong smell of onions, and so also of garlick and leek, is quite taken away from offending the head or eyes, by the eating of parsley leaves after them." Onions are also said to be "good for those that are troubled with coughs, shortness of breath and wheezing." Leeks, boiled and applied warm, are said to give relief when applied to swollen and painful hemorrhoids. Raw or boiled leeks are also recommended as a cure for hoarseness while garlic is praised as "the poor mans Treakle, that is a remedy for all diseases. It is never eaten raw of any man that I know..." Despite the presumed health-giving attributes of the alliums, and their mention in the Bible, "our dainty age now refuseth [their usage] wholly, in all sorts except [by] the poorest" (Parkinson, 1656).

Robert Thornton (an M.D. and member of Cambridge University) in his 1814 herbal says of garlic that "applied externally it acts successively as a stimulant, rubefacient [causes redness of the skin], and blister. Internally, from its very powerful and diffusible stimulus, it is often useful in diseases of languid circulation and interrupted secretion... [However] in cases in which...irritability prevails, large doses of it may be very hurtful... Taken in moderation it promotes digestion; but in excess, it is apt to produce headach, flatulence, thirst, febrile heat, and inflammatory diseases, and sometimes occasions a discharge of blood from the haemorrhoidal vessels. In fevers of the typhoid type, and even in the plague itself, its virtues have been much celebrated... It is much recommended by some as an anthelmintic [eliminating parasitic worms], and has been frequently applied with success externally as a stimulent to indolent [benign; slow to heal] tumours, in cases of deafness...and in retention of urine" (Thornton, 1814).

Thornton indicates different methods of administering garlic: "several cloves may be taken at a time without inconvenience, or the cloves cut into slices may be swallowed without chewing... The expressed juice, when given internally, must be rendered as palatable as possible, by the addition of sugar and lemon juice. In deafness, cotton moistened with the juice is introduced within the ear, and the application renewed five or six times in one day. Infusions in spirit, wine, vinegar and water, although containing the whole of its virtues, are so acrimonious as to be unfit for general use; and yet an infusion of an ounce of bruised garlic in a pound of milk, was the mode in which Rosenstein exhibited it to children afflicted with worms. By far the most commodious form for administering garlic is that of a pill or bolus conjoined with some powder... Garlic made into an ointment with oils, *etc*., and applied externally is said to resolve and discuss indolent tumours, and has been by some greatly esteemed in cutaneous disease. It has likewise sometimes been employed as a repellent... garlic applied to the soles of the feet... [can be used] in the confluent small-pox, about the eighth day, after the face begins to swell; the root cut in pieces, and tied in a linen cloth, was applied to the soles, and renewed once a day till all danger was over." With regard to onion, Thornton says: "The chief medicinal use of onions in the present practice is in external applications, as a cataplasm [poultice] for suppurating [pus discharging] tumours, *etc*." (Thornton, 1814). While many of the above treatments seem foolish based upon current medical knowledge, Thornton's astute observations on the chemistry of garlic, as well as his several cautionary statements concerning garlic, make him a very creditable observer of early nineteenth century medical practices in the pre-antibiotic era. Indeed, in 1916, a treatment for whooping-cough appeared in the *British Medical Journal* which is very similar to Thornton's treatment for small-pox, involving "cutting [garlic cloves] into thin slices, and wearing them under the soles of the feet between two pairs of socks ... the garlic can usually be smelt in the breath within half an hour ... and the whoop and spasm usually disappear within forty-eight hours" (Hovell, 1916).

In China, onion and garlic tea have long been recommended for fever, headache, cholera and dysentery (Corzo-Martínez, 2007). Legend has it that some followers of Huang Di, the so called Yellow Emperor (2697–2595 BCE), were cured of intestinal poisoning by eating "wild" garlic, possibly *A. tuberosum* or Chinese chives. In the Chinese Pharmacopoeia, garlic is suggested for controlling parasites and scalp ringworms, for dysentery and gastrointestinal disorders, as an expectorant and diuretic, and as a poultice for sores (Moyers, 1996). In India, garlic has been used as an antiseptic lotion for washing wounds and ulcers (Corzo-Martínez, 2007).

Ramsons (*A. ursinum*) or wild garlic grows in profusion in the wild. In Ireland, to treat a cold or cough, strips of *A. ursinum* are worn under the soles of the feet or inside shoes and socks. This treatment is not quite so far-fetched as it sounds, because it is known that allicin, formed by crushing garlic, will rapidly penetrate the skin and cause garlic breath upon crushing a garlic clove with bare feet (Allen, 2004).

Medical uses of garlic and onion at the beginning of the twentieth century, before the discovery of allicin and alliin and introduction of penicillin and modern antibiotics, are described in the *British Pharmaceutical Codex* and the *United States Dispensatory*, both of which provide extensive listings of existing and discontinued drugs. The former source states: "Garlic has antiseptic, diaphoretic, diuretic and expectorant properties. Large doses, continued, are stated to cause an increase in blood pressure. The juice has been used in laryngeal tuberculosis, and poultices made from pulped garlic are stated to be useful in accessible tuberculosis lesions. The juice, diluted with four parts of water, has been used externally for treating suppurating wounds, and by inhalation for pulmonary tuberculosis. The fresh juice is extremely irritating when applied to broken surfaces...Cases have been reported where its use internally has proved fatal to children" (Pharmaceutical Society of Great Britain, 1934). The *United States Dispensatory* states, concerning garlic and onion: "[the use of garlic] as a medicine is of the highest antiquity. The oil may be given with great advantage in obstinate bronchitis, and catarrhal pneumonia of young children. The bruised garlic cloves are often applied as poultices to the lungs, and similar applications were formerly used to the feet for nervous restlessness, or even the convulsions of young children...[Onion] is used as a stimulant, diuretic, expectorant, and rubefacient, and the juice is occasionally given made into syrup in infantile catarrhs and croup. Roasted and split the onion is sometimes applied as an emollient cataplasm to suppurating tumors" (Hall, 1918).

5.2 GARLIC DIETARY SUPPLEMENTS: MARKETING AND REGULATION

Garlic has long enjoyed a reputation for having remarkable medicinal qualities, both preventative and curative, effective for a wide range of human ailments. True believers tout "the miracle of garlic," while skeptics mock "the garlic cult" whose claims go far beyond what can be proven by scientific studies. Headlines of tabloids in the 1980s and 1990s trumpeted the health benefits of garlic: "From world's top university doctors – amazing garlic cures: heart disease + cancer + infections + arthritis + cholesterol + stroke + acne + asthma + stress" (*National Examiner*, July 18, 1995). Such publicity, along with the appearance of a profusion of garlic-promoting health books, websites and commercials featuring television personalities, helped make garlic the top-selling single-herb supplement, with more than 7 000 000 adult users and annual U.S. sales exceeding $160 000 000 in 2006 (Barnes, 2004; Weise, 2007). However, since the front pages of these same tabloids also featured stories on Elvis sightings, "UFO bases found on Mount Everest" and "Baby lives after three months in freezer," the legitimacy of their health reporting is suspect.

The popularity of garlic supplements (Figure 5.2) is part of a growing phenomenon reflecting a public desire for self-empowerment in health matters as well as distrust of pharmaceutical companies and regulatory agencies. This trend has been fueled by the escalating cost of drugs, well-publicized failures of major medicines, allegations of regulators being too cozy with drug companies, as well as "the trend for people to seek milder, more natural products" (Ipsen, 1994; Marcus, 2002; Halsted, 2003). In the United States, this

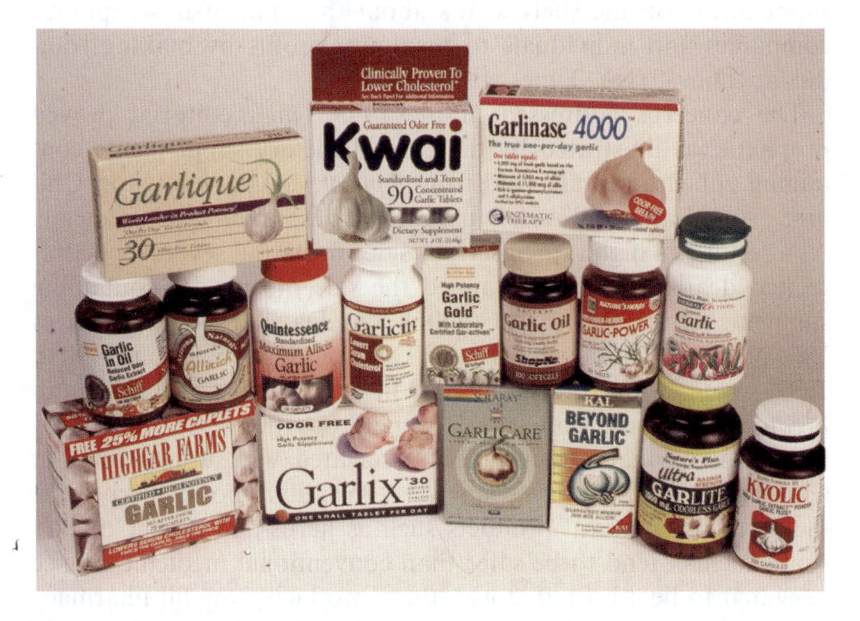

Figure 5.2 Examples of garlic health supplements. (Courtesy of Larry Lawson).

mindset led to passage in 1994 of the Dietary Supplement Health Education Act (DSHEA). A dietary supplement is defined as "an ingested product, intended to supplement the diet, which bears or contains one or more of the following dietary ingredients: a vitamin; a mineral; an herb or other botanical; an amino acid; a dietary substance for use by man to supplement the diet by increasing total dietary intake; or a concentrate, metabolite, constituent, extract; or combination of any ingredient described above" [U.S. Code (2003), Title 21 § 321 (ff)]. A garlic supplement is categorized as an herb or botanical, that is, a plant or plant part (root, flower, leaf, fruit) used for its medicinal or therapeutic properties.

Supporters of DSHEA called it the "Health Freedom Act" while detractors, including *The New York Times* editorial page, termed it "The 1993 Snake Oil Protection Act." A *Times* editorial, arguing against passage of DSHEA, said that "the legislation would lower the new labeling standard to allow health claims for supplements supported only by unconfirmed preliminary studies not subjected to any meaningful scientific peer review. It would also make it harder for the FDA to seize products promptly, or demand proof of safety, where there is mounting evidence that a product may be dangerous. This fight, in other words, really isn't about keeping supplements on the shelves. It's about the right of unscrupulous companies and individuals to maximize profits by making fraudulent claims" (NY Times, 1993).

The concerns addressed in the 1993 *Times* editorial proved prophetic, as illustrated by the case of dietary supplements containing ephedra, an alkaloid from *Ephedra sinica*, known in Chinese as *ma huang* and widely used in Chinese traditional medicine. As a result of a high rate of serious side effects and several ephedra-related deaths, the FDA banned the sale of ephedra-containing supplements in 2006 (FDA, 2006). Dr. David Kessler, Dean of Yale Medical School and former FDA Commissioner from 1990 to 1997, wrote in an editorial in 2000 in the *New England Journal of Medicine* in connection with a paper reporting cases of kidney failure and urinary tract cancer associated with use of another Chinese herb, *Aristolochia fangchi*: "Bestselling books perpetuate the myth that natural products such as herbs and other "dietary supplements" tend to be safer than conventional medicines. Once relegated to health food stores, these products now fill pharmacy and supermarket shelves. So-called natural substances are more popular than ever. Fueled by congressional passage of the Dietary Supplement Health and Education Act of 1994, which deregulated the industry by limiting the role of the Food and Drug Administration (FDA), the popularity of dietary supplements has created a $15-billion-a-year industry... Congress has shown little interest in protecting consumers from the hazards of dietary supplements, let alone from the fraudulent claims that are made, since its members apparently believe that few of these products place people in real danger. Nor does the public understand how potentially dangerous these products can be." Kessler argued that "Congress should change the law to ensure the safety and efficacy of dietary supplements before more people are harmed" (Kessler, 2000).

A New York State task force on safety of dietary supplements concluded: "Although the therapeutic effect of dietary supplements depends on their potency, there are no federal standards for dosage and purity, and the dose-finding studies that are mandatory for pharmaceuticals are rarely, if ever, performed. For many products, active ingredients have not been identified and the quantity needed to derive an effect has not been determined. Inferior manufacturing practices can lead to inaccuracies in product labeling (products may actually contain greater or lesser amounts of ingredients listed on their label) and the concentrations of active ingredients can vary among and within brands. Consumers may not know how much of any particular ingredient they consume. Dietary supplement manufacturers are not legally required to use any standardization processes to ensure batch consistency of their products, nor is there any legal or regulatory definition for dietary supplement standardization" (New York State, 2005).

In response to concerns such as those noted above, in 2006 the United States Congress modified DSHEA with the Dietary Supplement and Nonperscription Drug Consumer Protection Act, mandating supplement manufacturers to report all serious adverse events of over-the-counter drugs to the FDA within 15 business days of receipt and to maintain all adverse event reports received for six years. In 2007 the FDA announced a rule establishing current good manufacturing practice requirements (CGMPs) for dietary supplements. The FDA indicated that these regulations provide more accountability in the manufacturing process so that consumers can be confident that the products they purchase contain what is on the label (FDA, 2007, 2008).

To address problems of identification of active ingredients, journals publishing studies on botanical dietary supplements are requiring specific information in these articles, for example as indicated by Dr. Christine A. Swanson of the NIH Office of Dietary Supplements writing in the *American Journal of Clinical Nutrition*: "A study cannot be evaluated or replicated unless the test materials are properly identified and characterized. Investigators must provide an accurate and complete description of the botanical test material regardless of whether it is a finished product, commercial ingredient, extract, or single chemical constituent...[for plant extracts] the method of extraction must be described and the ratio of crude plant to plant extract in test preparations must be provided. Research cannot be replicated without this information. Chemical fingerprints of plant extracts (usually determined by HPLC or liquid chromatography–mass spectrometry) are informative and provide both qualitative and quantitative information. Plants have characteristic chemical profiles of secondary metabolites. These chemical fingerprints can be used to identify the source material and provide an indication of purity. Unexpected peaks, for example, may indicate contamination or adulteration" (Swanson, 2002).

In Europe, garlic and other herbal dietary supplements are regulated by the Directive on Traditional Herbal Medicinal Products (Directive 2004/24/EC of the European Parliament), which went into effect in 2005 (Silano, 2004; Kroes, 2006; Eberhardie, 2007). The Directive requires traditional, over-the-counter herbal remedies to be made to assured standards of safety and quality and for regulations to be standardized throughout the European Union. There will be a transitional period for products legally on the market since 30 April 2004, giving them protection until 2011. The Directive demands proof that a traditional herbal medicinal product has been in use for 30 years in the European Union (or at least 15 years in the European Union and 15 years elsewhere) for it to be licensed and obtainable over the counter.

The quality of website information on the health benefits of garlic varies, since many sites are affiliated with manufacturers or paired with online order catalogs. A study of the sale of garlic and other common herbal remedies found more than 400 Internet sites of which 149 claim that their products would treat, prevent or even cure particular diseases (Morris, 2003). Abstracts of the primary and secondary scientific and medical literature can be freely accessed at *PubMed* (United States National Library of Medicine – National Institutes of Health): http://www.ncbi.nlm.nih.gov/sites/entrez. In addition, general information on the health benefits of garlic (as well as onion), interpreted by expert panels, can be obtained from the following impartial sources:

- National Center for Complementary and Alternative Medicine (NCCAM: NIH, U.SA.), *Publication No. D274*, March 2008: http://nccam.nih.gov/health/garlic/index.htm
- Health Canada, Natural Health Products, Monograph on Garlic, May 2008: http://www.hc-sc.gc.ca/dhp-mps/prodnatur/applications/licen-prod/monograph/mono_garlic-ail-eng.php
- European Scientific Cooperative on Phytotherapy, Allii Sativi Bulbus – Garlic, *ESCOP Monographs*, 2nd Ed., 2003, 14–25, ESCOP, Exeter, UK and Georg Thieme Verlag, Stuttgart and New York.
- Agency for Healthcare Research and Quality Evidence Report, Garlic: Effects on Cardiovascular Risks and Disease, Protective Effects Against Cancer, and Clinical Adverse Effects, *Technology Assessment Number 20*, (AHRQ Publication No. 01-E023; 2000): *http://www.ahrq.gov/clinic/ep*csums/garlicsum.htm
- Full Evidence Report online on the U.S. National Library of Medicine Bookshelf: http://www.ncbi.nlm.nih.gov/books/bv.fcgi?rid=hstat1.chapter.28361
- M. Blumenthal, A. Goldberg and J. Brinckman, ed., Garlic, *Herbal Medicine: Expanded Commission E Monographs*, Lippincott Williams & Wilkins, Newton, MA, 2000, pp. 139–148.
- World Health Organisatioon, *World Health Organization (WHO) Monographs on Selected Medicinal Plants*, Vol. 1, 1999: http://whqlibdoc.who.int/publications/1999/9241545178.pdf
- PDR (Physicians' Desk Reference) for Herbal Medicines, Thompson Healthcare, Montvale, NJ, 4th edn, 2007.

The above referenced 2008 NCCAM summary on garlic states: "Some evidence indicates that taking garlic can slightly lower blood cholesterol levels; studies have shown positive effects for short-term (one to three months) use. However, an NCCAM-funded study on the safety and effectiveness of three garlic preparations (fresh garlic, dried powdered garlic tablets, and aged garlic extract tablets) for lowering blood cholesterol levels found no effect. Preliminary research suggests that taking garlic may slow the development of atherosclerosis (hardening of the arteries), a condition that can lead to heart disease or stroke. Evidence is mixed on whether taking garlic can slightly lower blood pressure. Some studies suggest consuming garlic as a regular part of the diet may lower the risk of certain cancers. However, no clinical trials have examined this. A clinical trial on the long-term use of garlic supplements to prevent stomach cancer found no effect."

The 2008 Health Canada web-document on garlic, cited above, states that it is: "traditionally used in Herbal Medicine to help relieve the symptoms associated with upper respiratory tract infections and catarrhal conditions....to help reduce elevated blood lipid levels/hyperlipidaemia in adults...to help maintain cardiovascular health in adults" (Health Canada, 2008).

The above cited 2003 ESCOP monograph on garlic gives the following therapeutic indications: "prophylaxis of atherosclerosis, treatment of elevated blood lipid levels insufficiently influenced by diet. Also used for upper respiratory tract infections and catarrhal conditions, although clinical data to support this indication is not available" (ESCOP, 2003). The recommended dosage for the different indications ranged from the equivalent of one to four cloves per day, with a cautionary statement about usage immediately following surgery.

The 1999 World Health Organization (WHO) monograph entitled *Bulbus Allii Cepae* states under principle uses for onion supported by clinical data that onion can be used: "to prevent age-dependent changes in the blood vessels, and loss of appetite." The monograph provides only one study involving rats as documentation. It also notes uses of onion in traditional medicine for "treatment of bacterial infections such as dysentery, and as a diuretic...to treat ulcers, wounds, scars, keloids, and asthma... [and] an adjuvant therapy for diabetes." Uses described in folk medicine, not supported by experimental or clinical data, include onion "as an antihelminthic, aphrodisiac, carminative, emmenagogue, expectorant and tonic, and for the treatment of bruises, bronchitis, cholera, colic, earache, fevers, high blood pressure, jaundice, pimples, and sores" (WHO, 1999).

The 1999 WHO monograph on garlic, *Bulbus Allii Sativi*, states under uses supported by clinical data that garlic can be used: "as an adjuvant to dietetic management in the treatment of hyperlipidemia and in the prevention of atherosclerotic (age-dependent) vascular changes. The drug may be useful in the treatment of mild hypertension." It also notes uses of garlic in traditional systems of medicine for the "treatment of respiratory and urinary tract infections, ringworm and rheumatic conditions. The herb has been used as a carminative in the treatment of dyspepsia." Uses described in folk medicine, not supported by experimental or clinical data, include garlic "as an aphrodisiac, antipyretic, diuretic, emmenagogue, expectorant, and sedative, to treat asthma and bronchitis, and to promote hair growth" (WHO, 1999).

Interpretation of human research on the effects of treatment with commercial garlic supplements is complicated by the great variety of preparations, high variability and inadequate definition in content of bioactive ingredients, and availability of these constituents after ingestion (Wolsko, 2005; Ruddock, 2005; Linde, 2001; Ackermann, 2001; Lawson, 2001). Four types of garlic-derived dietary supplements are sold: garlic powder (mainly alliin and alliinase), aged garlic extract (*S*-allyl cysteine), distilled oil of garlic (diallyl and allyl methyl polysulfides in an edible oil) and garlic macerate (ajoene and dithiins diluted in an edible oil).

5.3 GARLIC AS MEDICINE: A LEGAL RULING

A European importer sought to sell in Germany garlic pills in packages with a label devoid of medical claims showing only a picture of a garlic bulb and garlic capsules. German authorities required that: (1) the importer first provide documentation based on pharmacological tests; and (2) the product could only be sold in pharmacies. The importer filed a claim against Germany with the European Court of Justice. In 2007, the Court ruled against the Federal Republic of Germany, arguing that garlic extract powder capsules are not medicinal products and that "requiring marketing authorization as a medicinal product for garlic capsules constitutes an obstacle to the free movement of goods not justified on health protection grounds" (European Court of Justice, 11/15/2007, *Case C–319/05*). The Court argued that the European Community recognizes a product as medicinal, in protecting human health and life, either by presentation or function. While a capsule could be medicinal by its presentation, a capsule form is not exclusive to medicinal products, and while "physiological effect" could constitute a medicinal function, such an effect is also seen with food supplements. To qualify as a medicinal, a product must have the function of preventing or treating disease, rather than just a beneficial effect for health in general. Apart from an excipient, the garlic powder capsules do not contain any substance other than natural garlic and have no additional effects other than those associated with consumption of garlic in its native state.

The court held that (1) the garlic capsules fail to correspond to the definition of medicinal products either by presentation or by function and therefore cannot be classified as medicinal; (2) the requirement for marketing authorization as a medicinal product unfairly creates an obstacle to intra-EU trade in products legally marketed as foodstuffs in other Member states; and (3) Germany has failed to fulfill its Treaty obligations concerning the free movement of goods. The Court found that, instead of making garlic pills subject to an authorization procedure, they simply could have prescribed suitable labeling to warn consumers of potential risks associated with taking the product. "With regard to the pharmacological effects, the Commission does not dispute the fact that the product in question may serve to prevent atheriosclerosis, but points out that that effect may be achieved by taking a dose equivalent to 4 g of raw garlic each day. Therefore, where a product which is claimed to be a medicinal product does nothing more than a conventional foodstuff, it is clear that its pharmacological properties are insufficient for it to be accepted as medicinal products. In this case, the Federal Republic of Germany cites cases of spontaneous post-operative bleeding occurring after excessive consumption of garlic as a foodstuff or in the form of a preparation, the suppression of the effects of certain anti-retroviral drugs and an interaction with some anticoagulants. In that connection, it must be observed, first of all, that those risks arise from the absorption of garlic in general and not specifically from the ingestion of the disputed preparation."

5.4 HEALTH BENEFITS OF GARLIC AND OTHER ALLIUMS: EVIDENCE-BASED REVIEW SYSTEM FOR SCIENTIFICALLY EVALUATING HEALTH CLAIMS AND ITS APPLICATION TO EVALUATING RESEARCH

Modern medicine requires rigorous proof of efficacy for all drugs and medicines. The FDA provides guidelines for evaluating the scientific evidence for health claims (for example garlic as a food or as a dietary supplement). In considering studies supporting the claims by evaluating the substance–disease relationship, the FDA requires identification of a specific substance that is measurable, such as allicin, and a specific disease or health-related condition, such as cardiovascular disease. The FDA also provides the following detailed evaluation of different types of studies (CFSAN, 2007):

Randomized, placebo-controlled, double-blinded intervention studies (RCTs or randomized controlled trials) are considered to provide the strongest evidence of whether or not there is a relationship between a substance and a disease in a particular population. In intervention studies, subjects in a population are provided the substance in the form of a conventional food, food component or dietary supplement, or they receive a placebo or otherwise serve as controls. The substance is provided in a RCT to avoid selection bias, under conditions of "double blinding," in which neither the subjects nor the researcher who assesses the outcome knows who is in the intervention group and who is in the control group. The nature of the substance given should be thoroughly controlled and established.

The quality of a RCT is judged based on criteria, such as, whether: (1) appropriate control groups were included; (2) the relevant baseline data in the study differed significantly between the control and intervention group; (3) appropriate biomarkers were measured; (4) the study was conducted for a sufficiently long time period; (5) the specific geographic location of the study was relevant to a broader population; (6) the study subjects were healthy rather than having the disease that is the subject of the health claim; (7) the independent role of the substance in reducing the risk of a disease was measured; (8) the thoroughness of randomization of subjects into the control and treatment groups. When the substance is a food component, it may not be possible to accurately determine its independent effect when whole foods or multi-nutrient supplements are provided to the intervention group. Furthermore, food matrix physiological effects can affect the bioavailability and bioactivity of individual food components. It may not be possible to adequately blind trials involving garlic, due to the tell-tale garlic odor. However, in studies involving garlic supplements, minute amounts of garlic have been introduced into the bottles contain garlic-free placebos to cover up the absence of significant levels of garlic.

Prospective or retrospective observational studies lack the controlled setting of intervention studies but can still be useful.

In prospective studies, investigators recruit subjects and observe them prior to the occurrence of the disease outcome. The most reliable prospective studies are *cohort studies*, which compare the incidence of a disease in subjects who receive a specific exposure of the substance that is the subject of the claim with the outcome of otherwise similar subjects who do not receive that exposure. In cohort studies, the diets and other data from large groups (cohorts) of people who are assumed to be healthy are assessed, and the group is followed over a period of time. During the follow-up period, some members of the group will develop and be diagnosed with cancer (or other diseases), while others will not, and comparisons are then made between these two groups. Cohort studies, such as the European Prospective Investigation into Cancer and Nutrition (EPIC) study discussed below, need to involve tens or hundreds of thousands of participants to have sufficient statistical power to identify factors that may increase (or decrease) cancer risk by 20 to 30%.

The least reliable type of observational studies are retrospective studies, such as *case-control studies*, in which subjects with a disease (cases) are compared to subjects who do not have the disease (controls). In observational studies, biological samples should be used to establish intake of a substance only if a dose–response relationship has been demonstrated between intake of the substance and the level of a metabolite of the substance in the biological sample (for example, selenium intake and nail selenium concentration). A difficulty in observational studies involving a substance in food is estimating the amount of the substance actually consumed, factoring in changes associated with cooking, storage, variations in plant composition under differing growing conditions, and interactions of the substance in question with other components in the individual's diet. Due to the above complications, the FDA states that "scientific conclusions from observational studies cannot be drawn about a relationship between a food component and a disease. Observational studies, however, can be used to measure associations between a whole food and a disease."

Another part of the search for proof of efficacy involves studies with laboratory animals, who are given whole extracts or other plant preparations, presumed to contain the active ingredients, as well as pure samples of the various individual presumed active ingredients. Animal (*in vivo*) studies are generally done as a follow up to *in vitro* studies of the biological activity of individual compounds or mixtures toward cell cultures or other bioassays. Animal and *in vitro* studies are useful in providing background information and generating hypotheses on mechanisms that might be involved in any relationship between the substance and the disease, but do not provide information from which scientific conclusions can be drawn regarding a relationship between the substance and disease in humans. This is because the physiology of animals is different from that of humans. For example, platelet inhibitory response in dogs to an onion extract is higher than in humans (Briggs, 2001). In addition, *in vitro* studies are conducted in an artificial environment that cannot account for a multitude of normal physiological processes such as digestion, absorption and metabolism that affect how humans respond to the consumption of foods and dietary substances. Finally, *in vitro* experiments are usually conducted using concentrations of phytochemicals that are several orders of magnitude higher than those achieved with oral ingestion (Betz, 2001).

Disease risk determination associated with a substance can present difficulties because many diseases develop over a long period of time. Therefore, it may not be possible to carry out the study for a long enough period of time to a see a statistically meaningful difference in the incidence of disease among study subjects in the treatment and control groups. For this reason it is often necessary to use *surrogate endpoints*, that is, risk biomarkers that have been shown to be valid predictors of disease risk that can be used in place of diagnosis of disease. Examples of surrogate endpoints accepted by the FDA/NIH include: (1) serum low-density lipoprotein (LDL) cholesterol concentration, total serum cholesterol concentration and blood pressure for cardiovascular disease; (2) adenomatous colon polyps for colon cancer; and (3) elevated blood sugar concentrations and insulin resistance for type 2 diabetes.

In all of the above, the final feature is publication following rigorous peer-review by a respected journal. It is important that null (no effect), negative (harmful effect) as well as positive trials be published to give a properly balanced picture. Nothing less than the above is expected of *Allium*-containing products that are purported to be beneficial. There has been a surge of reviews on health-related properties of garlic including those appearing in issues of *Molecular Nutrition and Food Research* (2007, Volume 51, Issue 11) and *Medicinal and Aromatic Plant Science and Biotechnology* (2007, Volume 1, Issue 1). There is even interest in applications for human and animal health of more exotic alliums of the subgenus *Melanocrommyum*, which grow wild in the mountainous regions of Central and Southwest Asia. However, the active principles of these plants are still unknown and further investigation is needed (Keusgen, 2006).

Given the vast number of publications on health-related properties of alliums and the availability of specialized reviews, only representative human studies and a summary of *in vivo* (animal) and *in vitro* studies will be presented here.

5.5 ANTIMICROBIAL ACTIVITY OF *ALLIUM* EXTRACTS AND SUPPLEMENTS: *IN VITRO*, *IN VIVO*, DIETARY AND CLINICAL STUDIES

For centuries, garlic and onion have been used in traditional medicine against bacterial, fungal, parasitic and viral infections. The antibiotic properties of onion were recognized by Louis Pasteur in 1858 (Pasteur, 1858). Scientific studies documenting the antibiotic activity of garlic began to appear early in the twentieth century (Reuter, 1996). In 1937, seven years before Cavallito's isolation of allicin, University of Southern California bacteriologist Carl Lindegren reported that "garlic vapors have an active bactericidal effect...the bactericidal substance is much more volatile at 37.5 °C than at 10 °C...there was no evidence that resistance of organisms to the bactericidal substance could be developed...boiled garlic and autoclaved garlic are non-germicidal" (Walton, 1936).

The antibiotic activity of allicin from fresh garlic juice is described in this simple demonstration (Figure 5.3). Filter paper disks containing dilutions of fresh garlic juice were placed on a Petri dish containing an agar–agar growth medium innoculated with the *E. coli* bacterium. After incubation, the diameters of the inhibition zones were compared with similar data from an experiment using the conventional antibiotic ampicillin instead of garlic juice. It was found that as little as 84 μg of allicin gives a clear inhibition zone. While allicin on a molar basis is only a quarter to one-fifth as active as the conventional antibiotics ampicillin and kanamycin, given that the average allicin content of fresh garlic is 3. 9 mg g^{-1} of fresh weight, its antimicrobial potential is impressive

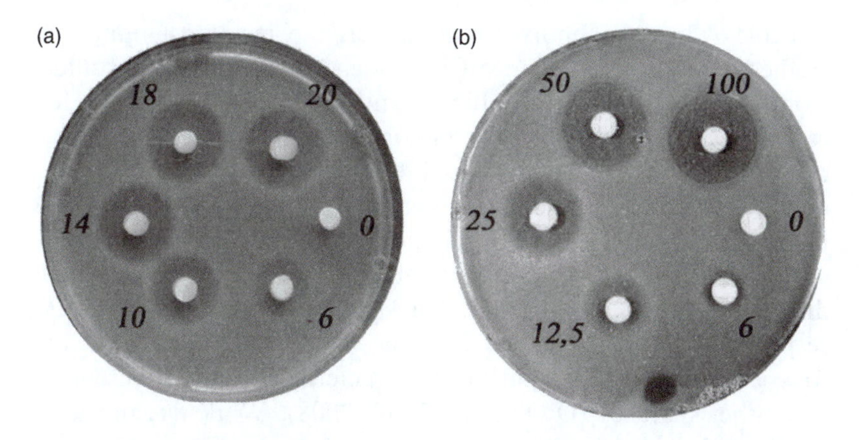

Figure 5.3 Standardized *E. coli*-seeded Petri plate assays comparing the respective antimicrobial activity of: (a) a garlic extract (20 μL of crude extract containing ~280 mg allicin, or 20 μL of dilutions containing 18, 14,10, 6 or 0 μL of extract), and (b) ampicillin (100, 50, 25, 12.5, 6 or 0 μg), a conventional antibiotic. (Image from Curtis, 2004).

indeed (Curtis, 2004; Slusarenko, 2008)! Similar tests using different strains of yeast (*Candida albicans*, *etc.*) showed that in many cases the yeast were more sensitive to a garlic extract (prepared by grinding garlic cloves with twice their weight of water) than to the conventional antifungal agent Nystatin (Arora, 1999).

A related study used the technique of atomic force microscopy (AFM) to compare the antimicrobial activities of a fresh aqueous extract of garlic and ampicillin against the Gram-negative and Gram-positive bacteria *E. coli* and *S. aureus*, respectively (Perry, 2009). In general, Gram-negative bacteria are more sensitive to garlic than Gram-positive bacteria, as was the case in this study. Treatment of *E. coli* with a garlic solution (25 mg mL^{-1} of a solution prepared by crushing fresh garlic with an equal weight of water and filtering, with storage at –20 °C if not used immediately) for 12 hours causes the bacteria to lose their bacillus shape, cluster together and show leakage of cell contents. Similar studies with *S. aureus* showed no changes in surface morphology compared with the untreated bacterium. The authors note that *S. aureus* has a thicker cell wall, lower cell membrane lipid content and higher cell membrane polysaccharide content, compared to *E. coli*, which may effect the relative cell membrane permeability of allicin and other constituents of fresh garlic extracts.

Table A2 in Appendix 1 summarizes the results of numerous studies demonstrating that allicin, ajoene and diallyl polysulfides from garlic, all possess significant antibotic activity. While a few reports have appeared on the biological activity of extracts from other alliums, such as onion, shallot, scallion, *A. bakeri*, *A. odorum* and *A. tuberosum*, these extracts are much less active than those from garlic (Yu, 1999). Thus, onion oil inhibited the growth of various Gram-positive bacteria (Zohri, 1995) and pyridine *N*-oxide disulfides and alkaloids from *A. neapolitanum* and *A. stipitatum*, respectively, are active at the 2 to 20 $\mu g\,mL^{-1}$ range against various *Mycobaterium* strains, pathogenic bacteria causing tuberculosis and other diseases (O'Donnell, 2007, 2008). While organosulfur compounds are thought to be the most active *Allium* antibiotics, saponins and phenolics may also contribute to this activity. Because of their antimicrobial activity, both garlic and onion have been recommended as natural preservatives to control microbial growth in food products (Corzo-Martínez, 2007). A group of pediatric clinicians urge that allicin be more thoroughly investigated as an orally administered drug for gastroenteritis involving antibiotic-resistant campylobacter strains (De Wet, 1999).

5.5.1 Antifungal Activity of *Allium* Compounds

Fungal pathogens are associated with clinical conditions ranging from irritating superficial infections to life-threatening systemic disease in immuno-compromised patients. While amphotericin B remains the treatment of choice for invasive fungal infections, it can cause negative side effects, in particular, nephrotoxicity. Drug resistance limits the effectiveness of "safer" antifungal agents such as fluconazole and other azoles. Garlic-derived compounds have been investigated for use as topical and intravenous (i.v.) therapies based on encouraging *in vitro* studies, summarized in Table A2 in Appendix 1 (Ankri, 1999; Davis, 2005).

Since 1973, diallyl trisulfide (DAT) in the form of a proprietary drug (Dasuansu, Allidridium, allitridium, allitridum, allitridi or allitridin) has been widely used in China for clinical treatment of patients infected with parasites, bacteria or fungi, *e.g.*, *C. neoformans*, which causes cryptococcal meningitis (Davis, 1990; Lun, 1994; Shen, 1996). Commercially available DAT is stable, can be easily synthesized and is said to be non-toxic (this assumption has been questioned; Iciek, 2009). It is administered orally or intravenously at a level of 60 to 120 mg per adult diluted in 500 to 1 000 mL saline or 5 to 10% glucose. *In vitro* studies of DAT showed that it had a mimimal inhibitory concentration (MIC) of 100 μg mL^{-1} toward *C. neoformans* and acted synergistically with amphotericin B (Shen, 1996). The pharmokinetics of DAT has been studied by Lawson (Lawson, 2005a). The data on anticandidal activity of diallyl polysulfides reveals an interesting trend, also seen with antibacterial activity: antibiotic activity (lower MICs) increases with the number of sulfur atoms in the polysulfide, at least up to diallyl tetrasulfide (Tsao, 2001b; O'Gara, 2000). This effect is presumably related to the weakening of the S–S bonds with increasing numbers of sulfur atoms, enhancing reactivity toward –SH groups.

Allicin-containing extracts of garlic are also active against *C. neoformans* and exhibit synergistic fungistatic activity with amphotericin B (Fromtling, 1978; Davis, 1994; An, 2008). Solutions of purified allicin show good antifungal activity against *Aspergillus* both *in vitro* and *in vivo* in mice (Shadkchan, 2004). Polybutylcyanoacrylate nanoparticles of allicin were approximately twice as active as fungicides *in vitro* as allicin itself (Luo, 2009). Compared to aqueous extracts of onion, aqueous extracts of garlic were 32 to 128 times more potent toward the dermatophytes (fungi causing skin disease) *Trichophyton rubrum* and *T. mentagrophytes* in an agar dilution assay, but only one-fifth to one-thirtieth as active as the conventional fungicide ketoconazole (Shams-Ghahfarokhi, 2004, 2006). Onion oil (200 ppm) completely inhibited the growth of *T. simii* (Zohri, 1995).

Garlic preparation components allyl alcohol and diallyl polysulfides, as well as an aqueous garlic extract, possess significant anticandidal properties toward *Candida* spp. as growth inhibitors and fungicides (Lemar, 2005, 2007; Choi, 2005; Moore, 1977; Barone, 1977). Compared to aqueous extracts of onion, aqueous extracts of garlic were four to eight times more potent toward *Candida* in an agar dilution assay, but only one-tenth as active as the conventional fungicide ketoconazole (Shams-Ghahfarokhi, 2006). The anticandidal activity of diallyl polysulfides may be due to their ability to elicit oxidative stress by depleting thiols and impairing mitochondrial function (Lemar, 2007), while the activity of fresh garlic extract is thought to be due to its ability to suppress hyphae production (Low, 2008).

Ajoene compounded into a gel at 0.6% and 1.0% wt/wt was compared to a similar 1% formulation of the antifungal drug terbinafine in a double-blind, three-armed trial involving 47 male volunteers, all presenting with tinea pedis (mostly due to *Trichophyton rubrium* or *T. mentagrophytes*) and using one of the three treatments twice daily for one week. Thirty days after ending the treatment, 60% of the patients treated with 0.6% ajoene, 100% of the patients treated with 1% ajoene, and 88% of the patients treated with 1% terbinafine had achieved a mycological cure. No adverse effects were observed during the study (Ledezma, 1996, 2000). Both ajoene and DAT show good *in vitro* activity toward *Scedosporium prolificans* (Davis, 1993). As indicated by data in Table A2 in Appendix 1, (*Z*)-ajoene shows higher antibiotic activity than (*E*)-ajoene.

5.5.2 Antibacterial Activity of *Allium* Compounds: Bad Breath – Disease or Cure?

Periodontitis, a severe form of gum disease caused by oral bacteria, is often treated with common antibiotics such as amoxicillin, but antibiotic resistance is a growing problem. *Pophyromonas gingivitis* and other oral bacteria causing periodontitis have been found to be particularly sensitive to allicin in fresh garlic extracts (see Appendix 1, Table A2; Bakri, 2005). This sensitivity is thought to be due to the very rapid reaction of allicin with the free thiol groups in bacterial enzymes such as cysteine proteases and alcohol dehydrogenases. Because these enzymes are essential for bacterial nutrition and metabolism, it has been suggested that development of resistance to allicin arises a thousand fold less easily than it does to other more conventional antibiotics. A five week trial involving 30 subjects examined the antimicrobial activity of a mouth rinse containing 2.5% garlic toward oral microorganisms. The rinse showed good antimicrobial activity. Maintenance of reduced salivary levels of streptococci was observed after two weeks at the end of mouthwash use. However, the unpleasant taste and halitosis almost universally reported by the trial participants, suggests that garlic mouthwash is unlikely to be a winning consumer product (Groppo, 2002, 2007).

Heliobacter pylori, a Gram-negative bacterium inhabiting regions of the stomach and duodenum, is strongly linked to development of chronic gastritis and gastric and duodenal ulcers.

H. pylori-infected populations have an increased risk of stomach cancer. Clinical evidence shows that eradication of *H. pylori*, through therapy with a combination of antibiotics and a proton pump inhibitor, results in significant relief from gastritis and ulcers, although drug-resistant strains of *H. pylori* are starting to appear. *H. pylori* has been found to be sensitive to various garlic-derived compounds including allicin, ajoene, diallyl trisulfide and diallyl tetrasulfide, suggesting the possible use of one or more of these compounds for treatment (Sivam, 1997, 2001; Ohta, 1999; O'Gara, 2000, 2008).

Epidemiological reports have revealed that high intake of *Allium* vegetables, including garlic, is associated with a reduced risk of gastric cancer (You, 1998; Bulatti, 1989; see Section 5.6.2 below). However, a pilot study of five patients with positive serology for *H. pylori*, who were given garlic oil capsules for 14 days, failed to show evidence of either eradication or suppression of *H. pylori* or symptom improvement while taking the garlic oil (McNulty, 2001). The results were similarly negative with two other small trials involving garlic oil (Aydin, 1997) or fresh garlic (Graham, 1994). In addition, a large randomized, double-blind trial examining different treatments to reduce the prevalence of precancerous gastric lesions found no correlation with long-term "garlic" supplementation. The "garlic" supplement used in this study was "a capsule that contained 200 mg of aged garlic extract...and steam-distilled garlic oil (1 mg) or placebo... two such capsules were to be taken twice daily" (You, 2006; Gail, 2007). The absence of a precise description of the chemical composition of the "aged garlic extract" makes it impossible to judge whether the participants in fact received the equivalent of fresh garlic, which is thought to have played a role in the earlier study (You, 1998). Others have faulted the "use of garlic products of undefined composition" in biological studies (Sivam, 2001; Ruddock, 2005; Koch, 1996). Recent work suggests that chronic *H. pylori* infection may also be beneficial by limiting the development of other important diseases. This suggests that *H. pylori* is actually an amphibiont: a microbial species that functions as a pathogen or a symbiont, depending on the specific context (Peek, 2008).

A literature review of controlled clinical trials on treatment of bacterial infections with herbal medicines including garlic concludes that "the most striking result of this review is the extreme paucity of controlled clinical trials testing herbal antibiotics...One obvious reason is the lack of patent rights on herbal medicines. Another reason could be that traditionally, herbal medicine has been hesitant to embrace modern methods for efficacy testing. . .and negative trials may not be published at all" (Martin, 2003). Based on the promising results from years of clinical experience in China treating diseases with garlic preparations, it is a pity that these treatments in China are not validated using controlled trial protocols, even if blinding is impossible due to the tell-tale garlic odors, and the results published in mainstream journals.

5.5.3 The Garlic Mask Treatment for Tuberculosis

Garlic has long been recommended for treatment of tuberculosis, an ancient scourge of mankind, previously called "consumption," or "phthisis." The utility of garlic preparations in the treatment of tuberculosis is the subject of monographs published in 1912 and 1927, before the advent of streptomycin and other modern antibiotics, by William C. Minchin, M.D., of the Kells Union Hospital in Ireland (Minchin, 1912, 1927). In his book, Minchin describes an inhaler mask containing a sponge soaked in garlic juice for treatment of the disease (Figure 5.4). He also recommends topical application to diseased areas of minced fresh garlic mixed with lard and chewing raw garlic (or chopped garlic and bread) and use of a fresh garlic spray for the throat for cases of laryngeal tuberculosis (Hall, 1913). The idea behind this rather bizarre device is that inhaled garlic juice would function as a volatile germicide which would destroy the "tubercle spheres in the infected tissue of the human lung." A review of Minchin's book in the *British Medical Journal* observes: "his success in some of the cases quoted appears to have been remarkable, and photographs of patients before and after treatment are given. It has long been known that salts [sic] derived from *Allium sativum* have bactericidal power, but the intense aversion which so many people have to the all-pervading smell of garlic has no doubt rendered their use unpopular. Such cases as those recorded by Dr. Minchin may fairly be claimed as proof that in certain individuals the drug may practically annihilate the activity of tubercle, but he has found that in others, where a strong antipathy to garlic exists, its administration in any form has failed to give relief" (Anonymous, 1912). However, the author of a

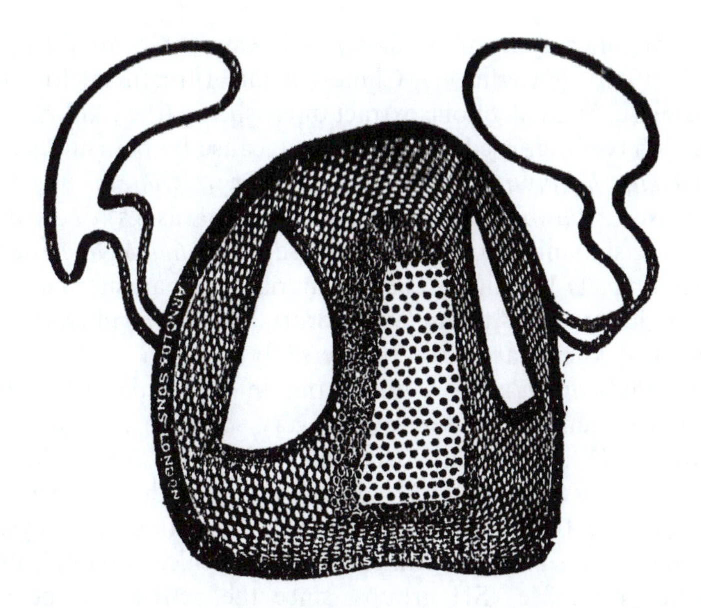

Figure 5.4 A face mask for inhalation of garlic oil for treatment of tuberculosis. (Image from Minchin, 1927).

1925 article in the same journal is skeptical about garlic treatments for tuberculosis, writing: "It is now well recognized that a patient suffering from tuberculosis who is placed under the charge of a doctor will, for a time at least, improve in health no matter what drugs, vaccines, or specialty of treatment may be employed. The beneficial result is due in this, as in other diseases, to efficient nursing, to the regulation of food, exercise, and sleep, and to light and fresh air... Essential oils have been the basis of many systems of "cure"; thus the oil of cinnamon and oil of garlic treatments have each had its day; it were [sic] idle to describe them in detail, as it is universally recognized that, like other specifics, they fail in the hands of the skeptical physician" (Dixon, 1925).

5.5.4 Antiparasitic Activity of *Allium* Compounds: *In vitro* Studies

As noted above, the antiparasitic effects of freshly crushed garlic were known by ancient cultures. Albert Schweitzer treated people in Africa suffering from dysentery or intestinal worms with garlic (Block, 1985). A traditional Chinese medical treatment for intestinal diseases is an alcoholic extract of crushed garlic (Ankri, 1999). Alliums have been used to treat disease cause by parasites such as *Entamoeba histolytica* (amoebic dysentery), *Giardia intestinalis* (giardiasis), *Plasmodium* species (malaria parasites), *Leishmania* species (leishmaniasis or sandfly disease), *Acanthamoeba castellanii* (*Acanthamoeba* keratitis), *Loa loa* filariasis (a parasitic nematode causing loiasis or African eye worm) and *Trypanosoma brucei* (sleeping sickness carried by the tsetse fly).

Entamoeba histolytica, a ubiquitous intestinal parasite causing amoebiasis (amoebic dysentery), is very sensitive to allicin, with growth totally inhibited at $30\,\mu g\,mL^{-1}$ (Mirelman, 1987). At lower concentrations ($5\,\mu g\,mL^{-1}$), allicin inhibits the virulence of *E. histolytica* trophozoites by strongly inhibiting their cysteine proteinases. It is likely that allicin acts by chemically modifying the cysteine proteinase –SH groups, since the activity of the latter enzyme can be fully restored by treatment with the thiol reagent dithiothreitol (Ankri, 1997). The LD_{50} of diallyl trisulfide for *E. histolytica* is $59\,\mu g\,mL^{-1}$. Since the cytotoxicity of diallyl trisulfide is $25\,\mu g\,mL^{-1}$, the latter may be too toxic for treatment purposes if levels at or above $25\,\mu g\,mL^{-1}$ are required (Lun, 1994).

The parasitic protozoon *Giardia intestinalis* is the most commonly diagnosed cause of waterborne diarrhea, a major problem worldwide. Current treatments of choice are metronidazole or other nitroimidazoles, nitrofurans, quinacrine or paromomycin, although all of these treatments are reported to have unpleasant side effects, and resistant strains of *G. intestinalis* are on the increase. Whole garlic and several compounds found in garlic preparations inhibit the growth of *G. intestinalis*, as shown by the following IC_{50} data (after 24 h exposure): whole garlic, $300\,\mu g\,mL^{-1}$; diallyl disulfide, $100\,\mu g\,mL^{-1}$; 2-propenethiol, $37\,\mu g\,mL^{-1}$; allyl alcohol, $7\,\mu g\,mL^{-1}$. Mechanisms have been proposed for the antigiardial activity of the garlic compounds, including the suggestion that it stimulates production of nitric oxide synthase (NOS). Nitric oxide is cytotoxic to *G. intestinalis* (Harris, 2000). The LD_{50} of diallyl trisulfide for *G. lamblia* is $14\,\mu g\,mL^{-1}$, which can be compared to the cytotoxicity of $25\,\mu g\,mL^{-1}$ (Lun, 1994).

Each year more than 1 500 000 people die from malaria, a major infectious disease throughout the world. Close to half of the world's population live in malaria-endemic areas and are at risk for contracting the disease. Infection is initiated when *Plasmodium* sporozoites are injected into the skin of vertebrate hosts by an infected mosquito. Allicin has been used as a cysteine protease inhibitor to inhibit processing of the *Plasmodium* circumsporozoite protein (CSP), in turn preventing invasion of host cells *in vitro* and *in vivo* (Coppi, 2006).

Leishmaniasis is a disease caused by intracellular parasitic haemoflagellates which infest macrophages of the skin and viscera of the host. Seventy-two hour incubation of filtered, centrifuged and lyophilized homogenates of fresh onions with five different leishmanial strains, gave an average LD_{50} of 376 ppm (Saleheen, 2004). Garlic extracts are more active, showing an LD_{50} of 5 ppm toward *Leishmania major* (Khalid, 2005). Garlic-induced nitric oxide production may play a role in parasite killing (Gamboa-Léon, 2007).

Acanthamoeba castellanii is a free-living protozoon that causes the sight-threatening infection, *A.* keratitis. Garlic extract components diallyl polysulfides and allicin are effective against this protozoon without showing cytotoxicity toward cornea cells (Polat, 2008). Typical of traditional medical usage of garlic and onion is treatment for the disease called loiasis or African eyeworm, a skin and eye disease caused by the nematode worm, *Loa loa*, afflicting as many as 13 million individuals living in swampy and forested areas of Cameroon in West Africa. The most common traditional treatment for eye worm is garlic or onion juice, which is dripped into the affected eye (Takougang, 2007). *Trypanosoma* are a class of unicellular, parasitic protozoa which include organisms such as *T. brucei* causing the fatal disease, sleeping sickness, in humans and cattle disease. The LD_{50} of diallyl trisulfide for *T. brucei* and related protozoa is 0.8 to 5.5 $\mu g\,mL^{-1}$, which is significantly lower than the cytotoxicity of 25 $\mu g\,mL^{-1}$ (Lun, 1994).

5.5.5 Antiviral Activity of *Allium* Compounds

Some garlic-derived compounds show *in vitro* activity against a number of human viruses, including the human cytomegalovirus (human herpes virus), although cytotoxicity is also observed (Zhen, 2006; Ankri, 1999; Esté, 1998; Guo, 1993; Weber, 1992). For example, while *Z*-ajoene inhibited spreading of the cytomegalovirus

(Terrasson, 2007), it was concluded that ajoene had no "antiviral activity at a concentration that was significantly lower...than the cytotoxic concentration" (Esté, 1998). A randomized clinical trial reported the use of a garlic supplement containing stabilized allicin, for the prevention and treatment of the common cold (Josling, 2001). Overall, 146 volunteers participated for 12 weeks; the group treated with one allicin-containing capsule per day had significantly fewer colds (24 *versus* 65, $P < 0.001$) of shorter duration (1.52 *versus* 5.01 days, $P < 0.001$) than those in the placebo group. Unfortunately, this trial suffers from several weaknesses including subjective data; absence of detailed information on dosing and characterization of the allicin preparation used (PDR, 2007); a small number of participants; and absence of financial disclosure or conflict of interest statement. Recent reviews of the use of garlic to treat the common cold conclude that not enough data is available for clinical recommendations (Pittler, 2007; Lissiman, 2009).

5.6 ALLIUMS AND CANCER: DIETARY, *IN VITRO* AND *IN VIVO* STUDIES

Cancer is a group of more than 100 diseases characterized by uncontrolled cellular growth as a result of changes in the genetic information of cells. Normally the division, differentiation, and death of cells are carefully regulated. All cancers start as a single cell that has lost control of its normal growth and replication processes. The transformation of a normal to cancerous cell involves several distinct phases – initiation, promotion, progression, and further evolution to malignant tumors – each of which can be modified by dietary habits. Indeed, it has been estimated that dietary factors account for approximately one third of cancer deaths, similar to the impact of smoking (Doll, 1981; Boivin, 2009). Furthermore, it has been consistently shown that individuals who eat five servings or more of fruits and vegetables daily have about half the risk of developing various cancer types, particularly those of the gastrointestinal (GI) tract (World Cancer Research Fund/American Institute for Cancer Research, 2007). Fruits and vegetables contain a wide variety of chemical compounds, some of which are potent modifiers of chemical carcinogenesis. One of the first plants reported to possess antitumor compounds was garlic, used for treatment of tumors by ancient civilizations. Cancer chemoprevention with garlic and its constituents has been the subject of numerous recent reviews (Nagini, 2008; Powolny, 2008; Stan, 2008; Shukla, 2007; Ngo, 2007; Siess, 2007; Herman-Antosiewicz, 2007, 2004; Stan, 2007; El-Bayoumy, 2006; Milner, 2006; Ariga, 2006; Khanum, 2004; Sengupta, 2004; Ejaz, 2003; Reuter, 1996).

A recent study evaluated the inhibitory effects of 34 commonly consumed vegetables on the proliferation of eight different tumor cell lines, derived from stomach, kidney, prostate, breast, brain (two types), pancreatic and lung cancer (Figure 5.5). The extracts from cruciferous and particularly genus *Allium* vegetables inhibited the proliferation of all test cancer cell lines. On the other hand, extracts from vegetables most commonly consumed in Western countries were much less effective. Furthermore, the anti-proliferative effect of these vegetables was specific to cells of cancerous origin – there was a negligible effect on the growth of normal

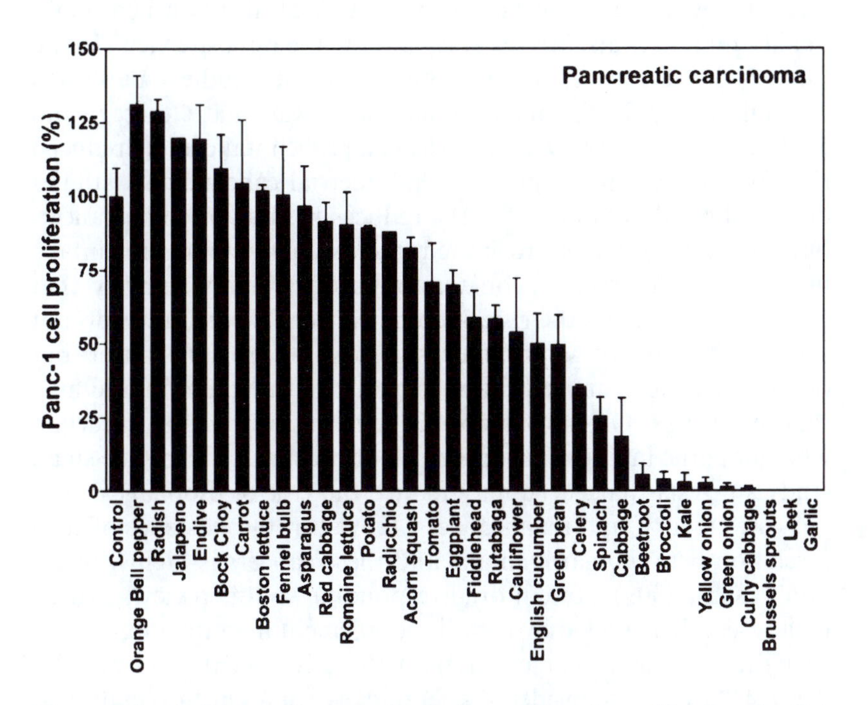

Figure 5.5 Inhibition of pancreatic carcinoma (Panc-1) proliferation, by 48 h incubation with vegetable extracts corresponding to 166 mg mL^{-1} of raw vegetable. (Image from Boivin, 2009).

cells. The study notes that "among all the vegetables tested in this study, the extract from garlic was by far the strongest inhibitor of tumour cell proliferation, with complete growth inhibition of all tested cell lines. Leek, immature (green) and mature (yellow) onions were also highly inhibitory against most cell lines." The study concludes that "cruciferous, dark green and *Allium* vegetables are endowed with potent anticancer properties...inclusion of cruciferous and *Allium* vegetables in the diet is essential for effective dietary-based chemopreventive strategies" (Boivin, 2009). Of course, it is a bit misleading to compare equal weight servings of all vegetables, as this study does, since a typical serving of broccoli, for example, would be much larger than a typical serving of garlic.

5.6.1 Evidence-based Review of Garlic Intake and Cancer Risk

While the above *in vitro* study provides a positive view of the cancer-preventive potential of dietary garlic, a review of all of the scientific evidence, focusing on the results of all published, high-quality human trials, arrives at essentially the opposite view. Using this evidence-based, review system, from studies published throughout 2007, the major conclusion was that "there is no credible evidence for a relation between garlic intake and a reduced risk for gastric, breast, lung, or endometrial cancer ...Six studies do not show that intake of garlic reduces the risk of colon cancer, but three weaker and more limited studies suggest that garlic intake may reduce this risk ... [on this basis] it is highly unlikely that garlic intake reduces the risk of colon cancer. Three studies do not show that intake of garlic reduces the risk of prostate cancer, but one weaker more limited study suggests that intake of garlic may reduce this risk. On the basis of these studies, it is highly uncertain whether garlic intake reduces the risk of prostate cancer. One small study suggests that garlic may reduce the risk of esophageal, larynx, oral, ovary, and renal cancers. However, the existence of such a relation between garlic intake and these cancers is highly uncertain" (Kim, 2009). An editorial responding to this review, written by a researcher funded by a garlic supplement manufacturer, notes that the number of clinical trials with active agents and placebo controls "that are considered scientifically sound in this analysis is remarkably few, and the number of subjects involved is generally small. Thus, the very strict criteria required to make a health claim may not be met by the limited number of studies conducted to date that are currently available" (Rivlin, 2009).

In a double-blind, placebo-controlled intervention study conducted in China and involving 5000 subjects (not included in the previous review), each person of the intervention group took 200 mg of synthetic diallyl trisulfide ("allitridum") every day, and 100 μg of sodium selenite every other day, orally for one month per year for two years. At the same time, people in the control group were given two placebo capsules. In the first five-year follow-up, the morbidity from gastric cancer was 10 and 19 persons for the intervention and control group, respectively. The study demonstrated a moderate reduction in incidence of gastric cancer for men; no significant protective effect was found for the female subgroup. The lack of an effect for the female subgroup could be due to a smaller female study size, or to a higher incidence of gastric cancer or a higher incidence of risk factors in men compared to women. The dose of diallyl trisulfide given was equivalent to eating 100 to 200 g of raw garlic each day for an adult, which is unreasonably high. Furthermore, the premise that supplementation with diallyl trisulfide for only one month per year can influence the incidence of cancer within five years after supplementation has been challenged (Iciek, 2009). The authors conclude that the preventive effect of diallyl trisulfide, is more likely due to inducing apoptosis in cancer cells, than by inhibition of formation of nitrosamines or growth of *H. pylori* in the stomach (Li, 2004).

A conclusion of the European Union study "Garlic and Health" (2000–2004) is that, from a human intervention study with 92 subjects, parameters involved in the initiation of cancer were not significantly influenced during a period of three months (Kik, 2004). Neither the World Cancer Research Fund report discussed in the following section nor a review by Gonzalez (2006) associate *Allium* consumption with a reduction in the risk of prostate cancer. Another author, reviewing much the same data on prostate cancer, comes to a different conclusion, recommending that men at risk for benign prostatic hyperplasia (BPH, benign enlarged prostate) or prostate cancer consume more garlic in their daily diets, and those diagnosed with BPH or prostate cancer should consume garlic to support their medical therapies (Devrim, 2007).

5.6.2 Epidemiological Studies

An extensive report by the World Cancer Research Fund examined the relationship of food and nutrition on the prevention of cancer, based on published cohort studies, case-control studies and ecological studies (WCRF/AICR 2007). With regard to dietary alliums, the report concludes: "The evidence, though not copious and mostly from case-control studies, is consistent with a dose–response relationship. There is evidence for plausible mechanisms. *Allium* vegetables probably protect against stomach cancer...garlic probably protects against colorectal cancer." These conclusions are based on the review of two cohort, 27 case-control, and two ecological studies of *Allium* vegetables, and one cohort, 16 case-control and two ecological studies involving garlic. A summary of all of the literature reviewed, concludes that there is evidence of "convincing decreased risk" of stomach cancer associated with the consumption of *Allium* vegetables, and of colorectal cancer associated with the consumption of garlic (WCRF/AICR 2007).

The European Prospective Investigation into Cancer and Nutrition (EPIC) study, specifically designed to investigate the relationship between diet and cancer, is a multicenter prospective study carried out in 23 centers from ten European countries: Denmark, France, Germany, Greece, Italy, the Netherlands, Norway, Spain, Sweden and the United Kingdom, including 519 978 subjects (366 521 women and 153 457 men), most aged 35 to 70 years. Among the findings, the study reveals that consumption of onion and garlic probably reduces the risk of intestinal and stomach cancers, but is probably not associated with cancer of the lung, prostate and breast (Gonzalez, 2006). Similar conclusions were reached by smaller European studies (Galeone, 2006, 2008). Another recent study concluded that diets rich in fruit and deep-yellow vegetables, dark-green vegetables, onions and garlic are modestly associated with reduced risk of colorectal adenoma, a precursor of colorectal cancer (Millen, 2007).

5.6.3 Clinical Trials: Use of Ajoene in Treating Nonmelanoma Skin Cancer and Allicin in Gastroscopic Treatment of Gastric Carcinoma

A clinical trial has been conducted on the use of topically applied ajoene for treatment of basal cell carcinoma (BCC), a type of nonmelanoma skin cancer that is the most common skin malignancy in

Caucasians (Tilli, 2003; Hassan, 2004). Nonsurgical treatment is often favored because of better cosmetic results. In this trial 21 patients with a total of 25 lesions were treated topically with a 0.4% ajoene cream. Before treatment, the surface area was determined for each tumor, and a biopsy was taken. The size of each lesion was measured once a month. After six months, the BCCs were completely removed by surgery, and the activity of the tumor cells examined. Most BCCs showed a positive response to ajoene with respect to size reduction, as well as a decrease in the expression of the so-called *B-cell lymphoma 2* (Bcl-2) gene, which protects the tumor cells against programmed cell death (apoptosis). Of the BCC lesions examined following treatment with ajoene, three showed an increase in size, one showed no change in size, while 21 showed a reduction in size, corresponding to an 84% reduction in size after six months. At the same time, no adverse reactions were experienced at the tumor sites, as was reported for another topical drug, imiquimod. Typical results are shown by photographs of BCC lesions on the forehead and on the abdomen, before and after treatment (Figure 5.6) and in graphical form for all 25 lesions (Figure 5.7). An effort was made to elucidate the underlying biological effect of ajoene on the tumor cells. It was concluded that tumor size reduction is best explained by the apoptosis-inducing ability of ajoene, whereby the number of cells expressing Bcl-2 is reduced. When directly added to cultured BCC cells, ajoene induced apoptosis in both a time- and dose-dependant manner (Tilli, 2003).

A clinical trial in China involved 80 patients with progressive gastric adenocarcinoma, whose diagnosis was confirmed by gastroscopy and pathological examination, and who were assigned to two groups, 40 in each group. Forty-eight hours before operation, allicin was infused *via* gastroscopy to the lesion region of patients in the allicin group, and normal saline was infused instead to those in the control group. It was concluded that local application of allicin *via* gastroscopy can inhibit the cell growth and proliferation of progressive gastric carcinoma, and can also promote gastric carcinoma cell apoptosis (Zhang, 2008). Concern has been expressed about the harmful effects of allicin on gastrointestinal mucosa (Lang, 2004).

5.6.4 *In vitro* and *In vivo* Mechanistic Studies

The first laboratory studies of the antitumor activity of the thiosulfinate analogs of allicin were reported in 1958, by Weisberger and

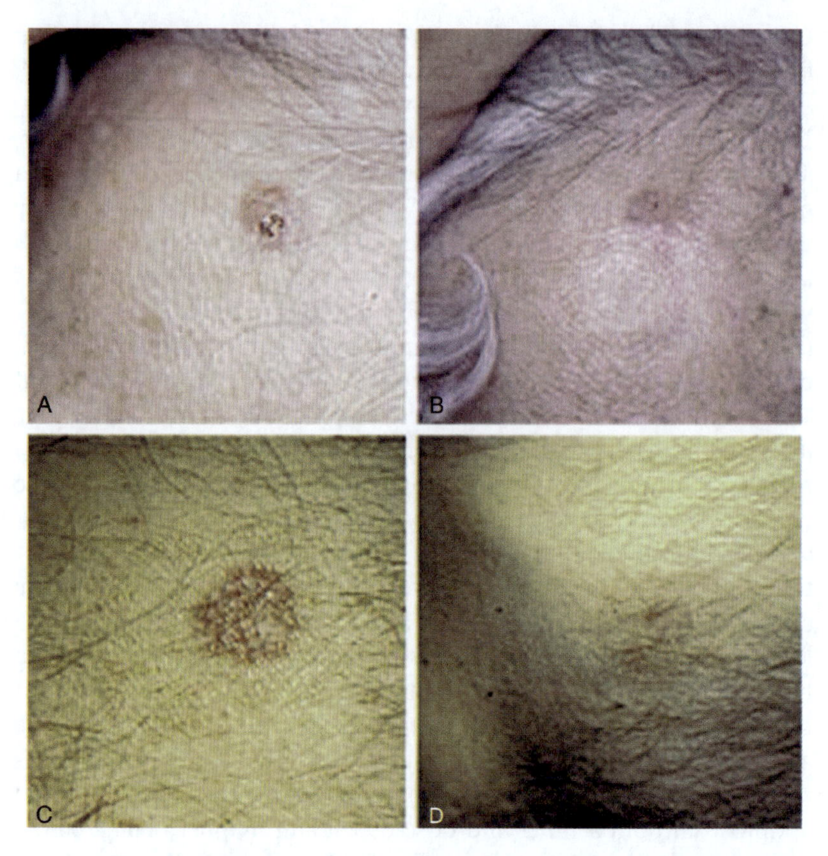

Figure 5.6 **(A–D)** Effect of ajoene on BCC tumor size after topical application. BCC located on the forehead (**A**, **B**) and on the abdomen (**C**, **D**), before (**A**, **C**) and after (**B**, **D**) topical treatment with ajoene. The tumor surface areas are: **A** 63 mm^2, **B** 42 mm^2, **C** 195 mm^2, **D** 24 mm^2. (Image from Tilli, 2003).

Pensky, using ethyl ethanethiosulfinate (EtS(O)SEt) and related aliphatic thiosulfinates. The antitumor activity of these compounds was attributed to their reactivity toward –SH groups (Weisberger, 1958). Interestingly, thiosulfinates from *A. tuberosum* are reported to induce apoptosis in human PC-3 prostate and other types of cancer cells (S.-Y. Kim, 2008a; Park 2007). *A. tuberosum* is the major ingredient for leek *kimchi* in Korea, which has long been used as a medicinal food for treatment of abdominal pain, diarrhea, hematemesis, snakebite and asthma in folk medicine. *A. tuberosum* compound methyl methanethiosulfinate (MeS(O)SMe), but not

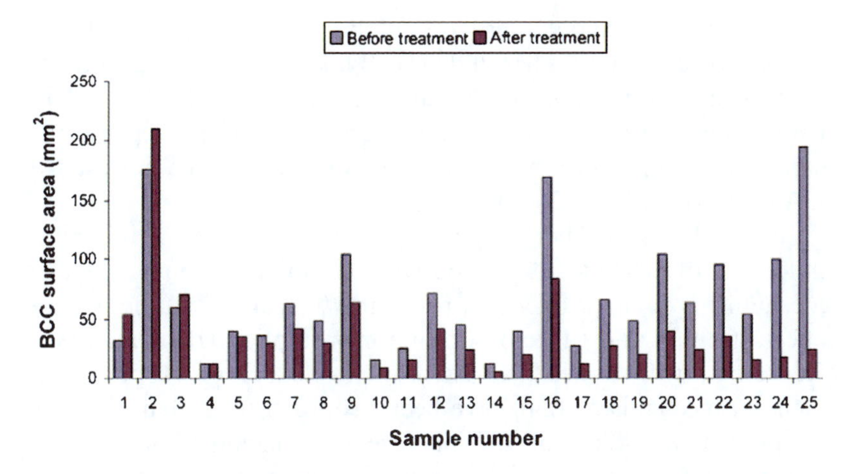

Figure 5.7 Effect of ajoene on tumor size after topical application in the individual samples (tumors no.1-3 increased in size after treatment, tumor no. 4 was without change in size, tumors no. 5–25 decreased in size after treatment). (Image from Tilli, 2003).

allicin, inhibits the *in vitro* growth of human acute myeloid leukemia cell lines with an IC_{50} of 2 μM (Merhi, 2008).

The potent antibiotic and cytotoxic effects of allicin have thus far been limited to topical or *in vitro* applications since allicin reacts rapidly with free thiol groups (*e.g.*, glutathione, abundant in plasma and in cells) and thus disappears from the circulation shortly after injection (in mice, injected allicin shows an LD_{50} of $60\,mg\,kg^{-1}$ body weight; Cavallito, 1944a). A new technique to circumvent this limitation in antitumor therapy involves *in situ* generation of allicin on the targeted cells. Alliinase enzyme is first targeted to a tumor, accomplished by conjugating the enzyme to a carrier, a monoclonal antibody (mAb) specific to a tumor-associated surface antigen. Alliin is then introduced into the circulation, resulting in formation of cytotoxic allicin only at the site of alliinase localization on the tumor cell surface. This strategy for drug delivery, based on the concept of antibody-directed enzyme prodrug therapy (ADEPT; Miron, 2003), has been used to kill human lymphocytic leukemia tumor cells *in vivo* (Arditti, 2005).

Allicin is a potent, microtubule-disrupting reagent that interferes with tubulin polymerization by reaction with tubulin SH groups and is thus able to inhibit cell polarization, migration and division (Prager-Khoutorsky, 2007). A somewhat related type of drug delivery system involves use of polybutylcyanoacrylate (PBCA) nanoparticles of diallyl trisulfide (DAT), shown to have antitumor activity in a mouse model of hepatocellular carcinoma. The DAT–PBCA nanoparticles (mean diameter: 113.4 nm), prepared by emulsion polymerization with PBCA as carrier, slowly release the DAT *in vivo* (Zhang, 2007).

A very large number of papers have appeared describing the influence of specific *Allium*-derived compounds on molecular mechanisms in carcinogenesis, following the earliest studies on the effect of onion and garlic oil on tumor promotion (Belman, 1983). Representative studies are listed in Table A3 of Appendix 1. General mechanisms that have been proposed to explain the cancer-preventive effects of *Allium* vegetables include inhibition of carcinogen formation, which can involve depression in nitrosamine formation. Thus, inhibition by garlic of *Helicobacter pylori* growth in the gastric cavity may result in reduced conversion of nitrate to nitrite in the stomach, thereby decreasing formation of carcinogenic *N*-nitroso compounds (You, 2006). Gastric tract reduction of nitrate to nitrite may be rendered harmless by scavenging of nitrite by garlic sulfur compounds, forming *S*-nitrosothiols. Other proposals for the mode of activity of *Allium* compounds include: free radical scavenging and antioxidant action (both discussed in Section 5.7.3); inhibition of genotoxicity and/or mutagenicity of carcinogenic agents; reduction in carcinogen bioactivation through modulation of the carcinogen metabolizing and detoxifying enzymes (cytochrome P450-dependent monooxygenases and phase II enzymes, such as glutathione transferases and quinone reductase); the effect on cell proliferation and apoptosis; inhibition of metathesis (Edwards, 2007) and angiogenesis (Herman-Antosiewicz, 2004; Siess, 2007; Nagini, 2008); and improving the chemotherapy of other drugs such as docetaxel (Howard, 2008).

A number of these studies use ajoene as an experimental antileukemia drug (Hassan, 2004) or garlic oil component diallyl trisulfide (Hodge, 2008). These and related compounds are thought to alter carcinogen metabolism, as well as suppress growth of cancer cells in culture and *in vivo* by causing cell cycle arrest, apoptosis induction and the suppression of angiogenesis and metastasis (Dirsch, 1998a, 2002). Cancer cell studies of various alkyl and allyl sulfur compounds have shown that diallyl sulfur compounds are more active than allyl alkyl sulfur compounds having only one allyl group, which in turn are more reactive than dialkyl sulfur compounds lacking allyl groups (Siess, 2007). Furthermore, diallyl polysulfides with three to five sulfurs, are more potent than diallyl disulfide for inhibiting cancer cell proliferation. Thus, the inhibitory effect increases with the number of sulfur atoms in the polysulfide (Nishida, 2008; Ariga, 2006). In addition, lipophilic sulfides are more effective than hydrophilic sulfides, such as *S*-allylcysteine.

The antitumor activity of garlic sulfur compounds is due, in part, to their ability to induce phase II detoxification enzymes (Powolny, 2008). Briefly, foreign chemical substances (xenobiotics) in an organism are detoxified mainly by two enzyme systems, so-called phase I and II, and then cleared from the body. Phase I enzymes oxidize the xenobiotics *via* the cytochrome P450 enzyme system, which can produce cytotoxic reactive intermediates. On the other hand, phase II enzymes eliminate the xenobiotics by conjugating the intermediates generated by phase I enzymes to glutathione or glucuronic acid. Diallyl trisulfide and other *Allium* polysulfides upregulate (cause a marked increase in) phase II enzymes glutathione *S*-transferase (GST) and quinone reductase (QR) in rats. This effect was not seen with diallyl sulfide or disulfide, but was seen with allicin, suggesting the need for compounds with weak S–S bonds. Diallyl trisulfide did not significantly affect phase I enzyme activity.

Natural organoselenium compounds found in garlic, such as γ-glutamyl-*Se*-methylselenocysteine, may also contribute to the protective effect of garlic, although dietary levels of such compounds are extremely low (Dong, 2001). Given the amount of daily garlic consumption, garlic hardly contributes to dietary selenium intake. However, the form in which the selenium is found in alliums is ideal for metabolic intake, since selenium-enriched garlic, where selenium appears as γ-glutamyl-*Se*-methylselenocysteine, was more effective in suppressing mammary carcinogenesis than supplementation with selenium in the form of inorganic selenite involving an equal number of moles of selenium (Ip, 1996; Zeng, 2008). Se-enriched garlic was found to lead to more desirable lower levels of accumulation in rat liver and kidney compared with Se-enriched kale or broccoli (Yang, 2008).

Despite the encouraging *in vitro* results on the anticancer effects of selenoamino acids, two recent publications show that there is *no protective association between plasma selenium (or selenium supplementation) and prostate cancer.* Results from the European Prospective Investigation into Cancer and Nutrition (EPIC) conclude that "plasma selenium concentration was not associated with prostate cancer risk in this large cohort of European men" (Allen, 2008). A second report, on the "Selenium and Vitamin E Cancer Prevention Trial" (SELECT), which used 200 μg day^{-1} selenomethionine as the sole selenium source in a randomized placebo-controlled trial (RCT) of 35 533 men for a median of 5.5 years, concludes that "selenium or vitamin E, alone or in combination at the doses and formulations used, did not prevent prostate cancer in this population of relatively healthy men" (Lippman, 2009). In an accompanying editorial, it is noted that, given this is the largest individually randomized cancer prevention trial ever conducted, it is unlikely to have missed detecting a benefit of even a very modest size. The editorial concludes that without a "firmer basis for casual hypotheses...physicians should not recommend selenium or vitamin E – or any other antioxidant supplements – to their patients for preventing prostate cancer" (Gann, 2009).

Allium components lacking sulfur or selenium are also of interest. Quercetin, an antioxidant flavonoid abundant in onions, is considered to be a promising agent for prevention of lung and colon cancers (Murakami, 2008), while an oligosaccharide from garlic (of molecular weight 1800) has been shown to be cytotoxic *in vitro* to several types of cancer cells and to significantly suppress the growth of murine colon adenocarcinoma cells *in vivo* (Tsukamoto, 2008).

5.7 THE EFFECT OF DIETARY GARLIC AND ONIONS AND GARLIC SUPPLEMENTS ON CARDIOVASCULAR DISEASE

Cardiovascular disease (CVD) is one of the leading causes of morbidity and mortality in industrialized countries. The CVD benefits of garlic and its constituents have been the subject of several recent reviews (El-Sabban, 2008; Espirito Santo, 2007; Gorinstein, 2007; Omar, 2007; Rahman, 2006; Banerjee, 2002b; Brace, 2002; Reuter, 1996). The reported effectiveness of garlic and onion in prevention of CVD is associated with the lipid-, cholesterol- and homocysteine-lowering (hypolipidemic, hypocholesterolemic, hypohomocysteinemic, respectively), anti-hypertensive, antidiabetic and antithrombotic (antiplatelet aggregatory) effects claimed of these plants and their derived supplements. The claimed effects will each be discussed in the following section, highlighting the most recent work. The reviews should be consulted for full details.

5.7.1 Epidemiological Studies

A recent epidemiological study from Mediterranean countries suggests that a diet rich in onion, but not in garlic, may have a favorable effect on the risk of acute myocardial infarction (Galeone, 2008). A Finnish study reached similar conclusions regarding onions (Knekt, 1996).

5.7.2 *In vitro* and *In vivo* Studies Relevant to the Cardiovascular Benefits of *Allium* Consumption

Several garlic-derived organosulfur compounds, including diallyl disulfide and trisulfide and 2-propenethiol, prevent cholesterol synthesis in cell cultures by inhibiting lanosterol 4α-methyl oxidase. Diallyl disulfide was active at a level of 15 $\mu mol\ L^{-1}$, while the other compounds were somewhat less active. *S*-Allylcysteine was only active at much higher millimolar concentrations (Singh, 2006). Earlier studies implicated ajoene and allicin as inhibitors of cholesterol synthesis (Ferri, 2003; Gebhardt, 1996). Allicin has been suggested to affect atherosclerosis not only by acting as an antioxidant, but also by modifying lipoproteins and inhibiting LDL uptake and degradation by macrophages (Gonen, 2005).

While hydrogen sulfide (H_2S) and nitric oxide (NO) were once both considered solely as metabolic poisons, they are now recognized as important, endogenously produced, biological signaling molecules ("gasotransmitters") with key roles in cardiovascular processes, including dilation of blood vessels, hypertension and myocardial infarction. Garlic-derived compounds are presumed to stimulate formation of H_2S, as has already been discussed (Section 3.10.3), and can either inhibit NO formation in macrophages by inhibiting inducible NO synthase (iNOS; Chang, 2005; Dirsch, 1998b, Kim, 2001) or promote NO formation by activating neuronal and endothelial NO synthases (nNOS; Aquilano, 2007; eNOS: Kim, 2001).

5.7.3 *Allium* Antioxidants

It has been inferred from epidemiological studies showing an inverse correlation between the intake of fruits and vegetables and the occurrence of cardiovascular disease, cancer and aging-related disorders, that the dietary antioxidants in these foods are the effective nutrients that prevent these and other oxidative stress related diseases (Huang, 2005). Antioxidants are compounds capable of counteracting the damaging effects of oxidation in animal tissue. More specifically, they are substances that may protect cells from damage, and resulting disease conditions, caused by free radicals. Antioxidants interact with, and stabilize free radicals and thus may prevent some of the damage they might otherwise cause. Antioxidants, in general, are widely used as ingredients in dietary supplements in the hope of maintaining health and preventing diseases. However, five large-scale clinical trials of antioxidants in the 1990s (involving β-carotene and vitamin E or A or aspirin) reached differing conclusions on their health benefits. While one trial, involving β-carotene, vitamin E and selenium, found a reduced incidence of gastric cancer, two of the trials demonstrated an *increase* in lung cancer rates among smokers, while two other trials found no change in cancer and/or cardiovascular disease rates (National Cancer Institute, 2004). There is another concern regarding the use of antioxidants by those undergoing chemotherapy. It is thought that the environment of cancer cells causes high levels of oxidative stress, making these cells more susceptible to the further oxidative stress induced by treatments (Trachootham, 2006; Schumacker, 2006). By reducing the redox stress in cancer cells, antioxidant supplements could decrease the effectiveness of radiotherapy and chemotherapy, although this subject is controversial (Seifried, 2003; Lawenda, 2008; Block, 2008).

In vitro measurement of antioxidant activity for alliums and other foods is not a straightforward process, due to the diversity of compounds with differing reactivities to different reactive oxygen species. The results of such assays cannot be applied to biological systems since "these assays do not measure bioavailability, *in vivo* stability, retention of antioxidants by tissues, and reactivity *in situ*" (Huang, 2005). In food science, the oxygen radical absorbance capacity (ORAC) has become the current industry standard for assessing *in vitro* antioxidant activity of foods (Prior, 2005). By use of this method, it is found that, on a fresh weight basis, garlic ranks very high among common vegetables and spices in antioxidant activity against peroxyl radicals, ROO· and Cu^{2+} (Cao, 1996), as explained above in Section 4.9.

Numerous studies have appeared on the ability of *Allium* preparations to increase the activity of protective enzymes, such as quinone reductase and glutathione transferase, as well as lower lipid peroxidation (Jastrzebski, 2007; Gonen, 2005; Pedraza-Chaverrí, 2004; Higuchi, 2003; Banerjee, 2003, 2002a; Lawson, 1996). Di-*n*-propyl trisulfide from onion oil has an ameliorative effect on memory impairment in the senescence-accelerated mouse. This effect is attributed to the antioxidant activity of this compound toward brain lipid hydroperoxide, thought to be involved in the pathophysiology of Alzheimer's disease (Nishimura, 2006).

5.7.4 The Effect of Dietary Garlic and Garlic Supplements on Cholesterol Levels: The Stanford Clinical Trial

Garlic supplements (Figure 5.2), seeking to package the benefits of raw garlic in more palatable forms, are promoted as cholesterol-lowering agents. Despite promising *in vitro* studies on inhibition of cholesterol by garlic-derived sulfur compounds, and a strong plausibility of effect demonstrated in numerous animal studies (85% significant positive effects; Koch, 1996), the evidence from clinical trials supporting a hypocholesterolemic (cholesterol-lowering) effect of various forms of garlic is highly inconsistent (Ackermann, 2001; Gardner, 2007). Thus, pre-1995 clinical trials with garlic powder tablets at doses of 0.6 to 1.2 g suggested a modest beneficial effect of garlic on lipids in substantially hypercholesterolemic adults, but these trials were criticized for serious design and conduct limitations. Post-1995 trials at similar doses consistently reported no significant effects on plasma lipids in similar populations (Gardner, 2007; Neil, 1996; Superko, 2000; Warshafsky, 1993; Turner, 2004; Zhang, 2006b). Almost all commercial garlic supplements used in these trials yield unexpectedly low amounts of the putative garlic active agent, allicin, under physiologically relevant dissolution conditions (Lawson, 2001a,b). Therefore, the effectiveness of garlic and garlic supplements has remained ambiguous. Furthermore, the bioavailability of the important sulfur-containing constituents differs significantly between raw garlic and the specific garlic supplement formulations: the precise amount of allicin released in the body from garlic supplements is never specified.

The objective of the Stanford Clinical Trial was to compare the effect of raw garlic and two garlic supplements with distinctly different formulations, on plasma lipid concentrations in adults with moderate hypercholesterolemia (elevated blood cholesterol levels) over a six-month period (Gardner, 2007, 2008). It was felt that if there were problems with the bioavailability of active ingredients from commercial supplement formulations, raw garlic should be superior. In this trial, 192 adults with low-density lipoprotein cholesterol (LDL-C) concentrations of 130 to 190 $mg\,dL^{-1}$ were randomly assigned to one of four treatment arms: raw garlic, powdered garlic supplement (Garlicin; Nature's Way Products Inc., Springville, Utah), aged garlic extract supplement (Kyolic; Wakunaga of America Co., Mission Viejo, California), or a placebo. The commercial supplements are representative of those most widely used. This sample size was substantially larger than almost all previous trials, and was designed to detect even modest effects on plasma lipid concentration; retention was uniformly good (87 to 90%). Garlic product doses, equivalent to one 4 g garlic clove, were consumed six days per week, for six months. The primary study outcome for the trial was LDL-C concentration. Fasting plasma lipid concentrations were assessed monthly, to reveal whether garlic exerted a moderate, albeit transient, cholesterol-lowering effect. Extensive chemical characterization of study materials conducted during the trial, showed that the chemical stability of study materials remained high throughout (Lawson, 2005).

The trial showed that *there were no statistically significant effects of the three forms of garlic on LDL-C concentrations*. The six-month mean changes in LDL-C concentrations for raw garlic, powdered supplement, aged extract supplement, and placebo were $+0.4\,mg\,dL^{-1}$, $+3.2\,mg\,dL^{-1}$, $+0.2\,mg\,dL^{-1}$ and $-3.9\,mg\,dL^{-1}$, respectively. There were no statistically significant short-term or longer-term effects on high-density lipoprotein cholesterol (HDL-C), triglyceride levels or total-cholesterol : HDL-C ratio. It was concluded that none of the forms of garlic used in the study, including raw garlic, when given at an approximate dose of a 4 g clove per day, six days per week, for six months, had statistically or clinically significant effects on LDL-C or other plasma lipid concentrations in adults with moderate hypercholesterolemia. Based on the results of these and other recent trials, it was concluded that physicians can advise patients with moderately elevated LDL-C concentrations that garlic supplements or dietary garlic in reasonable doses are unlikely to produce lipid benefits (Gardner, 2007).

A review of garlic and CVD by European experts states: "The investigations within the [European] Garlic & Health project as well as the majority of independent studies provided compelling evidence against a beneficial influence of garlic powders or garlic constituents on risk factors and pathological aspects of cardiovascular disease in animals and humans" (Espirito Santo, 2007). Reviews of statin alternatives in cardiovascular therapies and a meta-analysis of garlic supplementation and serum cholesterol by authors from Asia reach similar conclusions. These reviews conclude: "The negative result from the Stanford study, together with the absence of any evidence of clinical event reduction, means that at present garlic treatment can not be recommended for hyperlipidemia or for patients at risk of cardiovascular events" (Ong, 2008) and "the available evidence from randomized controlled trials does not demonstate any beneficial effect of garlic on serum cholesterol" (Khoo, 2009).

The Stanford trial did not specifically test garlic preparations containing distilled garlic oil or ajoene from garlic macerates. Ajoene, given at a dosage of 100 mg kg^{-1} ("polar fraction of garlic oil") to rats made hyperlipidemic by feeding them coconut oil and 2% cholesterol, was reported to counteract the effects of the dietary lipids (Augusti, 2005). However, this level of ajoene used with rats corresponds to 7.5 g of pure ajoene for a 75 kg human, substantially higher than the doses recommended by supplement manufacturers and the levels of garlic products (13–15 mg allicin) used in the Stanford trial. Indeed, using a typical concentration of 0.1 mg g^{-1} ajoene in oil-macerate, 75 kg of oil-macerate, equal to the weight of a human, would have to be consumed! The majority of earlier clinical trials employing steam-distilled garlic oil with both normolipidemic and hyperlipidemic subjects showed, at best, small effects on serum cholesterol levels after 12 weeks; the studies showing small effects can be faulted for lacking diet control (Barrie, 1987).

5.7.5 Antithrombotic Activity of Dietary Garlic and Other Alliums; Garlic Supplements and their Effect on Platelet Biochemistry and Physiology

The use of antiplatelet therapies decreases the incidence of mortality in persons prone to serious cardiovascular events. While several *in vitro* studies suggest that garlic or its individual components may decrease platelet aggregation, recent clinical trials suggest otherwise. For example, a small randomized, double-blind, placebo-controlled, crossover research study was performed, involving 14 healthy volunteers and the use of a garlic macerate preparation containing ajoene and dithiins. Four hours after consuming one large dose of garlic oil, the study showed that there was *little or no effect in the reduction of platelet aggregation.* This study used well prepared and characterized garlic-in-oil capsules, containing ajoene and dithiins. In particular, each dose capsule contained 55 mg of a garlic-derived oil dissolved in 445 to 500 mg of soybean oil. The garlic oil itself contained 22 to 26% 1,3-vinyldithiin and 17 to 21% of combined (*E*)- and (*Z*)-ajoene, according to HPLC analysis. Each 55 mg portion of garlic oil was derived from 9.9 g of fresh garlic. The fresh garlic was thoroughly minced at high speed with an electric mixing wand, and the resultant homogenate was found to contain 4.38 mg g^{-1} of fresh garlic (Wojcikowski, 2007).

A second recent study investigated whether standard commercial garlic supplements have an effect on platelet function *in vivo.* This investigation also sought to provide clinical scientific evidence of the safety of their use in the perioperative period. Thus, in a double-blind study, garlic supplements were administered to ten adult volunteers at the manufacturer's recommended dose for two weeks. At the end of the two-week period, *in vivo* platelet function was quantified using the PFA-100 assay. After a two-week "washout" period, the protocol was repeated using aspirin. It was found that *in vivo* platelet function was not affected by the administration of garlic supplements, but was markedly inhibited with the administration of aspirin. It was concluded that garlic supplements do not affect platelet function *in vivo.* Neither this experiment nor a review of the literature, supports the concern of perioperative bleeding in users of garlic (Beckert, 2007). Unfortunately, the above study did not specifically describe the type of garlic supplement tested, nor did it validate the extent of allicin formation from this supplement under simulated digestion conditions.

A third, small study involved randomized, crossover, observer-blinded and placebo-controlled research, in which 18 healthy volunteers were each given Greek tsatsiki (a dish made with Greek yogurt, cucumber, dill and garlic), with or without 4.2 g of raw garlic. Whole blood from all of the subjects was tested before, and five hours after, ingestion of the dish. In addition, the potential long-term effects of garlic were investigated in five volunteers after daily ingestion of 4.2 g of raw garlic over a period of one week. The study showed that *baseline values of platelet function were within the normal range in all volunteers and that platelet function was not impaired by single and repeated oral consumption of Greek tsatsiki containing raw garlic in any point-of-care monitoring test used.* The authors of this study conclude that consumption of "dishes containing socially acceptable doses of raw garlic are unlikely to increase the risk of perioperative bleeding" (Scharbert, 2007; also see: Beckert, 2007; Rahman, 2007).

A small open, clinical trial, funded by the U.S. National Institutes of Health/National Center for Complementary and Alternative Medicine (NIH/NCCAM), examined the effect of both short- and long-term ingestion of elevated levels of garlic on platelet aggregation in healthy volunteers. Six people consumed the equivalent of up to 12 grams of raw garlic, containing 72 mg of allyl thiosulfinates. Blood was drawn 4.5 and 7 hours after consumption, and collagen and ADP-induced aggregation was determined both for platelet-rich plasma (PRP) and whole blood. Small, inconsistent effects were found with PRP, but no effect at all was found with whole blood. In the long-term study, nine people were given the equivalent of 8 grams of raw garlic per day (35 mg allicin content) or an equivalent amount of garlic boiled for 10 minutes (84 mg alliin content; no allicin), both in the form of sandwiches. After 4 weeks, it was found that, while boiled garlic consumption had no effect, raw garlic consumption decreased PRP aggregation with all agonists to a small extent (5–8%), but increased collagen-induced whole blood aggregation by 10–13%. It was concluded that "the results are method-dependent, but considering the large dose used, it appears unlikely that consumption of normal levels of raw or boiled garlic (1–3 g) will have an effect on *ex vivo* aggregation in healthy individuals" (Lawson, 2007).

An Egyptian study (Ebid, 2006) examined the effects of concomitant intake of aspirin and garlic preparations on the coagulation profile in cardiac patients. A significant difference was evident in those who received both garlic and aspirin when compared with those who received aspirin alone. Results indicated that 31 out of 35 patients who received 100, 200 or 300 mg garlic concomitantly with aspirin had positive hemoccult tests indicating fecal occult blood. In addition, there was a significant association between garlic intake with aspirin and the incidence of gastrointestinal bleeding. While the authors raised the concern that garlic could increase the risk of bleeding during surgical procedures, this study dealt only with the combined action of garlic and aspirin, not the effect of garlic by itself.

In vitro testing of baboon blood-derived PRP showed that platelet aggregation with ADP (20 or 40 μM) was inhibited completely in a dose-dependent manner by ajoene at $\geq 75\,\mu g\,mL^{-1}$ (320 μM). Aggregation with collagen ($12.5\,\mu g\,mL^{-1}$) was inhibited from 66 to 11% by ajoene at $25\,\mu g\,mL^{-1}$ (101 μM), but higher concentrations had little additional effect (Teranishi, 2003). *In vitro* testing with human PRP gave similar IC_{50} values of 166 and 213 μM for inhibition of ADP-induced platelet aggregation by (*Z*)- and (*E*)-ajoene, respectively (Block, 1986). *In vivo* testing with baboons showed that ajoene in soybean oil at a dosage of $25\,mg\,kg^{-1}$ given intravenously either once, or repeatedly at intervals of 2 to 3 hours, to animals weighing 6 to 14 kg, completely inhibited platelet aggregation. Repeated doses at 2 hour intervals maintained complete inhibition of aggregation, but did not do so when the interval between doses was extended to 2.5 or 3 hours (Teranishi, 2003). It should also be noted that the baboon dosage used, translates to a level of *ca.* 1.9 g for a 75 kg human, which seems unacceptably high for a drug, particularly one with an apparently very short-lived effect.

In vitro testing showed that allicin and propyl propanethiosulfinate (PrS(O)SPr) rapidly interact with –SH groups on intracellular cysteine-proteins, especially calpain (a key calcium-dependent enzyme), both at the platelet surface and in its interior, inhibiting aggregation at 10 to 20 μM levels (Badol, 2007; Rendu, 2001). On the other hand, the smaller methyl methanethiosulfinate (MeS(O)SMe) enters the platelet more rapidly than allicin and reacts predominantly inside the cell, without effect at the membrane surface. These results explain the earlier observation

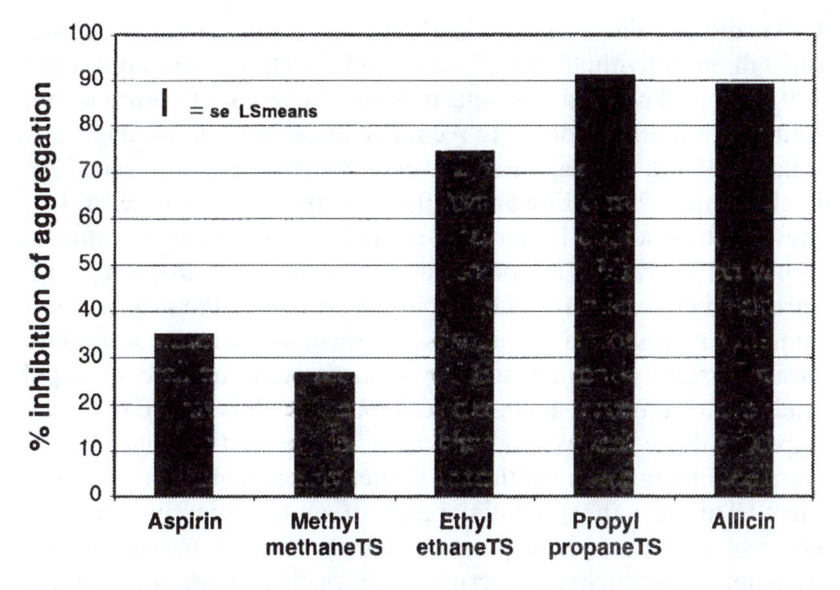

Figure 5.8 Inhibition of platelet aggregation by 360 μM aspirin and 400 μM thiosulfinates. The IC_{50} values for platelet inhibition for MeS(O)SMe, EtS(O)SEt, PrS(O)SPr and AllS(O)SAll are 550, 320, 270 and 270 μm, respectively. (Image from Briggs, 2000).

regarding the differential platelet inhibitory activity for thiosulfinates (Briggs, 2000). Inhibition of free cysteine and glutathione in whole blood may also contribute to their anti-aggregant properties. *In vitro*, allicin, PrS(O)SPr and EtS(O)Set, but not MeS(O)SMe, are more potent platelet inhibitors than aspirin at nearly equivalent concentrations, as illustrated in Figure 5.8; this relative order of potency may not be true *in vivo*, where rapid deactivation of thiosulfinates is likely in blood plasma (Briggs, 2000). Both thiosulfinates and diallyl trisulfide are potent inhibitors of calcium ion mobilization, which also accounts for their ability to inhibit platelet aggregation (Qi, 2000). Alcoholic extracts of *A. ursinum* and *A. sativum* both showed similar antiaggregatory activity toward human platelets (Hiyasat, 2009).

5.7.6 Antihypertensive Activity of Dietary Garlic and Garlic Supplements

Hypertension [systolic blood pressure (SBP) $\geq$ 140 mm Hg; diastolic blood pressure (DBP) $\geq$ 90 mm Hg] is a known risk factor for cardiovascular morbidity and mortality, affecting an estimated one billion individuals worldwide (Ried, 2008). A meta-analysis in 1994 reported promising results in mildly hypertensive subjects, but found insufficient evidence to recommend garlic for clinical therapy (Silagy, 1994); similar results were reported in 2001 and 2005 (Ackermann, 2001; Edwards, 2005). A problem with several of these meta-analyses is that the majority of the results from the published literature on potential blood pressure lowering with garlic come from trials where lipid lowering was the *primary* outcome of interest, and blood pressure was assessed as a secondary measure, simply because it is easy to measure. In many of the lipid trials, an important inclusion criteria was that participants be hyperlipidemic at baseline. However, many of these individuals were normotensive, resulting in limited opportunity for improvement. This has the potential result of underestimating the true effect of garlic on blood pressure, which is better measured with hypertensive participants recruited specifically for a blood pressure lowering trial.

Two more recent meta-analyses (literature reviewed through mid-2008) suggest that garlic is associated with blood pressure reductions in patients with an elevated systolic blood pressure (SPB), although not in those without elevated SBP. When compared with placebo in patients with elevated SBP, garlic reduced SBP by an average of 8.4 mm Hg in one study (Ried, 2008) or 16.3 mm Hg in a second study (Reinhart, 2008). The author of one these recent studies concludes that the effects of garlic on blood pressure are "comparable to the hypotensive effects of commonly prescribed blood pressure drugs" (Ried, 2008). The blood-pressure reducing properties of garlic have been linked, in part, to its hydrogen sulfide-producing ability (Benavides, 2007).

Studies with neuronal cell cultures have shown that the pungency of allicin and other garlic-derived compounds is associated with activation of a member of the so-called transient receptor potential family of cation-selective sensory ion channels, TRPA1. These pungent sulfur compounds also induce vasorelaxation (vasodilation) through a TRPA1-dependent mechanism involving release of special peptides (CGRP) from the activated nerve terminals. It remains to be seen whether this vasodilation mechanism modeled *in vitro* contributes to systemic *in vivo* hypertensive activity of garlic (Bautista, 2005).

A study examined the frequency of garlic usage in hypertensive populations and then clinically evaluated the acute effects on blood pressure in patients with hypertension. Garlic supplements were found to be used by 53.3% of 7703 hypertensive patients responding to a questionnaire. In a clinical study, 75 hypertensive patients (blood pressure ≥ 140 mm Hg systolic or ≥ 90 mm Hg diastolic) were randomly separated into three groups: a placebo group, a group taking one clove of garlic orally "after slightly chewing" and a group taking 12 tablets of a commercial garlic tablets orally "after slightly chewing." Blood pressure was measured before garlic or placebo consumption, and then every ten minutes for a total of 70 min. No significant effect on blood pressure was observed in any of the three groups (placebo, fresh garlic or garlic tablets; Capraz, 2007).

A prospective and uncontrolled clinical study attempted to evaluate the effects of short-term supplementation with a garlic macerate formulation on lipid metabolism, glucose level and antioxidant status, in patients suffering from primary arterial hypertension. Seventy subjects aged 30 to 60 years with primary arterial hypertension took part in the study. They each took two capsules of garlic preparation after meals, three times daily, for a period of 30 days, at the same time continuing their hypotensive pharmacotherapy as recommended by their physician. The daily dose of garlic corresponded to 1.62 mg of allicin-derived compounds (dithiins) in a rapeseed oil base. The garlic formulation was found to significantly lower lipid levels and the level of lipid peroxidation products in the blood. It markedly increased the vitamin E concentration in the serum, whereas the increases in the levels of other antioxidant vitamins and glutathione peroxidase activity proved insignificant. The product did not affect arterial blood pressure in the study subjects (Duda, 2008).

A randomized placebo-controlled trial involving 62 subjects, examined the effect of dried garlic powder tablets (10.8 mg d^{-1} of alliin, corresponding to about one-quarter of a garlic clove) on blood pressure and arterial stiffness (assessed by pulse wave velocity), and concluded that garlic powder tablets have no clinically relevant lipid-lowering or blood pressure-lowering effects in middle-aged, normo-lipidemic individuals (Turner, 2004). The lack of an effect may be due to the low dosage of alliin used in this study.

5.7.7 Dietary Garlic and Inflammation

In vitro studies using epithelial cells showed anti-inflammatory activity for allicin (Lang, 2004). However, in a twelve-week, double-blind, randomized, placebo-controlled clinical trial with 90 overweight subjects, aged 40 to 75 years, who smoked more than ten cigarettes per day, a chemically well-characterized, enteric-coated garlic preparation (2.1 g garlic powder per day, equivalent to 9.4 mg allicin per day) had no significant effect on inflammatory biomarkers, endothelial function, or lipid profile in normolipidemic subjects with risk factors for cardiovascular disease (van Doorn, 2006). A review of the medical literature concludes, "Even though part of the experimental work done *in vitro* points to some chances for defined effects of garlic components on cardiovascular and inflammatory parameters, it remains doubtful whether these effects may be of any significance with respect to the human situation" (Espirito Santo, 2007).

5.7.8 Garlic and Hyperhomocysteinemia

Elevated homocysteine (hyperhomocysteinemia) is a risk factor linked to cardiovascular disease. While a commercial aged garlic extract (AGE) was found to reduce homocysteine levels in animal studies, these studies were criticized because "the doses of AGE applied are extremely high and probably without relevance for the human situation" (Espirito Santo, 2007).

5.7.9 Garlic and High-Altitude Sickness and Hepatopulmonary Syndrome

Use of garlic to prevent altitude sickness is noted in the medical literature (Basnyat, 2004). In the early nineteenth century during Argentina's fight for independence, San Martin observed that administration of garlic to his soldiers while they crossed the Andes countered the effects of high altitude (Harris, 1995). Among the beneficial effects attributed to garlic is the ability to relax vascular smooth muscle. One explanation for the beneficial effect of garlic ingestion in preventing altitude sickness might be amelioration of hypoxic pulmonary vasoconstriction (HPV or narrowing of pulmonary blood vessels due to oxygen deprivation). Recent experimental studies have demonstrated several effects of garlic, including activation of endothelial nitric oxide synthase and smooth muscle cell membrane hyperpolarization, which could decrease pulmonary vascular tone. Together, these observations suggest that garlic administration might block the development of HPV by inhibiting pulmonary vascular smooth muscle contraction (Fallon, 1998). Clearly, further testing is needed to verify that garlic has some prophylactic activity against high-altitude sickness (also known as acute mountain sickness).

Hepatopulmonary syndrome is a syndrome of shortness of breath and hypoxemia (low oxygen levels in the blood of the arteries) caused by vasodilation (broadening of the blood vessels) in the lungs of patients with liver disease. Several pilot studies and case reports suggest that, with some but not all patients, garlic may be useful in treatment of this syndrome by improving arterial oxygenation and diminishing shortness of breath on exertion (Abrams, 1998; Akyuz, 2006; Caldwell, 1992; Chan, 1995; Sani, 2006).

5.7.10 Summary of Cardiovascular Benefits of Garlic

The cardiovascular benefits of garlic summarized by Larry D. Brace, a pathology professor at the University of Illinois, Chicago, remain valid: "The vast majority of recent randomized, placebo-controlled studies do not support a role for garlic in lowering blood lipids. There also is insufficient evidence to support a role in reducing blood pressure. While there have been indications of antiatherosclerotic effects associated with garlic consumption, there are insufficient data in humans. Investigation of antithrombotic effects of garlic consumption appears to hold promise, but too few data exist to draw firm conclusions" (Brace, 2002).

5.8 DIETARY ALLIUMS AND DIABETES

In 2006, according to the World Health Organization, at least 171 million people worldwide suffer from diabetes. Its incidence is increasing rapidly, and it is estimated that by the year 2030, this number will double. It is therefore not surprising, that the possible role of garlic as an antidiabetic agent has been examined. A recent, very thorough review of the use of garlic in the treatment of diabetes concludes that, while promising results have been obtained from animal studies on the use of garlic as an antidiabetic agent, "an understanding of the antidiabetic effects of garlic is still in its early stages.... The antioxidative, anti-inflammatory and anti-glycative effects of garlic in diabetic subjects or in animal models of diabetes warrant further investigation. Large-scale, well-designed clinical investigations are needed to establish the antidiabetic efficacy of garlic... the active compounds in garlic, and the doses thereof, that can effectively provide antidiabetic effects ... remain to be established" (Liu, 2007). Earlier, brief reviews of the literature came to a similar conclusion (Morelli, 2000; Mulrow, 2000; Ackermann, 2001).

Studies with alloxan-diabetic rats, showed that treatment with alliin from garlic improved the diabetic condition to the same extent as glyburide and insulin (Augusti, 1996). In a separate study, alloxan-diabetic rats fed 12.5% of their body weight of fresh garlic for 15 days, showed reduced fasting blood glucose (182.9 mg dL^{-1}) compared to the diabetic untreated group (258.7 mg dL^{-1}), but greater than the normal health group (67.2 mg dL^{-1}). These results suggest that, at the above dosage, fresh garlic has hypoglycemic activity in rats. There was no difference in fasting blood glucose between the group which received 12.5% of their body weight of fresh onion (269.4 mg dL^{-1}) for 15 days, and the diabetic untreated group (Jelodar, 2005). A single clinical trial examining the effect of garlic consumption on type 2 diabetes has been published. This double-blind, randomized trial of 20 patients with type 2 diabetes performed in Thailand, showed that there were no glucose-lowering effects for garlic when compared with the placebo (Sitprija, 1987). A significant limitation of this study was the use of a spray-dried garlic preparation, which was likely to contain little active ingredients (*e.g.*, little allicin potential).

5.9 GARLIC SULFUR COMPOUNDS: ANTIDOTES FOR CYANIDE, ARSENIC AND LEAD POISONING

The effect of garlic and allicin on the acute lethality of cyanide was studied in rats (Aslani, 2006). It was found that the lethality of cyanide intoxication was markedly reduced in rats that received dietary garlic or allicin, and that the rate of protection was dose-dependent. The effect of 1 000 ppm of allicin on cyanide lethality was equal to 20% of garlic in the diet. It is concluded that garlic sulfur compounds may have a protective effect on cyanide intoxication, presumably by facilitating conversion of cyanide to less toxic thiocyanate. Separate studies reported encouraging results involving diallyl and dialkyl disulfides and tetrasulfides, which provided protection against cyanide intoxication in mice (Baskin, 1999; Iciek, 2005, 2009).

Treatment of rats and sheep with garlic extracts or allicin was found to significantly reduce lead toxicity induced by administering lead acetate (Senapati, 2001). Sheep administered allicin at the level of 2.7 mg kg^{-1} body weight, twice daily for 7 days, showed significantly lowered blood lead levels, as well as kidney, bone and ovary lead contents, after having been given 80 mg kg^{-1} body weight of lead acetate for 5 days (Najar-Nezhad, 2008). Similar results were obtained with rats given lead acetate at the rate of 5 mg kg^{-1} body weight and garlic extract given at the rate of 400 mg kg^{-1} body weight, *versus* controls that were not given garlic (Senapati, 2001).

Chronic arsenic toxicity is a widespread problem, particularly in India and Bangladesh but also in various other regions of the world. Exposure to arsenic may occur from natural or industrial sources. Recent studies using rodents, as well as several different cell lines, suggest that aqueous garlic extracts coadministered with sodium arsenite attenuated the arsenite-induced toxicity (Chowdhury, 2007; Flora, 2009).

5.10 DIETARY ALLIUMS AS ANTI-ASTHMATIC AND ANTI-INFLAMMATORY AGENTS AND USE FOR STINGS AND BITES; TOPICAL *ALLIUM* EXTRACT APPLICATION AND SCAR HEALING

Ancient herbals describe topical application of *Allium* preparations for treatment of stings and bites. The Dakota and Winnebago Indians of North America reportedly applied bruised wild onion for the stings of bees and wasps with excellent results. Other *Allium* species have been used for rashes and pain relief (Lewis, 1977). A letter in an 1883 issue of the *Journal of the American Medical Association* observes that the juice of the common onion is excellent for stings, particularly if "thoroughly applied to the wound immediately after the sting has been received. It acts as a very perfect antidote to the poison, prevents swelling and speedily relieves the pain" (Ingals, 1883). Similarly, in 1932 in the *British Medical Journal* it was noted: "Rub site of sting gently with a raw cut onion for ten minutes or so. The effect is dramatic, and the swelling goes down under observation. An onion should be in every picnic basket" (Stobie, 1932). A recent report in the *Medical Journal of Australia* describes a remedy used in the Northern Territory of Australia for treating stingray as well as fish-spine wounds which involves bandaging half of an onion bulb over the wound. The scientist/victim of a painful foot wound reported that "after 30 minutes the pain had largely subsided" and that he "could walk easily, with only slight joint stiffness... The wound did not become necrotic or infected and healed without incident with only one application of antiseptic cream" (Whiting, 1998). Despite the abundance of anecdotal reports on the beneficial effect of raw onions on stings and bites, no scientifically rigorous studies have yet been published on this subject, although work has been done on the effect of onion preparations on scar healing, as will be described below.

The possible biochemical basis for the above described effects of *Allium* preparations on inflammation and pain follows from several studies. Extracts of dried garlic powder and diallyl disulfide modulate cytokine patterns in human whole blood, which, in turn, may result in a reduction of the inflammatory response by reducing the proinflammatory activity of NF-κB in adjacent tissue (Keiss, 2003). Cytokines are signaling compounds which are involved in a variety of inflammatory, immunological and infectious diseases. Thiosulfinates from onions are reported to inhibit histamine release, leukotriene and thromboxane biosynthesis *in vitro* and counteract allergen-induced bronchial obstruction *in vivo*. Thiosulfinates and cepaenes from onions also exhibit an anti-inflammatory effect by inhibiting chemotaxis of human polymorphonuclear leukocytes (Dorsch, 1988, 1990). Simply stated, alliums may act on stings and bites by inhibiting production of inflammation- and pain-causing prostaglandins which are produced in response to trauma.

Onions have traditionally been recommended to treat ulcers, wounds, scars and keloids, as well as earaches (WHO, 1999). The latter usage has been reviewed in an article humorously entitled, "An onion in your ear" (Brooks, 1986). An over-the counter (OTC) topical onion extract-based gel (Mederma®, Merz Pharmaceuticals, Greensboro, N.C., U.S.A.) is claimed by its manufacturers to "improve the appearance and feel of just about any kind of scar ... [including those] from surgery [www.mederma.com]." A similar European product (Contractubex gel, Merz Pharma, Frankfurt, Germany) contains both onion extract and heparin as active ingredients. The chemical composition of the onion extract is not specified, but may contain the flavonoid quercitin (Saulis, 2002) and/or cepaenes (Dorsch, 1988, 1990). Based on published clinical trials, the evidence for the efficacy of these products is mixed. Thus, a pilot study at the University of Texas M.D. Anderson Cancer Center, involving 17 patients with surgical scars, compared the effectiveness of Mederma to a petrolatum-based ointment over a period of one month. The conclusions of this study were that "no statistically significant difference between the pre- and post-treatment evaluations of scar erythema and pruritus in patients using topical onion extract gel were found" and that "topical onion gel extract was ineffective in improving scar erythema and pruritus in our patients." In contrast, "a statistically significant reduction in scar erythema was found in patients using a petrolatum-based ointment" (Jackson, 1999).

A Harvard Medical School study involving 24 patients in a randomized, double-blind, split-scar model, also compared Mederma to a petrolatum-based ointment over the course of eight weeks, with follow-up interviews conducted at least 11 months postoperatively. The overall conclusion of this study was that "the onion extract gel did not improve scar cosmesis or symptomatology when compared with a petrolatum-based ointment" (Chung, 2006). A prospective double-blind study (that is a study in which the subjects are identified and then followed forward in time), involving 99 patients with scars ranging in age from three weeks to eight years, found no difference in physician assessment of scar improvement between placebo-treated and onion extract gel-treated scars (Clarke, 1999). A prospective, randomized, double-blind, placebo-controlled study at the University of Miami, involving 30 subjects with a keloid or hypertrophic scar, compared the effectiveness of 0.5% hydrocortisone, onion extract gel and a placebo after 16 weeks. While the three study treatments were well-tolerated, the hydrocortisone treatment significantly improved more clinical parameters than did the onion extract gel or the placebo (Berman, 2007). A study at Northwestern University employing the rabbit ear model, found no significant reduction in scar hypertrophy or erythema using Mederma onion extract gel treated scars, compared with untreated scars. An improvement in collagen organization was noted in the

Mederma-treated scars, suggesting that Mederma "may have an effect on the pathophysiology of hypertrophic scar formation" (Saulis, 2002). A double-blind study at a private dermatology clinic in the United States, compared the treatment of postsurgical scars with an onion extract gel *versus* no treatment. It was concluded that the onion extract gel significantly improved scar softness, redness, texture and global appearance at the excision site (Draelos, 2008).

A six-month prospective study in Turkey involving 60 patients, compared the therapeutic activity of topical onion extract to silicon gel sheets only, to a combination of onion extract and silicon gel sheets, on hypertrophic and ketoid scars. While the onion extract gel was more effective in relation to scar color, it was statistically ineffective in improving scar height and itching. The most effective therapeutic results were obtained when the silicon gel sheet treatment was combined with onion extract (Hosnuter, 2007). A study in Germany comparing Contractubex gel to no treatment of post-thoracic surgery scars in 45 young patients, concluded that "Contractubex gel is useful in scar treatment after thoracic surgery" (Willital, 1994). A Hong Kong study compared the effect of treatment with Contractubex gel to no treatment, following Q-switched laser tattoo removal. It was found that the Contractubex group had a statistically significantly lower rate of scarring than the control group (Ho, 2006). In a multicenter, retrospective cohort study in Germany involving 771 hypertrophic scar patients, 553 received Contractubex and 216 received cortisteroid. Based on an assessment of the normalization of erythema and pruritus and on timely healing, the local administration of Contractubex was judged to be significantly more effective than corticosteroid treatment. Contractubex treatment was also associated with significantly fewer adverse events than topical corticosteroid application (Beuth, 2006). A 2006 review of topical treatments for hypertrophic scars (Zurada, 2006), concluded that although onion extract gels have had long-term use in Europe, are relatively low in cost with "botanical" ingredients, are easy to use and are readily available, clinical studies in the United States show no benefit in scar appearance.

5.11 ONION CONSUMPTION AND BONE LOSS

It has been reported that an onion-rich diet decreases bone loss in rats and that the active compound is γ-L-glutamyl-(*E*)-*S*-1-propenyl-

L-cysteine sulfoxide (GPCS; Wetli, 2005; Huang, 2008). The effectiveness of onion in preventing bone loss was confirmed by exposing osteoclast cells to parathyroid hormone to stimulate bone loss, and then exposing some of the treated cells to GPCS. Treatment with GPCS significantly inhibited the loss of bone minerals, including calcium, when compared to cells that were not exposed to GPCS. The minimal effective dose of GPCS was $\sim$2 mM. It is important to do the math to put this number in perspective. Fresh onions contain about 0.7 grams or 2.2 mM GPCS per kilogram (Lawson, 1991c). Since we have five liters of blood in our body, we would need to consume 4.5 kilograms, or about ten pounds of onions in one sitting, to achieve a concentration of 2.2 mM. That's a lot of onions! Since 3.2 grams of GPCS would be needed to reach this blood concentration, it might prove possible to supplement the diet with such an amount of the pure compound, but GPCS would then be considered a drug and would have to undergo thorough testing, at considerable cost and, because it is a naturally occurring compound, without the benefit of patent protection for the manufacturer. Despite the above concerns, a recent study found that women, 50 years and older, who consumed onions once a day or more, had an overall bone density that was 5% greater than individuals who consumed onions once a month or less (Matheson, 2009).

5.12 ADVERSE EFFECTS AND HEALTH RISKS ASSOCIATED WITH *ALLIUM* FOODS

5.12.1 How much Garlic can be Safely Consumed per Day?

"Natural" does not necessarily mean "safe." Consider mushrooms that grow in the wild: some are safe to eat, while others are poisonous. A study was performed using rats to determine the optimal oral dose of garlic to avoid liver damage. The researcher, using fresh garlic homogenates, found evidence of liver damage after use of garlic for periods up to 21 days at levels $\geq$ 0.5 g of garlic per kg of body weight. They concluded that, based on a rat model, only daily doses of garlic $\leq$0.25 g kg^{-1} body weight ($\leq$19 g or $\sim$five cloves for an adult weighing 150 lbs or 75 kg) are without harmful effects on the liver. They also note that for doses as low as 0.1 g kg^{-1} body weight, given daily for 28 days, there was significant deterioration in liver function tests (Rana, 2006). With rats given 2 g kg^{-1} fresh garlic, 55%

mortality was observed after 15 days (Banerjee, 2002a). It should be noted that the average consumption of garlic, from all sources, in the United States is 4.2 grams per day, or slightly more than one clove (Lucier, 2000). Allicin administered to four rabbits intraveneously as a 0.5% solution at 1 mL min^{-1} caused immediate constriction of the pupils of the eyes, darkening of the blood, salivation and marked peripheral vasodilation. Death occurred at levels averaging 44 mg kg^{-1} (Sterling-Winthrop, 1984).

It is well known that garlic can cause breath and skin odor as well as heartburn and flatulence. The difficulties some individuals have with severe halitosis and body odor following consumption of garlic, or even allergies to garlic and other alliums, may be associated with impaired ability to oxidize sulfides to sulfoxides through the cytochrome systems (Scadding, 1988; Harris, 1986a,b). Consumption of raw garlic in quantities greater than one or two cloves at a time can irritate the lining of the digestive tract (Koch and Lawson, 1996). Taking garlic with food may reduce the chance of stomach irritation. Shallots and other alliums are rich in polyphenolic compounds which can inhibit iron absorption by forming iron complexes. *In vitro* studies showed that these alliums reduced iron availability in a dose-dependent manner (Tuntipopipat, 2008).

5.12.2 Garlic Consumption by Pregnant or Lactating Women: Infant Bad Breath

A garlic odor has been detected in the amniotic fluid of pregnant women who have ingested garlic (Mennella, 1995). Furthermore, garlic has been detected on the breath of newborns whose mothers had garlic-flavored meals prior to childbirth (Snell, 1973). In a study investigating the effects of garlic ingestion by a mother on the odor of her breast milk and the suckling behavior of her infant it was found that: (1) evaluation of the milk samples by a sensory panel revealed garlic ingestion significantly and consistently increased the perceived intensity of the milk odor; this increase in odor intensity was not apparent one hour after ingestion, peaked in strength two hours after ingestion, and decreased thereafter; (2) the nurslings detected these changes in mother's milk and remained attached to the breast for longer periods of time, sucked more when the milk smelled like garlic, and ingested more milk as well; (3) the infants who had no exposure to garlic volatiles in their mother's milk during the experimental period, spent significantly more time breast feeding after their mothers ingested garlic capsules compared to those infants whose mothers repeatedly consumed garlic during the experimental period. These results demonstrated an effect of prior experience with garlic in mother's milk. (Mennella, 1991, 1993).

Several studies have examined the possible use of garlic or garlic supplements to prevent pre-eclampsia (pregnancy-induced hypertension). Administration of a garlic supplement during the third trimester of pregnancy was found to be ineffective in preventing pre-eclampsia (Ziaei, 2001). A thorough review of the literature concludes that "there is insufficient evidence to recommend increased garlic intake for preventing pre-eclampsia and its complications" (Meher, 2006). While the above studies suggest that it is not necessary to avoid fresh garlic during pregnancy or lactation, there are no data on the safety of commercial garlic supplements for pregnant or lactating women.

5.12.3 How do Onions cause Gastric Reflux and Heartburn?

Raw onions can make the esophagus more acidic causing heartburn and acid reflux (Allen, 1990). Hypersensitive individuals can suffer noncardiac chest pain and spasms of the esophagus in as little as one minute after consuming green peppers or raw onions. In the latter case the symptoms, relieved by inhalation of the vasodilator amyl nitrite, were attributed to "certain ingredients in the ingested food" (Bajaj, 2004). Prostaglandins, which can regulate the contraction and relaxation of the lower esophagal sphinter muscle, are found in onions (Ali, 1990; Al-Nagdy, 1986). In addition, sulfur compounds from onions may inhibit prostaglandin activity (Block, 1992). Heartburn and gastric reflux occur when the sphinter muscle inappropriately relaxes during digestion allowing small amount of stomach acid to enter the esophagus.

5.12.4 *Allium*-Linked Botulism and Hepatitus

Several cases of botulism associated with consumption of garlic-in-oil, as well as onion products, have been reported (Morse, 1990; MacDonald, 1985). Since garlic is a bulb, spores of the soil-dwelling organism *Clostridium botulinum,* which can cause botulism, may be present. Garlic in olive oil in a sealed container at room temperature provides the perfect anaerobic growing conditions for this microorganism. Since the microorganism is sensitive to acid, the FDA recommends soaking garlic cloves for 24 hours in an acidic solution (citric or phosphoric acid, vinegar or wine). Thus treated, the drained garlic cloves, if covered with an edible oil and refrigerated, will keep for 6 to 10 months.

Another health-related problem associated with food use of alliums is hepatitis A, following consumption of uncooked, chopped green onions. A number of incidents in the United States, involving hundreds of infected individuals and several deaths, have been reported in which green onions imported from Mexico were identified as the most likely source of restaurant-associated hepatitis A (Wheeler, 2005; Dentinger, 2001). Green onions require extensive handling during harvesting and preparation for packing, including removal of outer skins and soil. Workers infected with hepatitis A virus could contaminate the onions during this process. Alternatively, use of contaminated water during irrigation, rinsing, processing, cooling and icing could expose green onions to the hepatitis virus (or to bacteria such as *E. coli*, which have also been involved in cases of food-related sickness from consuming tainted green onions). Viral or fecal particles are especially difficult to wash off layered onion surfaces, which can trap particles within the layers. While the hepatitis A virus cannot multiply in foods, it persists because of its adsorption capacity on the surface of the green onions. Adsorption protects so efficiently that washing, chlorination or even refrigeration does not affect the survival of several types of viruses in green onions. Hence, instead of cleaning the green onions by rinsing them with plain water, it is recommended that immediately before consumption they should be washed, and the outer layer subjected to another peeling (the first layer had been removed during harvesting). After that, the green onions should be slowly washed in chlorinated water and then rinsed with plain water, at which point they are ready for consumption (Vale, 2005).

5.12.5 Choking on Cloves

An unusual injury associated with consumption of raw garlic was reported in a medical journal (Kim, 2008c). A 60-year-old woman in Korea complained about severe and sustained chest pain after

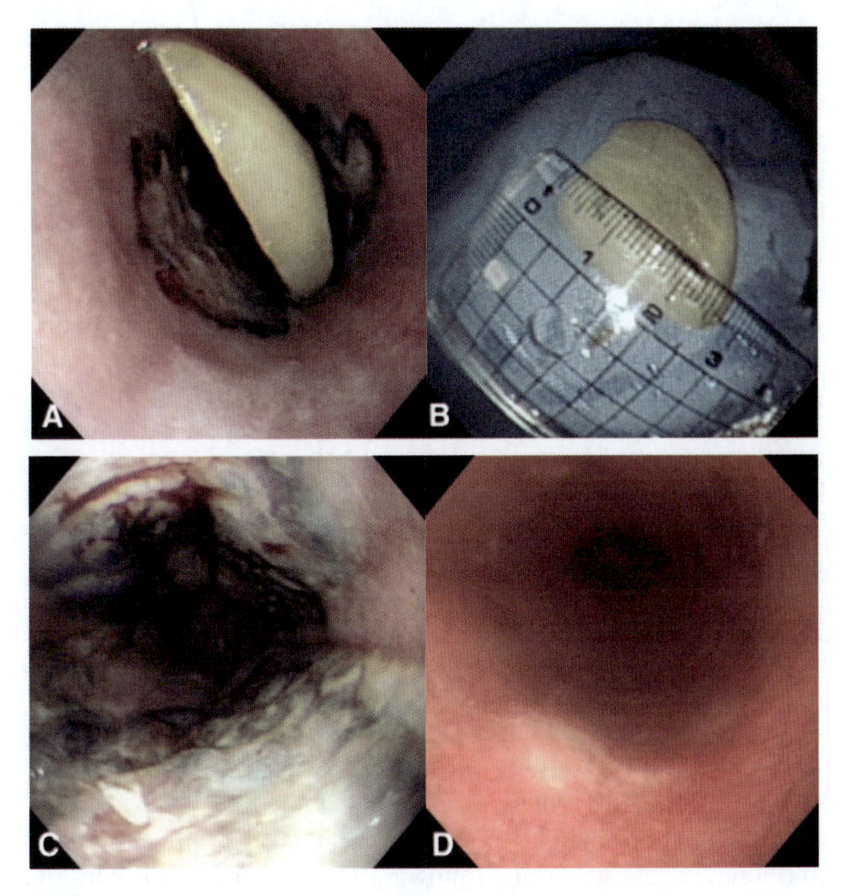

Figure 5.9 Garlic clove shown lodged in esophagus (**A**), after removal (**B**), esophageal injury (**C**), fully healed esophagus (**D**). (Image from Kim, 2008c).

eating sliced raw fish served with garlic. Endoscopic examination revealed the presence of an 2.7 cm long intact clove of garlic at mid-esophagus (Figure 5.9). The impacted clove was removed using retrieval forceps. After removal, a 4 cm segment of "whitish and bluish bullous necrotic change was noted at the site where the garlic had been impacted; the esophagus was otherwise normal." After three days of hospitalization, endoscopic examination revealed that the patient's esophagus had fully healed. Based on this incident, as well as on a report that "undamaged garlic (swallowed) had no lowering effect on lipid level of serum" (Jabbari, 2005), it would be unwise to swallow an intact clove as a possible means of realizing the health benefits of fresh garlic, as some recommend.

5.12.6 *Allium* Allergies and Contact Dermatitis

Contact dermatitis, allergic asthma and rhinitis (inflammation of the mucous membrane in the nose) from garlic and garlic dust among cooks and garlic workers, first reported in 1950 (Edelstein, 1950), is now well known (Falleroni, 1981; Lybarger, 1982; Añibarro, 1997; Jappe, 1999; Pires, 2002; Bassioukas, 2004; Hubbard, 2005). Garlic-sensitive patients showed positive tests to diallyl disulfide, allyl propyl disulfide, 2-propenethiol and allicin (Hubbard, 2005; Pappageorgiou, 1983). Patch testing with 1% diallyl disulfide in petrolatum is recommended when garlic allergy is suspected (Delaney, 1996). It is also recommended that diallyl disulfide be included in photopatch-test series to identify photocontact allergy to diallyl disulfide, even though it is rare (Alvarez, 2003). While gloves might offer a solution for allergic individuals handling garlic, it has been found that diallyl disulfide penetrates most commercially available glove types (Moyle, 2004). Furthermore, it is impractical for chefs to wear gloves. Some of the chefs with garlic allergies, who had to continue chopping garlic, responded well to Acitretin or topical psoralen-ultraviolet A after 12 weeks (Hubbard, 2005).

Cases of food allergies involving garlic (Perez-Pimiento, 1999) and anaphylactic reactions following ingestion of raw onion and garlic have been reported, although they are relatively rare (Valdivieso, 1994; Arena, 2000; Pérez-Calderón, 2002, Moyle, 2004). Alliinase from garlic (alliin lyase) was found to be a major allergen in a garlic-allergic group of patients in Taiwan. Skin tests showed that the purified protein elicited IgE-mediated hypersensitive responses in patients with garlic allergy. Carbohydrate groups are involved in the antigenicity, allergenicity and cross-reactivity. Garlic alliinase showed strong cross-reactivity with alliinases from other *Allium* species, namely leek, shallot and onion (Kao, 2004).

5.13 DON'T FEED YOUR PET ONION OR GARLIC!

Onions, whether fresh, cooked or dehydrated, similar forms of garlic, leek and Chinese chives, as well as wild alliums, including

A. canadense, *A. cernuum*, *A. validum* and *A. vineale*, are toxic to dogs, cats, monkeys, sheep, cattle, water buffalo, horses, pigs and other animals, particularly small or young animals, where the amount of *Allium* consumed could be large relative to their size (Cope, 2005; Borelli, 2009). The association of canine hemolytic anemia in dogs with the ingestion of onions has been known since 1930 (Spice, 1976). Similar results have been reported for cats fed baby food containing onion powder (Robertson, 1998) and horses fed freeze-dried garlic (Pearson, 2005). Unfortunately, several books suggest feeding garlic or onion to pets for treating or preventing intestinal worms – advice which could prove fatal to the pet!

A veterinary case report is informative. A four-year-old male Yorshire Terrier developed polyuria and depression, 36 hours after having been fed one-tenth of an onion soufflé that was made with one cup of dehydrated onion. The dog passed very dark urine and was diagnosed with hemolytic anemia. The dog fully recovered with conservative treatment after two weeks. The author notes that the toxic dose of onions for dogs is an amount equal to or greater than 0.5% of the animal's weight (Spice, 1976). Dog breeds such as Akitas and Shibas are especially susceptible to *Allium* poisoning. In adult dogs intentionally fed onions, a series of biochemical changes in plasma constituents could be followed, including a decrease in the level of reduced glutathione, as well as a decrease in erythrocyte membrane fluidity (Tang, 2008). A specific diagnosis of *Allium* poisoning in dogs is appearance in the blood of eccentrocytes, red cells with a ragged appearing, poorly hemoglobinized fringe of cytoplasm along one side of the cell (Yamato, 2005; Lee, 2000).

The situation with livestock is more variable. Onion poisoning can occur in dairy cows (Carbery, 1999). It is common practice in onion-growing regions around the world to feed cull onions to livestock. One report indicates that sheep and goats are able to tolerate onions in their diet due to the ability of the animal's rumen microflora to rapidly change to a population of organisms capable of reducing the sulfur compounds. Pregnant ewes are able to eat a diet of 90 to 100% cull (waste) onions without developing severe anemia, while up to 25 percent dry matter of onions can be successfully fed to cattle in a balanced ration (Knight, 2000). On the other hand, a more recent report describes severe onion toxicosis in the case of sheep which had grazed daily on onion fields for one month. Two ewes died, four aborted and many showed clinical signs of toxicosis, including pale mucous membranes, weakness and poor appetite. Hematology of 12 ewes revealed low erythrocyte number and hemoglobin, Heinz-bodies in erythrocytes, polychromasia and mild leukocytosis (Aslani, 2005). As can be seen, animals given free access to onions with other feed sources may prefer the onions and ingest toxic quantities. The severity of the toxicosis depends on the animal species and the quantity and variety of onions ingested, since mild varieties of onions contain lower levels of the toxic compounds than more pungent varieties (Parton, 2000).

The specific effect of consumption of onions and other alliums on animals, is to cause methaemoglobinaenia and hemolytic anemia with Heinz-body formation, that is, destruction of red blood cells, which become rigid and rupture, liberating hemoglobin into the urine and coloring it dark red-brown (hemoglobinuria). The animals act dull and depressed, and often their breath, urine or milk (if lactating) has the odor of alliums. The onion flavor disappears from the milk after the lactating animal has had onions removed from its diet for 24 hours. The toxic effects of the alliums are associated with oxidative damage to animal erythrocytes, due to compounds such as 2-propenylthiosulfate, $CH_2{=}CHCH_2SO_2SNa$, and related thiosulfates (Yang, 2003; Yamato, 1998, 1999, 2003; Hu, 2002). In particular, oxidation of exposed sulfhydryl (SH) groups on hemoglobin causes formation of disulfide bonds, distorting the tertiary structure of the hemoglobin molecule, resulting in precipitation of hemoglobin that coalesces, forming Heinz bodies. Various factors, other than *Allium* ingestion, can cause an increase in levels of Heinz bodies in blood, including diabetes, lymphoma and hyperthyroidism. In animals anemic from *Allium* poisoning, *Allium* feeding should be discontinued. Blood transfusions may be necessary in severely anemic animals.

5.14 ADVERSE EFFECTS AND HERB–DRUG INTERACTIONS FROM MEDICINAL USE OF GARLIC

There is a growing concern in the medical community about concomitant use of botanical supplements and conventional medications (Eisenberg, 1998; Ang-Lee, 2001; Gurley, 2005). Surveys indicate that 24 to 36% of all consumers use botanical supplements on a regular basis and that herbal supplement usage is prevalent among those taking prescription medications, many of whom are 65 years of age or older. Since those over 65 constitute the largest demographic with regard to prescription medications, they may also exhibit an increased risk for herb–drug interactions. The practice of self-medicating with botanical supplements is common among those over 65. Garlic is one of the products most frequently consumed.

5.14.1 Burns from Medicinal Use of Garlic

Burns associated with application of garlic to the body, mostly for treatment of various ailments, are the subject of a number of medical reports. It is well known that raw garlic along with its constituent allicin, is a potent skin irritant, particularly for young children (Garty, 1993; Parish, 1987). Topical burns, including blistering lesions, have been reported on children as young as three months following application of garlic to the skin (Rafaat, 2000). Other reports discuss contact dermatitis following rubbing the cut end of a fresh garlic clove on the skin; the practice of direct application of fresh garlic onto the skin for treating infections or other conditions should be discouraged (Lee, 1991). A recent book on garlic notes the following home remedy for corns: "Cut a sliver of garlic the same size as the corn, put it on top of the corn and hold it in place with a plaster or bandage. Replace the garlic sliver every day until the corn drops off" (Holder, 1996). What is suggested is risky and liable to lead to skin irritation more painful than the corn being treated (see below)!

There is a significant literature on self-inflicted garlic burns, for example as a way to get out of military service (Friedman, 2006). Medical papers have considered the etiology and management of the burn as well as the associated pathology that led to the treatment (Al-Qattan, 2008; Borrelli, 2007). One of the papers discussed three cases, both female and male, involving burns to the upper arm where the burn was less than 0.5% of total body surface. "Naturopathic physicians" had treated all three individuals for localized pain. Treatment involved application of fresh garlic under a firm bandage to the site of pain. It was stated that "all of the patients felt a burning sensation in the area but they initially did not remove the bandage when the burning sensation became severe. The presentation was erythema and small blisters in all cases." (Figure 5.10). In all three cases, the patients were treated with topical antibiotics ointments and the burns healed within ten days.

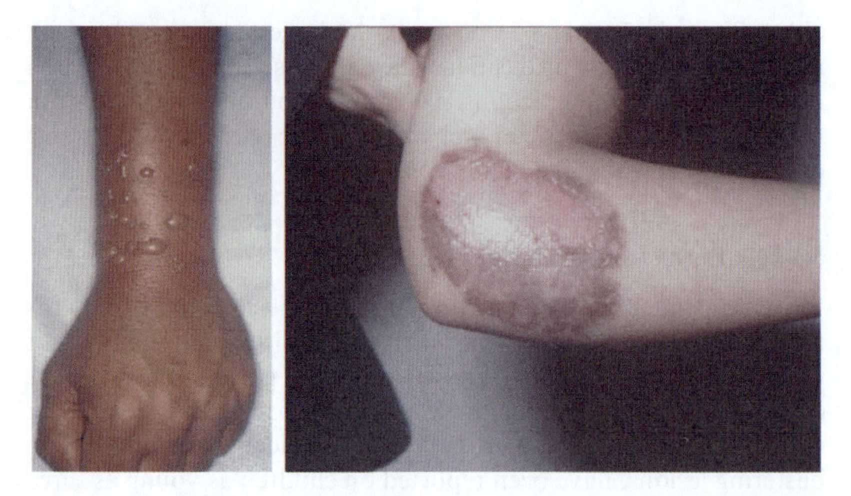

Figure 5.10 Right: a garlic burn to the lateral aspect of the elbow. Left: a garlic burn to the dorsal aspect of the wrist. (Image from Al-Qattan, 2008).

The type of treatment described previously is said to be a common practice of naturopathic physicians worldwide and is traditional in Arab medicine in the Middle East. As already noted, both diallyl disulfide and allicin are allergins and irritants, and can cause dermatitis. Occasionally, a deeper "chemical" burn may develop with long exposure periods. All of the above cases of garlic burns occurred when raw garlic was kept in contact with the skin for several hours. When such a burn is examined by a physician, it is recommended that the reason for garlic application should be investigated, as there is often an underlying pathology for the local pain. Reasons for topical application of garlic included fever, an acute asthmatic attack or localized pain. In the cases reported, garlic burns were sustained on the feet, wrists, forehead, legs, breasts and upper abdomen (Dietz, 2004; Baruchin, 2001; Roberge, 1997; Farrell, 1996; Canduela, 1995; Garty, 1993). Figure 5.11 shows burns to the feet of a 60-year-old man who applied crushed raw garlic to both feet and covered the garlic with bandages for 12 h (Dietz, 2004).

A case report (Ekeowa-Anderson, 2007) describes a 40-year old telephone engineer with a five-month history of worsening cheilitis (Figure 5.12), who over nine-months added several cloves of raw, chopped garlic to his meals because of its reported beneficial

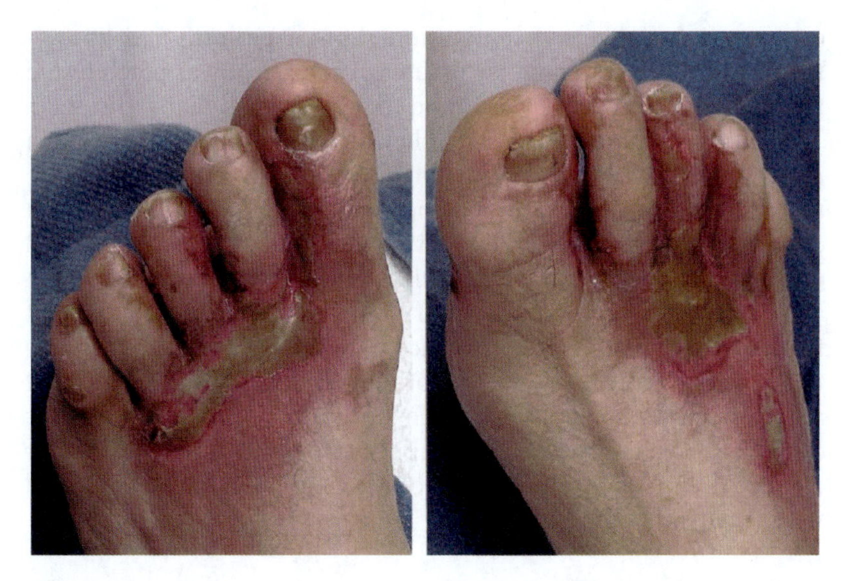

Figure 5.11 Garlic burns to the feet. (Image from Dietz, 2004).

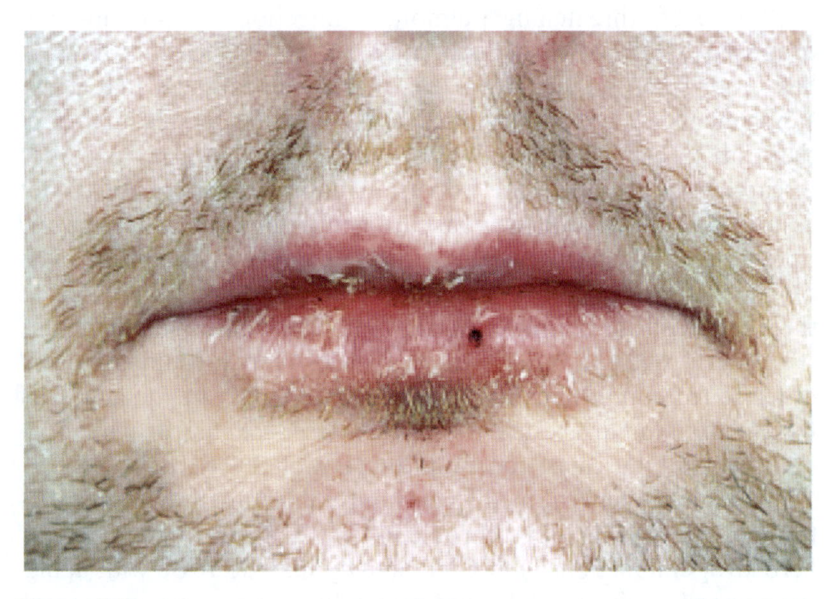

Figure 5.12 Lips with erythema marked scale and angular fissures. (Image from Ekeowa-Anderson, 2007).

cardiovascular effects. Examination showed perioral and lip erythema and scale with angular fissures. The patient was patch tested for contact allergies and was found to be allergic to diallyl disulfide. The patient was advised to shop chewing garlic, which resulted in a rapid and sustained resolution of his symptoms. In another case report, an individual sustained chemical burns of the oral mucosa caused by crushed garlic. To relieve a toothache, the patient placed crushed garlic cloves in the cheek ("buccal vestibule") overnight and developed garlic burn injury, including ulceration in that region (Bagga, 2008). The incidence of painful garlic burns is likely to increase if individuals inappropriately follow the suggestion that allicin from fresh garlic might constitute a useful treatment for canker sores (Jiang, 2008), or employ a risky "home remedy" for earache involving pinching a garlic clove and plugging the ear canal with it. Ouch!

5.14.2 Garlic–Drug Interactions

The confusing situation facing patients is well illustrated by the case of garlic and AIDS. On the one hand, it has been reported that garlic extracts inhibit the growth of *Mycobacterium avium*, which can cause life-threatening pulmonary infection in immuno-compromised AIDS patients (Deshpande 1993). On the other hand, it has more recently been reported by NIH researchers that garlic supplements ("caplets") significantly reduce the bioavailability of the anti-HIV protease inhibitor drug, saquinavir. The NIH researchers conclude that "clinicians and patients should not assume that dietary supplements are benign therapies – [they may] have potent pharmacological actions and may alter the blood levels of concomitant medications... Patients receiving saquinavir as their sole protease inhibitor should avoid using garlic supplements" (Piscitelli, 2002). In addition, garlic-derived thiosulfinates and other sulfur compounds may possibly interfere with the action of thiol-containing drugs such as Plavix®. Users of such drugs should consider limiting garlic consumption (Badol, 2007).

Since phytochemical-mediated changes in human drug metabolizing enzymes such as cytochrome P-450 (CYP) may underlie many herb–drug interactions, and CYP activity may decrease with aging, the effect of garlic oil (GO) supplementation on cytochrome activity was investigated. The study involved 12 healthy volunteers between the ages of 60 and 76 (mean age of 67 years). GO (characterized by GC-MS; 500 mg, 3× daily for 28 days, followed by a 30-day washout period) was found to inhibit CYP2E1 activity by 22%, compared to 40% inhibition in a separate study with young volunteers. The authors conclude that, based on their study, GO supplements pose a minimal risk for CYP-mediated, herb–drug interactions (Gurley, 2005).

5.14.3 Effect of Garlic on Platelets and Blood Clotting Processes

Excessive ingestion of garlic (four cloves daily) by an 87-year-old man has been reported to cause spontaneous spinal epidural hematoma, causing paraplegia secondary to a qualitative platelet disorder (Rose, 1990). While this anecdotal report has been used to suggest that there is an increased risk of bleeding associated with the use of garlic and garlic supplements, it has already been indicated in Section 5.7.4 that, apart from a possible synergistic effect of garlic and aspirin, clinical trials have failed to substantiate this claim. None-the-less, concerns continue to be raised that the use of garlic preparations increases the risk of bleeding during surgical procedures and with the use of anticoagulants, such as warfarin (Ang-Lee, 2001; Pribitkin, 2001; Heck, 2000; Argento, 2000; Evans, 2000; Burnham, 1995).

5.15 OVERVIEW OF THE CLINICAL EFFECTIVENESS OF GARLIC

The plural of anecdote is not data.

Roger Brinner

Garlic teases us. Garlic-derived compounds display remarkable *in vitro* activity and simultaneously show low toxicity to mammalian cells. *In vivo* testing of garlic preparations and garlic-derived compounds in rats and other laboratory animals reinforces the *in vitro* studies. Testimony on the medicinal benefits of garlic, much of it anecdotal, coming not only from our grandparents but also from great sages of ancient civilizations, is reinforced by a well-orchestrated advertising campaign by purveyors of garlic supplements, taking advantage of the relaxed regulatory atmosphere toward dietary supplements. To persuade us of the "miracle of garlic," supplement manufacturers make reference to the enormous number of published scientific studies on garlic and remind us that garlic must be safe, because it is not only natural, but edible. Of course, the large number of publications on the biological activity and health benefits of garlic should not be used to infer the extent to which the health benefits of garlic are valid, and may simply reflect the relative ease with which such research can be performed. Furthermore, the significance of the "placebo effect" should not be discounted in clinical and traditional uses of garlic.

More sobering, however, is a recent review of clinical trials on garlic, part of a 2007 issue of the journal *Molecular Nutrition and Food Research* devoted to the health effects of garlic, which comes to the conclusion that evidence for the efficacy of garlic in treatment of diseases is not convincing. In particular, the review states that "for hypercholesterolemia, the reported effects are small and may therefore not be of clinical relevance. For reducing blood pressure, few studies are available and the reported effects are too small to be clinically meaningful. For all other conditions, not enough data are available for clinical recommendations" (Pittler, 2007). Other reviews have voiced similar concerns (Tattelman, 2005).

What are we to believe? Shouldn't we accept only double-blind randomized placebo-controlled clinical trials to confirm hypotheses concerning clinical efficacy? Shouldn't clinical trials on dietary supplements involve supplements that are well characterized and standardized for content of the component(s) believed to be active? Shouldn't we require that the number of trial participants be large enough for meaningful statistical analysis, that the duration of the trial be long enough for an effect to be seen, that the dosage of the supplement be representative of labelling claims, that the participants' health status be appropriate for the hypothesis being tested, and that, to avoid possible conflicts of interest, clinical trials be funded independently of supplement manufacturers? Do we agree that dietary supplements should be held to the same high standards of proven efficacy and safety that apply to pharmaceuticals?

Glutathione and other biological thiols, abundant in our bodies, readily deactivate garlic-derived compounds, accounting for their low toxicity. A consequence of this easy deactivation is that ingestion of garlic leads to its rapid metabolic degradation. Thus, many *in vitro* active compounds are destroyed *in vivo*. The exceptions would be the effect of garlic on the oral cavity, the digestive tract and stomach, and topical (and gastroscopic!) applications, where allicin and other thiosulfinates, as well as allyl polysulfides, should survive long enough to be active. Future research may allow the active compounds in garlic and other alliums to be "packaged" in such a way that they can be delivered directly to a site within the body, bypassing metabolic degradation. In addition, it would be worthwhile to carefully scrutinize the clinical uses of garlic preparations in China as antibiotics, ideally through studies involving Western collaborators, since the information on the results of these treatments is tantalizing, even if not rigorously supported by data.

Finally, those choosing to use "home remedies" employing fresh garlic should fully recognize the risks, graphically described earlier in this chapter. Allicin from freshly crushed garlic is powerful stuff and can easily cause irritation and burns to sensitive parts of the body or elicit an allergic response. It is particularly unwise to use garlic remedies on young children or small pets. In addition, since compounds found in garlic can interfere with the action of drugs, it is important to fully discuss dietary supplement usage with health care providers, assuming the latter have also adequately educated themselves about dietary supplements.

CHAPTER 6

Alliums in the Environment: Allelopathy and *Allium*-Derived Attractants, Antibiotics, Herbicides, Pesticides and Repellents

A wild Rose-tree from the hedge brought he,
And planted it well in the mould,
Digging around, and making a mound,
To stand it up high, and bold.
Then a hole he made, at its back in the shade,
And an Onion deep tilled in;
For the Onion was bound to make roses sound,
And a fine rich perfume to win.

Mrs. Thomas Hardy

Plants live by their chemical wits.

Richard Schultes

Garlic and Other Alliums: The Lore and the Science
By Eric Block

Published by the Royal Society of Chemistry, www.rsc.org

6.1 IT'S A JUNGLE OUT THERE!

Almost since their first appearance on Earth, the 300 000 or so known species of plants have been under attack by the 400 000 or so known species of insects as well as by foraging animals. With insects consuming 10 to 15% of plant growth, plants have of necessity evolved ingenious defensive strategies to ward off herbivorous predators (Schultz, 2002). While some of these strategies are mechanical (*i.e.*, thorns), other strategies employ a range of chemical substances known as *secondary metabolites*. In contrast to *primary metabolites*, which are associated with essential functions, *secondary metabolites* are specialized substances considered to be dispensable, although their influence on the environment of the plant can be profound, allowing the plant to survive and thrive. An indication of the importance of secondary metabolites is the observation that 15 to 25% of each plant genome is associated with the metabolites (Pichersky, 2000) and that 800 to 1500 differentially regulated genes are involved in a plant's response to a single insect. It is said that "plants are not static, chemically defended fortresses – they respond to attack with rapid, long-lasting, variable, and often specific biochemical, physiological and developmental changes" (Schultz, 2002).

In this chapter we consider the environmental influence of *Allium* secondary metabolites. The same secondary metabolites can serve both as attractants for pollinators, seed-dispersal agents and entomophagous insects (*e.g.*, insectivorous insects feeding on other insects) as well as provide protection against pathogens, pests and other forms of environmental stress (Durenkamp, 2004). Secondary metabolites are abundant in roots and bulbs where they can act on soil-borne organisms or be exuded into the soil surrounding the root (rhizosphere) where they can contribute to the competitiveness of the species, *e.g.*, by acting as germination inhibitors for plant competitors. Some insect pests have adapted so that these same compounds attract them, for example larvae of the onion fly (*Delia antiqua*) are attracted by onion oil (Soni, 1979).

The foliage of living or withering plants can release volatile secondary metabolites that repel insects and animals or prevent growth of other plants. The term *allelopathy* (from the Greek words "allelon," meaning mutual, and "pathos," meaning harm or affection) refers to the effect of one plant (including microorganisms) on the growth of neighboring plant species, competing for the same growing space, by the release of chemical compounds into the environment. Both beneficial as well as inhibitory effects are included in this definition. We will focus primarily on the latter types of effect, involving phytotoxic biomolecules. As suggested above, all plant parts, including roots and bulbs, stems, leaves, flowers, fruits and seeds, can release chemicals into the environment, while plant residues can release chemicals following microbial degradation (Zeng, 2008). This release can involve volatile compounds from foliage as well as soil-borne agents from roots (Field, 2006). While the term allelopathy was first coined in 1937, ancient civilizations, including that of China, are known to have observed and recorded the phenomenon of plant chemical interference more than 2000 years ago, applying the knowledge to increase crop yield and to control weeds, disease and pests (Zeng, 2008).

Plant-released toxic molecules specifically targeting microorganisms are termed *phytoalexins* or *phytoanticipins*. Phytoalexins are low molecular weight, antimicrobial compounds that are both synthesized by and accumulated in plants *after* attack by microorganisms. Phytoanticipins are "low molecular weight, antimicrobial compounds that are present in plants *before* challenge by microorganisms or are produced after infection solely from preexisting constituents" (VanEtten, 1994). By the above definition, allicin and other thiosulfinates are phytoanticipins, protecting *Allium* species against plant pathogens, including bacteria, fungi and oomycetes (water molds; Curtis, 2004; Slusarenko, 2008).

6.2 HERBICIDAL AND PESTICIDAL ACTIVITY OF *ALLIUM* PLANTS IN THEIR NATURAL ENVIRONMENT

Defense-related secondary metabolites are commonly stored, for example, in vacuoles, as inactive precursors and then converted into biologically active agents by plant enzymes in response to challenge. Compartmentalization breakdown occurs upon tissue damage caused by pathogen or pest attack, often resulting in a cascade of chemical reactions releasing a burst of key defensive compounds. These defensive compounds can produce pain and inflammation in animal predators by activating TRPA1, an excitatory ion channel on primary sensory neurons of the pain pathway, to achieve chemical deterrence (see Section 3.5; Bautista, 2005).

Control of weeds and pests in agriculture is almost completely dependent on commercial herbicides and pesticides, many of which are toxic and harmful to the environment. The use of allelopathic and defensive compounds from plants represents a promising alternative to the agricultural chemicals presently in use. Selected examples of the natural defensive systems of *Allium* plants are presented below together with examples illustrating recent very promising agricultural and horticultural use of these systems.

6.2.1 Learning from Leeks

Great fleas have little fleas upon their backs to bite 'em,
And little fleas have lesser fleas, and so ad infinitum.

A. De Morgan, *A Budget of Paradoxes* (1872)

Auger, Thibout and coworkers in France have conducted a fascinating study of the chemical response of leek (*A. porrum*) to herbivore attack (Auger, 1979, 1989a,b; Dugravot, 2004–2006; Ferary, 1996; Thibout, 1997). The main pest for field-produced leeks is the specialist leek moth *Acrolepiopsis assectella*, which is attracted by the defensive sulfur compounds produced when leek is damaged (USDA, 2004). The leek moth, in turn, is preyed on by the pupal form of *Diadromus pulchellus*, an insect-eating insect. The leek can also undergo mechanical damage, *e.g.*, during harvesting, or can be damaged by the generalist moth *Agrotis ipsilon*. In contrast to *A. assectella*, the generalist *A. ipsilon* is highly mobile with a tendency to consume only leaf tips (the site where the concentrations of secondary sulfur metabolites is lowest) and then leave the plant. To model herbivore damage, *A. assectella* leek moth caterpillars were placed on leek plants in a greenhouse, with both the number of caterpillars and duration of exposure varied to model intensity of attack. Mechanical wounding consisted of ten 1 cm incisions on leek plants twice a day, for eight days, on a new leaf every day. As summarized in Scheme 6.1, it was found that:

1. The concentrations of the precursor compound propiin under natural conditions were almost three times higher in

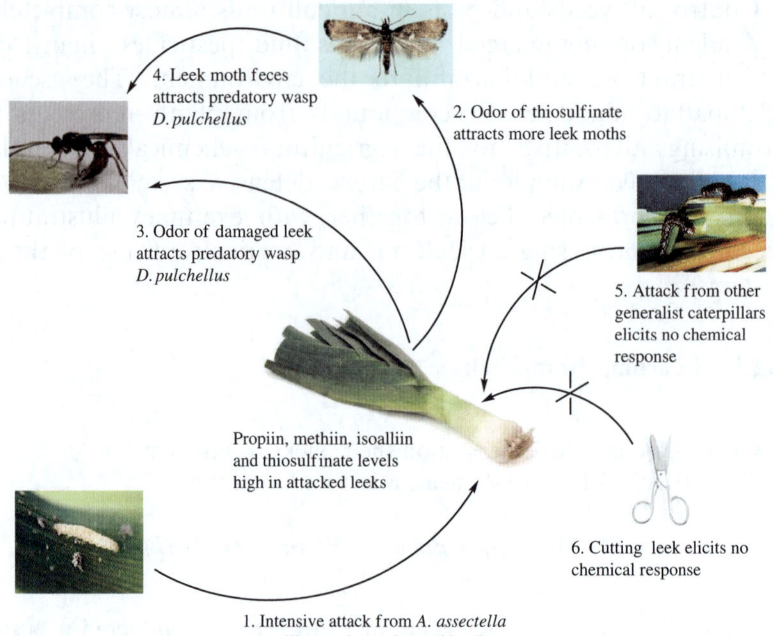

Scheme 6.1 Precursors propiin, methiin and isoalliin and thiosulfinate levels increase following attack by leek moth (*A. assectella*) larva, as it mines and bores through leek leaves and stems (1), but not following daily cutting (6) nor following attack by generalist caterpillar (5). Odor of leek moth-damaged leek attracts more leek moths (2). Odors of leek moth-damaged leek and leek moth feces attract predatory *D. pulchellus* wasp (3, 4), which parasitizes leek moth larva. (Insect images from Landry, 2007).

A. assectella-damaged leeks than in healthy leeks; similar increases were found in the concentrations of methiin and isoalliin on damaged leeks.

2. Cutting the leaves daily did not induce a significant change in propiin concentration; production of sulfur precursors occurs only if the stimulus from *A. assectella* caterpillar activity is sufficient, *e.g.*, leeks produce elevated levels of sulfur precursors only in cases of intensive attack.
3. Response induced in leek plants by *A. assectella* is systematic. Undamaged leaves from damaged plants showed higher concentrations of propiin than in leaves of undamaged plants, but significantly lower concentrations of

propiin than those in leaves directly damaged by caterpillars.

4. The concentration of propiin in undamaged leeks rose sharply immediately after the damage and remained elevated for at least one month. As a result, previously damaged plants can be better protected against herbivory and against the offspring of the insects that inflicted the first damage (*e.g.*, like an immune response).
5. Mean concentrations of propiin in leeks damaged by the generalist *A. ipsilon* were not significantly different from the same mean concentrations in undamaged leeks, suggesting that the elicitors of the induced response to *A. assectella* caterpillars in leeks are not present in *A. ipsilon* caterpillars.
6. Undamaged leeks gave off no detectable odor under the experimental conditions used, in contrast to leeks damaged mechanically or by *A. assectella*.
7. Direct analysis (using LC-APCI-MS) of odors (volatiles) from leek, garlic and onion reveal the presence of thiosulfinates (and the onion LF), but not disulfides or their rearrangement products. Similarly, ether extraction of damaged leeks followed by analysis established the presence of a series of mixed methyl-1-propenyl-*n*-propyl thiosulfinates and zwiebelanes, similar to what is observed on chopping the plant. Analysis of leek odor on a short GC column confirms that the main component is *n*-propyl *n*-propanethiosulfinate, the most attractive compound to the leek moth (Auger, 1989). Other studies confirm the gas-phase stability of thiosulfinates (Auger, 1990).
8. No immediate adverse effect on larval development of *A. assectella* is seen, but the duration of male development is significantly longer and an almost 20% decrease is observed in the number of oocytes that developed on attack-damaged leek plants produced by adult female *A. assectella*. Leek moths do not avoid damaged plants systematically.
9. The female of the parasitoid *D. pulchellus*, which feeds on *A. assectella*, is more strongly attracted to attacked leeks than to undamaged plants and mechanically damaged plants. In addition to leading to a greater release of thiosulfinates, the increase in the alk(en)ylcysteine precursors is probably accompanied by a greater release of disulfides

from *A. assectella* feces, and thus to further attraction of the parasitoid *D. pulchellus*. In summary, the concentration of propiin in leek plants increases in response to damage by the herbivore specialist *A. assectella*, and this increase is followed by a characteristic profile of sulfur volatile release.

6.2.2 An Assertive Plant: The Case of Bear's Garlic (*A. ursinum*)

Consider the fascinating example of the humble *A. ursinum*, also known as ramsoms, broad-leaved garlic or bear's garlic (presumably because of the fondness of bears for the bulbs of this plant). It forms dense populations in beech–fir forests in hollows and plateaus in areas as large as tens of hectares (Figure 6.1). Other forest species are either sparse or completely absent in populations where *A. ursinum* is dominant, pointing to inhibitory allelopathic effects. Free phenolic phytotoxins are abundant in the leaves of *A. ursinum*. At the beginning of summer the leaves, stems and inflorescences of *A. ursinum* wither and disintegrate, and the phenolics enter the surface soil layer, preventing germination of other other species (Djurdjevic, 2004).

At the time *A. ursinum* starts to wither away, the entire forest smells of the biogenic emissions of this wild garlic. In a fascinating study, ambient air sampling was performed in a forest one meter above ground and four meters from a stand of *A. ursinum*. The measurements were conducted at the fringe of the *A. ursinum* covered area in order not to damage plants (which could lead to false elevated levels). Air samples were collected using an "enclosure box" containing an air flow system using glass tubes filled with the absorbants Tenax TA and Carbotrap (carbon black).

Back in the laboratory the sample tubes were thermally desorbed and cryofocused for analysis by GC-MS. Thermal desorption was performed at 250 °C with sample trapping at –196 °C prior to introduction into the GC (Puxbaum, 1997; König, 1995). The following compounds were detected: allyl methyl sulfide, dimethyl disulfide, diallyl sulfide, methyl 1-propenyl disulfide, methyl *n*-propyl disulfide, bis(1-propenyl) disulfide, (*Z*)-1-propenyl *n*-propyl disulfide, and di-*n*-propyl disulfide. The predominant substance in these emissions was methyl 1-propenyl disulfide followed by bis(1-propenyl) disulfide. The calculated ratios of

Figure 6.1 *A. ursinum* growing wild in a forest. (Courtesy of University of Basel Botanical Database).

methyl : allyl : 1-propenyl : *n*-propyl groups in these volatiles of 28.6% : 11.1% : 49.6% : 10.7%, compare reasonably well with the measured methiin : alliin : isoalliin : propiin ratio of 35% : 28% : 37% : 0% in *A. ursinum* (Fritsch, 2006).

Several features of this work are noteworthy. First, in these air samples "classical" biogenic organic compounds such as isoprene and monoterpenes (*e.g.*, limonene) were present in very low concentrations or were undetectable. The emissions of C_2–C_6 organosulfur compounds from *A. ursinum* ranged from *ca.* 44 to 100 $\mu g\,S\,m^{-2}\,h^{-1}$ with disulfides making up 88% of the total sulfur emissions. The mean emission rate per unit of ground area per hour of *ca.* 62 $\mu g\,S\,m^{-2}\,h^{-1}$ was the highest ever reported for organic sulfur species emitted from a terrestrial plant.

While the results are impressive, given the very high temperature (250 °C) employed during thermal desorption into the GC, it is unclear if the compounds actually present in the forest air were only polysulfides or instead mixtures with thiosulfinate precursors. Thiosulfinates are known to be unstable in soil and would be expected to be unstable under conditions of thermal desorption. However, they are known to be stable in air (Auger, 1990; Arnault, 2004).

Detection of dimethylthiophene among the emissions directly captured in close proximity to *A. ursinum* suggests that the reported polysulfides may be, at least in part, artifacts of the analytical method since it is known that dimethylthiophene is a thermal decomposition product of bis(1-propenyl) disulfide, the major reported emission compound. The authors also comment on the unusual abundance of a compound they identify as 2-hexanal, stating: "the relatively high ambient mixture ratio of hexanal could be explained as coming from sources other than *A. ursinum* emissions." However, it is likely that 2-hexanal has been misidentified and that the actual compound present is in fact the isomeric 2-methyl-2-pentenal, a known condensation product of propanal, in turn formed from the onion LF, which might be expected to be produced by *A. ursinum* due to the significant concentration of isoalliin.

The organosulfur volatiles also inhibit seedling germination and growth of sensitive species. Insect visitors to *A. ursinum* are reported to be rare (Tutin, 1957). Further evidence for the proposal that the volatile emissions from *A. ursinum* are mainly thermally unstable thiosulfinates comes from analysis of volatiles from garlic, leek and onion using cryotrapping and direct injection of aqueous solutions of the samples into an HPLC-MS-APCI. These studies led to the conclusion that *no* disulfides or other rearrangement products were formed: *Allium* odors contain only thiopropanal

S-oxide and thiosulfinates as sulfur volatiles (Ferary, 1996). It has been shown that only extremely mild GC conditions can be used to separate thiosulfinates, and that even under the mildest of conditions many thiosulfinates do not survive analysis (Block, 1992).

An additional example of an allelopathic effect involves Chinese chive (*A. tuberosum*). Tomato plants were grown alone or intercropped with Chinese chive plants, with or without the inoculation of bacterial wilt agent *Pseudomonas solanacearum*. Chinese chive plants had no detrimental effects on the growth of tomato plants, but significantly delayed and suppressed the occurrence of bacterial wilt of tomato. The *P. solanacearum* population in bare soil was higher than that grown with Chinese chive; root exudates of Chinese chive collected with a continuously trapping system were inhibitory to multiplication of *P. solanacearum* (Yu, 1999). Examples of allelopathic effects involving onion and garlic are also known. At $<10\,g/350\,mL$ of ground onion bulbs in air, the onion bulbs stimulated cucumber germination in closed containers. At $>15\,g/350\,mL$ of onion, the germination was inhibited, and at 20 g/350 mL of onion in air, germination was completely suppressed. S- and O-containing compounds (in the case of the O-compounds probably lower aldehydes) were isolated from the onion volatile fractions (Rik, 1974).

Dried and milled residues of onion and garlic plants added to soil inhibited germination of seeds of several types of weedy annual plants; the herbistatic/herbicidal activity was more pronounced at higher soil temperatures (*e.g.*, 39 °C) and higher *Allium* residue concentrations. It is suggested that the germination-inhibiting activity of *Allium* residues could be exploited as a component of integrated weed management (Mallek, 2007). Research in Bangladesh funded by the U.S. Department of Agriculture shows that seeds treated with inexpensive garlic tablets dissolved in water show 95 to 100% germination, compared with 56 to 60% germination for untreated seeds (ABC News, Australia, 2005).

6.3 PESTICIDAL AND ANTIBIOTIC ACTIVITY AND INSECT REPELLENT PROPERTIES OF COMPOUNDS FROM *ALLIUM* PLANTS

Use of alliums as repellents has a long history, going back to the Egyptians who used onions to repel snakes (Manniche, 2006).

Roman naturalist Pliny the Elder wrote: "garlic has powerful properties . . . it keeps off serpents and scorpions by its smell." Household use of garlic as an insect repellent for preserving food grains has been described (Huang, 2000). Even today in Nigeria it is common practice to scatter garlic cloves around residential areas to drive away snakes (Oparaeke, 2007). Kipling bases a tale in his *Second Jungle Book* on the supposition that bees "hated the smell of wild garlic." With increasing legislative pressure upon chemical pesticides such as carbamates and organophosphorus compounds, it is timely to consider the scientific basis for the extensive lore on insect repellent (or biocidal and pesticidal) properties of *Allium* formulations.

6.3.1 Nematodes

Nematodes, microscopic roundworms typically inhabiting topsoil at levels in excess of one billion per acre (four out of five animals on the planet are nematodes), cause considerable economic loss, particularly to root crops, *e.g.*, as shown by occurrence of excessive root forking damage in carrots (Figure 6.2, left). Since several widely used nematicides have been removed from the marketplace due to high toxicity and environmental concerns, conditions are favorable for the introduction of effective plant-based nematicides (Block, 2008). Garlic extract has been reported to be repellent to the nematode *C. elegens* (Figure 6.2, right; Hilliard, 2004). Extracts of garlic have further been found to kill root knot nematodes *in vitro* within 40 minutes and considerably reduced the number of galls in potted plants inoculated with this nematode (Sukul, 1974).

The toxicity of aqueous and methanolic garlic extracts, steam-distilled garlic oil and synthetic diallyl disulfide (DADS) toward the mushroom and citrus nematodes *Aphelenchoides sacchari* (Hooper, 1958) and *Tylenchulus semipenetrans* (Cobb, 1913) has been demonstrated (Nath, 1982). The aqueous and methanolic extracts were diluted with water while the garlic oil and DADS were diluted with water containing 1% of an emulsifier (Triton X-100). The treatments involved 200 nematodes with four replicates; observations were recorded after 48 hours. While the aqueous and methanolic extracts completely killed *A. sacchari* and *T. semipenetrans* at relatively high concentrations of 0.5%, the garlic oil (GO) and synthetic DADS were much more toxic,

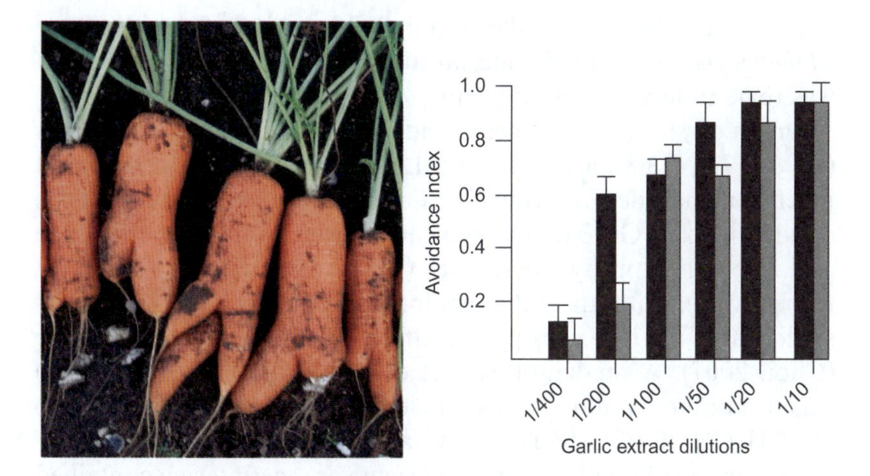

Figure 6.2 Left: Carrots showing nematode damage (*e.g.*, multiple roots) (Photo by Eric Block). Right: Bars represent the avoidance index of a population of 100 wild type (black bars) or mutant (grey bars) *C. elegens*, each tested with a single drop of garlic extract, at different dilutions, as repellent. (Image from Hilliard, 2004).

completely killing *A. sacchari* at 8 ppm (GO) and 25 ppm (DADS); similarly for *T. semipenetrans*, complete killing was observed at 16 ppm (GO) and 25 ppm (DADS). The authors state that the higher toxicity of GO compared to DADS suggest the presence of other constituents toxic to nematodes in the GO (Nath, 1982). Since all commercial DADS contains higher diallyl polysulfides, it is likely that the toxicity of pure DADS is lower than the values reported while the toxicity of the polysulfides is significantly higher than pure DADS.

In agreement with the above studies, of the various garlic oil components diallyl trisulfide (DAT) has been reported to be more potent than diallyl disulfide (DADS) toward the pine-wood nematode, *Bursaphelenchus xylophilus* (LC_{50} values of 2.79 *vs.* 37.06 $\mu L\ L^{-1}$ for DATS and DADS, respectively, toward juvenile nematodes) (Park, 2005). Diallyl sulfide (DAS) showed minimal activity in this assay. It is suggested from structure–activity studies that the higher molecular weight compounds such as DAT from garlic oil should be the most active (Park, 2005). Immersion of tomato roots in allicin solutions as a prophylactic measure was beset by problems of phytotoxicity and lack of nematotoxicity, but a 5 minute immersion in 25 $\mu g\ mL^{-1}$ allicin inhibited penetration of roots by juveniles by 50%

and was not phytotoxic. Allicin inhibited hatching of the nematode *Meloidogyne incognita* at concentrations as low as 0.5 mg mL^{-1} and was toxic to juveniles at 2.5 μg mL^{-1} (Gupta, 1993).

Onion oil showed good nematicidal activity against *B. xylophilus* (LC_{50} 17.6, 13.8, and 12.1 μg mL^{-1} against male, female and juvenile nematodes, respectively). Of the several components of the oil identified by GC-MS, the most active were di-*n*-propyl trisulfide and methyl *n*-propyl trisulfide (LC_{50} 5.0 and 22.9 μg mL^{-1}, respectively), with the other major components, di-*n*-propyl disulfide and methyl *n*-propyl disulfide, being only weakly active (Choi, 2007). A methanol extract of *A. grayi* (Japanese: nobiru) was found to contain a nematicidal thiosulfonate of structure (*E*)-$CH_3CH{=}CHSO_2SCH_3$, active against *M. incognita.* Similarly, the steam distillate of *Allium fistulosum* var. *caespitosum* contained nematicidal disulfides, trisulfides and a thiosulfonate. Related synthetic thiosulfonates and thiosulfinates had similar nematicidal and antibacterial activities (Tada, 1988).

The effect of garlic oil (GO) on the needle nematode *L. elongates*, the potato cyst nematode *G. pallida*, the stubby root nematode *P. anemones* and the root knot nematode *M. Hapla* was investigated. Garlic granules formulated from a combination of wood flour (55% wt/wt) and garlic juice concentrate (45% wt/wt) known to contain garlic oil polysulfides, were dissolved with water. All nematodes were killed after four hours exposure to the garlic granule supernatants at concentrations of 25 or 12 mg mL^{-1}. Repetition of this work using the garlic juice concentrate only showed 24 hour LD_{50} towards *L. elongates* at concentration levels of 0.025% wt/vol (Groom, 2004). In field trials involving carrot and parsnips, it was found that GO impregnated granules offered significant protection against damage from free-living nematodes (Figure 6.3; Groom, 2004, 2006) and that nematicidal effects attributable to the GO constituents in the granule can be achieved which are comparable to those of widely used carbamate and organophosphate nematicides.

The parasitic gastrointestinal nematode (GIN) *Haemonchus contortus* is considered the single most important pathogen in sheep and is a problem for goats as well. Goat kids naturally infected with GIN were administered fresh garlic bulbs, freshly squeezed garlic juice from the same number of garlic bulbs, a commercial garlic juice product or water (control). None of the garlic products

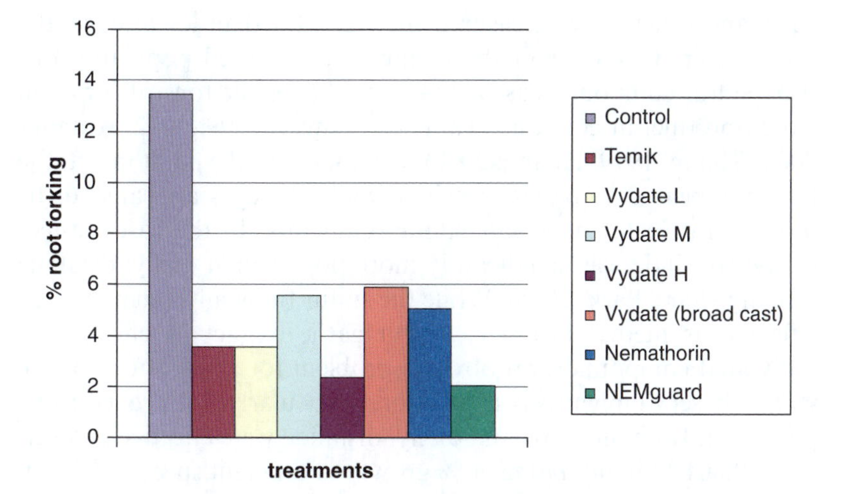

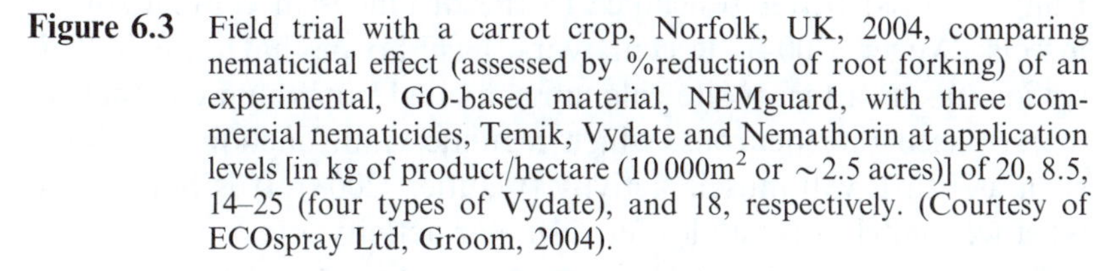

Figure 6.3 Field trial with a carrot crop, Norfolk, UK, 2004, comparing nematicidal effect (assessed by %reduction of root forking) of an experimental, GO-based material, NEMguard, with three commercial nematicides, Temik, Vydate and Nemathorin at application levels [in kg of product/hectare (10 000m^2 or ~2.5 acres)] of 20, 8.5, 14–25 (four types of Vydate), and 18, respectively. (Courtesy of ECOspray Ltd, Groom, 2004).

were effective in controlling the GIN compared to the control. The authors conclude that garlic "is not recommended as an aid to control GIN in goats or lambs" (Burke, 2009).

6.3.2 Coleoptera (beetles), Lepidoptera (moths, butterflies), Hemiptera (true bugs; includes aphids), and Diptera (true flies; includes mosquitoes), Hymenoptera (bees, wasps, ants) and Isoptera (termites)

Insect infestation, particularly from Coleoptera (beetles), Lepidoptera (moths) and Isoptera (termites), is the cause of major damage and food loss, particularly in tropical areas. While methyl bromide is very effective as a gaseous fumigant on insect eggs as well as on adults, its negative impact on stratospheric ozone led to its ban under the Montreal Protocol. Auger and coworkers in France suggest that *Allium* thiosulfinates and dimethyl and diallyl disulfides could be used as fumigants instead of methyl bromide (Auger, 1999).

They and others have measured the LD_{50} (median lethal dose: the dose required to kill half the members of a tested population) of these sulfur compounds as well as methyl bromide toward common stored-product insect pests (Table 6.1; Auger, 1999, 2002; Arnault, 2004; Chiam, 1999; Huang, 2000). Compared to the adult insect, the larvae appeared to be less sensitive while the eggs appeared to be more sensitive to the organosulfur fumigants. In the cases tested, diallyl trisulfide was significantly more potent than diallyl disulfide (Huang, 2000; Park, 2005). While the sulfur fumigants seem to be as effective as methyl bromide, and a garlic extract as effective as individual compounds, an obvious problem for applications of the *Allium*-based fumigants is their odor, particularly if it proved to be persistent. Both dimethyl and diallyl disulfide were effective at levels of *ca.* 0.001 M in inhibiting 90% growth of different species of fungi. Fungi showed better resistance to the disulfides in comparison to insects (Auger, 2004). Recent work has shown that dimethyl disulfide has as a target site calcium-activated potassium channels in insect pacemaker neurons, which regulate vital functions in insects such as heart wall muscle activity (Gautier, 2008). It is not known whether diallyl polysulfides have similar activity.

Extracts of garlic are toxic across all life stages to such pest species as Coleoptera (beetles), Lepidoptera (moths, butterflies), Heteroptera (true bugs) and Diptera (true flies, including mosquitoes). Thus, aqueous extracts of garlic inhibit egg hatch of mosquitoes (Jarial, 2001), and garlic extracts and steam distillates have toxic and antifeedant effects on stored product pests of species Coleoptera (Chiam, 1999) and Lepidoptera (Gurusubramanian, 1996) and repellent and toxic effects on Homoptera (Gurusubramanian, 1996). The toxicity of garlic extract toward the giant red bug (*Lohita grandis*) is due to the antibiotic action of garlic toward symbiotic intestinal bacteria, which provide the insect with essential cholesterol that the insect itself is unable to synthesize (Mandal, 1982).

A two-year study in Nigeria demonstrates the utility of a homemade garlic preparation for foliar application to cowpea (*Vigna unguiculata* (L.) Walp) to protect against *Maruca vitrata* (Lepidoptera), the legume pod borer moth, and *Claigralla tomentosicollis* (Hemiptera). Locally grown garlic was dried at 60 °C for 12 h, powdered and thoroughly mixed with water to make 5%, 10% and 20% solutions which, after standing overnight, were filtered,

Table 6.1 Comparative toxicities of *Allium* sulfur compounds and methyl bromide toward stored-product insect pests: LD_{50} in mg L^{-1} after 24 h exposure.

Insect	*MeS(O)SMe*	*AllS(O)SAll*	*MeSSMe*	*AllSSAll*	*Garlic extract*	*MeBr*
Ephestia kuehniella (Mediterranean flour moth)	0.04	0.02	0.2	0.02		–
Plodia interpunctella (Indian meal moth)	0.02					
Bruchidus atrolineatus (Cowpea bruchid beetle): adult	0.15	0.18	0.2	0.6		
B. atrolineatus: larvae L4			2.0			
Sitophilus granaries (Wheat weevil)	0.14					
Sitophilus oryzae (Rice weevil)	0.19		1.23			1.05
Callosbruchus maculates (Cowpea weevil): adult	0.25	0.16	1.1	0.5	0.37	
C. maculates: larvae L4			2.04			
C. maculates: eggs			0.17			
Dinarmus basalis (Solitary parasitic wasp): adult			0.31	0.35	0.14	
D. basalis: larvae L4				1.58		
Tineola bisselliella (Clothing moth)	Pr_2S_2: 1.26 Me_2Se_2: 0.002			0.013		
Tribolium castaneum (milled rice pest)[a]			4.32[c]	3.8 (0.83)[b]		
Sitophilus zeamais (maize pest)[a]			12.1[c]	20.0 (6.3)[b]		
Reticulitermes speratus (Japanese termite)[d]				<0.25 (~0.125)[d]		

[a]From Chiam (1999).
[b]In parentheses, for All_2S_3, from Huang (2000).
[c]For AllSSMe, from Huang (2000).
[d]In parentheses, for All_2S_3, from Park (2005).

mixed with starch and flaked bar soap and sprayed on cowpea six weeks after sowing, which corresponded to the onset of flowering.

A split plot design compared in triplicate the effect of garlic extract and a synthetic insecticide (sprayed four times weekly) with an untreated control, with subplots for garlic extract comprising 5, 10 and 20% concentrations with three different spraying regimes (two, four and six times weekly). The higher concentrations and more frequent applications of garlic extracts were superior to the lower concentrations and less frequent applications, and to the untreated control. The higher garlic concentrations were effective in reducing insect populations and were comparable in this regard to synthetic insecticides. While all garlic extract treatments were inferior to the synthetic insecticide treatment in both pod damage and grain-yield assessments in the two growing seasons, cost and environmental factors were thought to make the garlic extract a viable alternative to synthetic pesticides on cowpea fields managed by farmers in developing countries with limited resources (Oparaeke, 2007).

A study from Bangladesh comparing five botanical extracts – tobacco, neem, garlic, eucalyptus and mahogany – on aphid (*Aphis* spp.) populations on yard-long bean, *Vigna unguiculata*, showed that the garlic extract was less effective than both the tobacco (most effective) and neem extracts. None of the extracts were toxic to ladybird beetles (Bahar, 2007). Another study used a methanolic garlic extract on *Plutella xylostella* (cabbage moth). Contact toxicity bioassay results indicated 67% 48 hour mortality of first instar larvae with a 5.0% methanol extract (Samarasinghe, 2007).

The repellent and anti-feeding effect of garlic oil against the bite of female phlebotomine sandflies (*Phlebotomus papatasi*) was evaluated under laboratory conditions. Two different laboratory procedures were used: (1) topical application of garlic oil on five human volunteers, using the "standard cage test;" (2) feeding sandflies on artificial membranes treated with the compound. Topical application of garlic oil on the skin of volunteers afforded 97% and 40% protection, respectively, at 1% and 0.005% dilution. When tested on an artificial membrane feeding apparatus, garlic oil showed a dose-dependent anti-feeding effect of 100% and 38%, respectively, at 0.1% and 0.005% dilution (Valerio, 2005).

Mosquitoes are of great importance as disease transmitters, acting as vectors for malaria, dengue, West Nile virus, yellow fever, encephalitis and other infectious diseases. Garlic oil was shown to be toxic to late third-instar larvae of the mosquito *Culex pipiens quiquifasciatus* at levels ≥ 3 ppm. Diallyl disulfide (DADS) and diallyl trisulfide were as toxic as garlic oil, whereas diallyl sulfide and dipropyl disulfide/trisulfide were inactive (Amonkar, 1970, 1971; Kimbaris, 2008). Compared to larvae, pupae of *C. pipiens* show higher tolerance for DADS. Both larvae and pupas metabolize ^{35}S-DADS to inorganic ^{35}S-sulfate (Rawakrishnan, 1989); DADS inhibits larval protein synthesis (George, 1973). In an anecdotal report, an onion and garlic field in central Oregon, which was flooded for several months to kill a severe infestation of white rot disease and which had a slight garlic odor of DADS leaking from the decaying garlic, showed no signs of mosquitoes. There were many other insects and invertebrates present in abundance (Crowe, 1995). Garlic oil component DADS shows low environmental persistence, with exponential loss of ^{35}S-labelled DADS from treated water with respect to time; the environmental persistence of the less volatile diallyl polysulfides was not mentioned (Rawakrishnan, 1989). A fresh garlic extract retained 83% of its original activity toward mosquito larvae (LD_{50} *ca.* $0.1\,\mathrm{g\,L^{-1}}$) after 90 h (Thomas, 1999). Garlic oil components repel mosquitoes (Gries, 2009).

It is a common misconception that eating garlic affords protection against mosquitoes. The results of a recent study show that this is not so. Thus, using a randomized, double-blinded, placebo-controlled crossover protocol, subjects were asked to consume either garlic (one visit) or a placebo (the other visit). The subjects were then exposed to laboratory-reared mosquitoes (*Aedes aegypti*, L.). The numbers of mosquitoes that did not feed on the subjects, the number of mosquito bites, the weights of the mosquitoes after feeding and the amounts of blood ingested were determined. The data did not provide evidence of significant systemic mosquito repellence. A limitation of the study is that more prolonged ingestion of garlic may be needed to accomplish repellence (Rajan, 2005).

Evaluation of several earlier commercial garlic-based products used as pesticides uncovered serious problems with batch inconsistency (Flint, 1995; Liu, 1995). Such properties are at odds with pesticide registrations, *e.g.*, in the United Kingdom, which requires consistency in physical chemistry and efficacy to be demonstrated. The Advisory Committee on Pesticides has approved a new commercial garlic-based biocide. Most of the difficulties in eliminating batch-to-batch variability during the regulatory review process were removed since production of the juice follows strict food-grade quality level protocols to ensure chemical consistency. The same material also exhibits insecticidal properties and has been approved as an insecticide against cabbage root fly (*Delia radicum*) in Denmark. The LD_{50} across each of the three life stages of the cabbage root fly (*Delia radicum*) and the housefly (*Musca domestica*) were determined. For *D. radicum* LD_{50} values recorded were: eggs (7 day exposure), 0.8%; larvae (24 h exposure), 26.4%; larvae (48 h exposure), 6.8%; adults (24 h exposure), 0.4%. For *M. domestica* LD_{50} values recorded were: eggs (7 day exposure), 1.6%; larvae (24 h exposure), 10.1%; larvae (24 h exposure), 4.5%; adults (24 h exposure), 2.2%. Mortality rates caused by the garlic juice were comparable to those obtained with the organophosphate pesticide Birlane®, indicating parity of effect at various concentrations depending on life stage. It was concluded that this garlic oil product may provide an effective, naturally derived insecticide for use in agricultural systems against dipteran pests (Prowse, 2006).

When honeybees visit animal feeders with chopped sugar cane, the bees can prevent the cattle from eating. It was found that garlic extracts are efficient in repelling honeybees from these confined beef cattle feeders and that garlic extracts were more repellent than citronella (Nicodemo, 2004).

6.3.3 Arachnida (Acari: mites and ticks)

Northern fowl mites (*Ornithonyssus sylviarum*) are external parasites that can lower egg production and cause anemia and even death in laying hens. In an experiment conducted with laying hens with light and heavy mite infestations, each hen was sprayed with either water or 10% garlic juice in water, with spraying continued each week for three weeks. At the end of this period, it was found that there were significantly fewer northern fowl mites on the birds treated with garlic juice compared with controls. It was concluded that topical application of garlic juice may be an effective way to decrease northern fowl mites in laying hens (Birrenkott, 2000). A garlic juice preparation has been found to be effective *in vitro* against the poultry red mite, *Dermanyssus gallinae* (Maurer, 2009). In the United Kingdom, a garlic juice concentrate has been approved for use against the poultry red mite.

The parasitic mite *Tropilaelaps clareae*, also known as the Asian bee mite, is a serious threat to beekeeping, affecting both developing brood and adult honey bees. Parasitization by these mites can cause abnormal brood development, death of both brood and bees, leading to colony decline and collapse, and can cause the bees to abscond from the hive. The natural host of the mite is the giant Asian honey bee, *Apis dorsata*, but *T. clareae* can readily infest colonies of *A. mellifera*, the Western honey bee. Mites feed on the bee larvae inside the sealed honeycomb wax cells, which make their control and elimination difficult. It was found that a single spray of a 2% aqueous garlic extract resulted in significantly more dead *T. clareae* mites in hive debris than in the untreated group. In addition, significantly more worker adults and brood honeybees and fewer mites were present in the treated groups as compared to the untreated group. The extract was prepared by grinding garlic cloves with an equal weight of water followed by intermediate shaking for 48 hours, filtering through filter paper and then dilution with water to afford a 2% solution. A single spray of the garlic extract led to a significant increase in the number of worker bees and brood and fewer mites compared to the untreated group; it did not affect the organoleptic quality of the honey nor the bee colony strength (Hosamani, 2007).

6.3.4 Gastropods: Slugs and Snails

While most slugs are harmless to humans, a few species are pests of agriculture and horticulture, feeding on fruits and vegetables prior to harvest, making holes in the crop, rendering it more vulnerable to rot and disease and unsuitable for sale for aesthetic reasons. The efficacy of a garlic-oil containing preparation for slug and snail control was investigated using the slug *Deroceras panormitanum* (Lessona and Pollonera, 1882; *D. caruanae*) and the snail *Oxyloma pfeifferi* (Rossmässler, 1835), which are the most abundant slug and snail pest species found damaging hardy ornamental plants in commercial nurseries in the UK (Schüder, 2003). Garlic had irritant, antifeedant, physical barrier, chemical repellent, or molluscicidal effects, or showed a combination of more than one effect. Aqueous extracts of garlic as well as synthetic allicin had high time- and dose-dependent molluscicidal activity against the snails *Lymnaea acuminata* (Lamarck) and *Indoplanorbis exustus* (Singh 1993, 1995, 2008).

6.3.5 Plant Pathogenic Bacteria, Fungi and Oomycetes: Phytoanticipins

Use of antibiotics and fungicides to control plant pathogens is limited due to the residues that may occur both in the plants to which they are applied and to the soil. The residue problem can be avoided using antimicrobial substances already present, or naturally produced, in food plants such as allicin. Allicin is an example of a phytoanticipin: a low-molecular weight antifungal compound produced after infection solely from preexisting constituents. Such antifungal compounds protect garlic from fungal infestation (Durbin, 1971). To standardize testing of the antimicrobial activity of allicin from fresh garlic juice, it was shown that on a molar basis, allicin was approximately one quarter as effective in killing *E. coli* as the antibiotic kanamycin (see Figure 5.6, Chapter 5), and that allicin is bacteriocidal rather than bacteriostatic against *E. coli.*

Through use of an inverted Petri dish it was also established that allicin can be effective as an antibiotic *via* the vapor phase. The antibiotic action of fresh garlic juice containing allicin remained relatively constant when stored at 4 °C. Activity was rapidly lost after 10 min exposure to temperatures >80 °C. Growth of the plant pathogenic bacteria *Agrobacterium tumefaciens*, *Erwinia carotovora*, *Pseudomonas syringae* pv. *Maculicola*, *P.s.* pv. *phaseolicola*, *P.s.* pv. *tomato* and *Xanthomonas campestris* pv. *Campestris* was inhibited by allicin-containing garlic juice. In addition, growth of the plant pathogenic fungi *Alternaria brassisicola, Botrytis cinerea, Magnaporthe grisea* and *Plectosphaerella cucumerina* (*Fusarium tabacinum*) was similarly inhibited, as was *Alternaria* spore germination (Curtis, 2004; Slusarenko, 2008). Plant pathogens could also be controlled using an aqueous extract of garlic powder. The garlic powder preserved its antibiotic activity toward plant pathogens for up to three years when stored at room temperature in a closed container (Ark, 1959).

Reduction of rice blast, *Arabidopsis* downy mildew disease and potato tuber blight caused by *Phytophthora infestans* could be achieved by spraying the plants with a 1 : 20 dilution of garlic juice (0.7 $mg\,mL^{-1}$ allicin), applying the garlic extract to the innoculation site, or exposing the organism to allicin *via* the vapor state (Curtis, 2004). Promising preliminary results were also obtained encapsulating garlic juice and applying the capsules to soil around *Phytophthora* inoculated tomato seedlings (Slusarenko, 2008). Garlic juice containing allicin at a level of 50 $\mu g\,mL^{-1}$ inhibited the germination of sporangia and cysts of *P. infestans* both *in vitro* and *in vivo* on the leaf surface and reduced the severity of cucumber downy mildew caused by *Pseudoperonospora cubensis* by 50–100% (Portz, 2008). It should be noted that downy mildews are among the most destructive and economically important agricultural problems worldwide.

Garlic extracts inhibit mycelial development in pathogenic fungi (Bianchi, 1997), protect bambara groundnut (*Vigna subterranean* L.; a widely cultivated legume in Africa) against brown blotch caused by *Colletotrichum capsici* (Obagwu, 2003a), protect oranges and grapefruits from citrus green and blue molds, due to *Penicillium digitatum* and *P. italicum* (Obagwu, 2003b), control foot rot of the common bean, *Phaseolus vulgaris*, caused by *Fusarium solani* (Russell, 1977), and wilt disease in gram, *Cicer arietinum*, due to *F. oxysporum* (Singh, 1979). Stem and leaf blight in pigeonpea (*Cajanus cajan* (L.) Millsp.) due to *Phytophthora drechsleri* can be controlled with 2.5 ppm ajoene (Singh, 1992). Bacterial wilt of tomatoes from *Ralstonia solanacearum* can be controlled by applying an aqueous extract of garlic to the soil (Abo-Elyousr, 2008). Studies found solutions of a garlic extract or oil and diallyl disulfide to be fungicidal against a broad range of soil-borne fungal organisms (Murthy, 1974; Sealy, 2007).

Sclerotium cepivorum, a fungus that is the causal agent of white rot of onions, persists in soil in the form of sclerotia whose germination is triggered by volatile sulfur compounds released into the soil from roots of genus *Allium* plants. Treatment of the soil with diallyl disulfide (DADS) triggers germination of the sclerotia in the absence of the host crop, leading to death of the fungus. The effectiveness of DADS treatment is related to the rate at which the material disappeared from soil, which was itself dependent upon soil temperature (Coley-Smith, 1986). Diallyl disulfide was found to be superior to dipropyl disulfide. In field trials, survival of sclerotia decreased after three months exposure to DADS-treated soil. Single DADS applications (86% DADS injected into soil at 10 L ha^{-1} in 500 L ha^{-1} of water) reduced disease incidence on onions at harvest compared with the untreated checks at three of four sites, with only 1% of the fungal population remaining (Hovius, 2002).

6.4 USE OF ONION AS AN INSECT REPELLENT BY CAPUCHINE MONKEYS

White-faced capuchin monkeys (Figure 6.4; *Cebus capucinus*), highly intelligent primates found in Central and South America, are known to rub plants on their fur, often in unison. Similar activities have been observed with captive tufted capuchin monkeys (*Cebus apella*) (Meunier, 2007, 2008; Paukner, 2008). It is thought that this is done to repel insects and is considered to be a type of self-medicative behavior. Such behavior has been extensively studied in primates during the past 20 years. It has been found that items used for fur rubbing are pungent smelling or tactually stimulating, and contain secondary metabolites that are known to have anti-insect and/or medicinal benefits. In controlled experiments, captive white-faced or tufted capuchin monkeys were provided with cut up onion bulbs, or separately cut onion bulbs and apples, which were part of their regular diet. The monkeys were observed rubbing their fur with the onions in each of 30 experiments in which they were provided with onions. They often performed this action in unison. Fur rubbing did not occur when the monkeys were provided with cut up apples. Monkeys were also observed eating an onion, manipulating it by tearing it apart, sniffing it or biting into it, but spitting it out again. While the above explanation of fur rubbing is interesting, another explanation has been offered, namely, that the intent of the monkeys when applying these plants may not be for medicinal purposes. The scents of these plants may serve to establish a group scent, much the way that urine washing and other forms of scent marking do for other primates. Social fur rubbing may be similar to grooming, in that it may serve to reinforce social ties. They may simply enjoy the interaction with the plant material as cats enjoy catnip (Baker, 1996).

Figure 6.4 A white-faced capuchin monkey. (Photo by T. Beth Kinsey).

6.5 BIRD REPELLENT ACTIVITY OF COMPOUNDS FROM *ALLIUM* PLANTS

Garlic oil (GO; diallyl di-, tri- and tetra-sulfide mixture) based products have been examined as environmentally benign, nonlethal bird repellents for crop damaging birds (Hile, 2004; Mason, 1997). It was found that European starlings significantly reduced their consumption of a food mixture that was 50% food-grade GO-impregnated granules ("ECOguard") compared to a control,

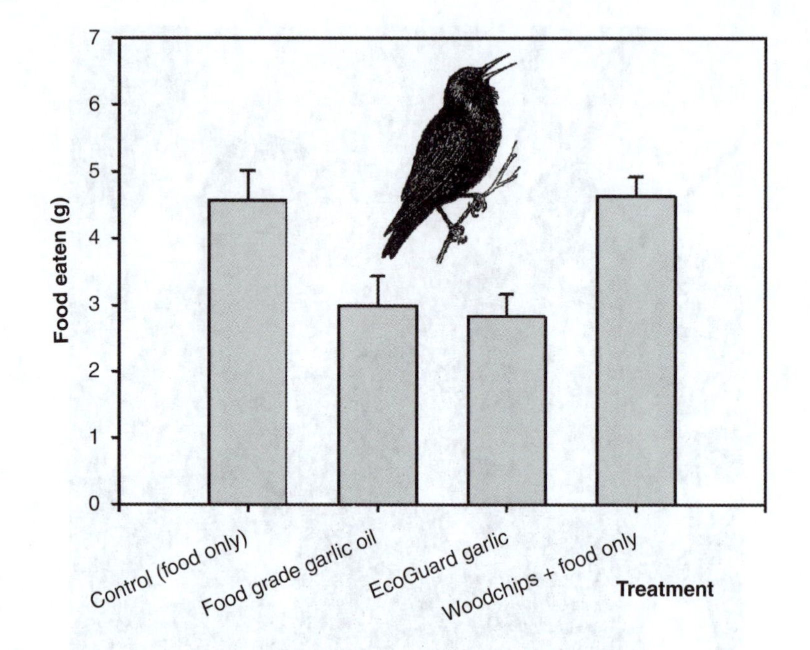

Figure 6.5 Effect of added garlic oil on food consumption by European starling. (Image from Hile, 2004).

even after overnight food deprivation (Figure 6.5). Food consumption during 3 h following overnight food deprivation was reduced by 61 to 65% compared to controls. By testing the same subjects with 25, 10 and 1% mixtures of granules in feed, it was shown that commercial ("ECOguard") GO granules were repellent to birds in lower concentrations, with more than a 50% decrease in feeding for birds presented with a 10% mixture of commercial GO granules in food and a 17% decrease for the 1% treatment. It was considered unlikely that the results were due to neophobia, since the same group of animals was used repeatedly, and repellency was consistent across experiments (Hile, 2004).

6.6 COMPANION PLANTING/INTERCROPPING WITH ALLIUMS

Polyculture (multispecies cropping) is agriculture using multiple crops in the same space and is the opposite of monoculture, which involves large stands of single crops. Monoculture is widely practiced in the mechanized agriculture of Europe, North America and parts of Asia, but is criticized for its negative environmental impact. Polyculture, which includes related methods of intercropping, companion planting, crop rotation, multicropping and beneficial weeds, avoids the susceptibility of monocultures to diseases while at the same time increases local biodiversity, among other advantages (Malézieux, 2008). Intercropping, cultivation of two or more crops in the same space at the same time, has been practiced for at least 2000 years. While a detailed consideration of polyculture, intercropping and companion planting is beyond the scope of this book, several examples involving *Allium* species merit consideration here.

A first century CE Chinese agricultural book describes the virtues of interplanting leek and cucurbits (cucumbers, squash) to reduce the diseases of the latter. A twelfth century CE Chinese book states that planting chive, garlic, or leek around flowering plants protects the latter against attack by the musk deer (Zeng, 2008). It is popularly suggested that companion planting of beets, carrots, celery, brassicas, chard, lettuce, cucumber, peppers, roses, squash or tomato with *Allium* species offers protection against insect attack due to release of repellent or odor-masking substances into the air by the *Allium* species (Web resources on companion plants, 2008). Various field trials involving intercropping alliums and study of the effect on the population dynamics of the companion crop pests and their main natural enemies have been reported, generally with encouraging results, *e.g.*, onion, garlic and potato in Egypt and Indonesia (Mogahed, 2003; Potts, 1991), garlic and sugar beet in Egypt (El-Shaikh, 2004), onion and cauliflower and cabbage in Turkey (Yildrim, 2005; Guvenc, 2006), garlic and winter wheat in China (Wang, 2008), onion and carrots in the United Kingdom (Uvah, 1984) and, as already discussed above, Chinese chive (*A. tuberosum*) and tomato in China (Yu, 1999). Garlic oil preparations applied as sprays were ineffective for control of the silverleaf whitefly (*Bemisia argentifolii*) in cotton (Flint, 1995).

Odor-mediated effects of leek, *A. porum*, and chives, *A. schoenoprasum*, on the host searching behavior of the aphid *Myzuspersicae* (Sulzer; Homoptera: Aphididae) were studied. In an olfactometer, odor of the host plant sweet pepper *Capsicum annuum* L. was significantly attractive, whereas odor of chives was significantly repellent. The combined odor of sweet pepper and chives was neither attractive nor repellent. When sweet pepper plants were exposed to volatiles from chives for five days, their odor subsequently became repellent to *M. persicae*. An extract of leek plants was significantly repellent to aphids in the olfactometer, as were sweet peppers sprayed with this extract. Because both leek and chives can disrupt host finding by the aphid, both plants have potential for intercropping with sweet pepper. If successful in the field, such a system could bring economic benefits for farmers in the wet zone of Sri Lanka and other parts of the world (Amarawardana, 2007).

It is informative to consider a representative trial, *e.g.*, mixed cropping of carrots, the "host" plant, and onions, the "non-host" plant. Release of repellent or odor-masking compounds into the air, in this case by the onion, confers some protection to the carrots. The researchers found that row-intercropping the carrots with onions caused fewer carrot flies, *Psila rosae* Fab., to enter the plots. Attack by *P. rosae* was reduced by the onion volatiles disturbing carrot-finding, particularly when the onions were young and emission of inhibitory volatiles was highest. *P. rosae* infestation and damage levels on carrots decreased progressively with increased density of the surrounding onions. In a parallel study, it was found that intercropping the French marigold, *Tagetes patula* L., was ineffective in protecting the carrots. Despite the above encouraging results, the researchers conclude that intercropping carrots with onion as a cultural method for managing carrot pests, and *P. rosae* in particular, is unlikely to be sufficient on its own to provide the high levels of control required for commercial crops, that is, 55 to 78% undamaged roots through intercropping *versus* 95% required for profitability in the primary wholesale market. In addition, intercropping is generally too labor intensive to be practical for large-scale farming operations. However, research suggests that intercropping may be useful for gardeners or growers concerned with limiting pesticide usage (Uvah, 1984).

The effectiveness of companion planting was evaluated under field conditions for protecting roses against the Japanese beetle (*Popillia japonica* Newman). Garlic chives were interplanted with roses in replicated garden plots. Numbers of beetles on these roses were compared with rose-only control plots on six days during beetle flight. There was no significant reduction in the numbers of beetles attacking the roses relative to the controls. The results suggest that the use of companion or reputedly repellent plants such as garlic chives probably will be ineffective for protecting roses or other highly susceptible ornamentals from the Japanese beetle. Use of such tactics in an effort to discourage other garden pests might even increase Japanese beetle damage in those plantings (Held, 2003).

6.7 CONCLUSIONS

Microbial and animal studies, intercropping and field trials confirm the ability of *Allium*-derived organosulfur compounds to repel predators and kill insect pests and plant pathogens, thereby protecting crops and serving as useful, non-toxic biocides, *e.g.*, for mite-infected chickens and bees. The biological activity of alliums in their natural environment logically follows from Darwinian theories of natural selection. Because the repellent and toxic properties of alliums have been naturally developed over tens of thousands of years, the chances for development of resistance to *Allium*-based pesticides is diminished, although not completely eliminated. If natural and/or synthetic *Allium*-derived compounds can be demonstrated to be sufficiently long lasting in the soil and at the same time cost-effective for the farmer, then the combination of an ancient natural and human history and years of scientific investigation, together with the endorsement of green agriculture, bodes well for widespread use of *Allium*-based pesticides and biocides. Only time will tell!

Bibliography

1 BOOKS AND MONOGRAPHS

1. G. J. Binding, *About Garlic, The Supreme Herbal Remedy*, Thorsons Publishers Ltd., London, UK, 1970.
2. R. Bird, *Growing Bulb Vegetables*, Lorenz Books, London, UK, 2004.
3. E. Block, *Reactions of Organosulfur Compounds*, Academic Press, New York, 1978.
4. J. L. Brewster, *Onions and Other Vegetable Alliums*, 2nd edition, CABI, Wallingford, UK, 2008.
5. M. Coonse, *Onions, Leeks, and Garlic. A Handbook for Gardeners.* Texas A & M University Press, College Station, TX, 1995.
6. D. Davies, *Alliums: The Ornamental Onions*, Timber Press, Portland, OR, 1992.
7. G. Don, *A Monograph of the Genus Allium*, Wernerian Natural History Society, Edinburgh, UK, 1832.
8. R. L. Engeland, *Growing Great Garlic: The Definitive Guide for Organic Gardeners*, Chelsea Green Publishing, 1991.
9. M. Gregory, *Nomenclator alliorum: Allium Names and Synonyms – A World Guide*, Royal Botanic Gardens, Kew, UK, 1998.

Garlic and Other Alliums: The Lore and the Science
By Eric Block

Published by the Royal Society of Chemistry, www.rsc.org

10. S. Fulder and J. Blackwood, *Garlic: Nature's Original Remedy*, Healing Arts Press, Rochester, VT, 2000.
11. S. Fulder, *The Garlic Book: Nature's Powerful Healer*, Avery Publishing Group, Garden City, NY, 1997.
12. L. Guangshu, Proceedings of the Fourth International Symposium on Edible Alliaceae, *Acta Horticulturae*, **688**, 2005 (published by the International Society for Horticultural Science).
13. L. J. Harris, *The Book of Garlic*, Aris Books, Berkeley, CA, 1979.
14. L. J. Harris, *The Official Garlic Lovers Handbook*, Aris Books, Berkeley, CA, 1986.
15. K. Holder and G. Duff, *A Clove of Garlic*, Reader's Digest Association, London, UK, 1996.
16. H. A. Jones and L. K. Mann, *Onions and their Allies*, Interscience, NY, 1963.
17. H. P. Koch and L. D. Lawson (ed.), *Garlic: The Science and Therapeutic Applications of Allium sativum L. and Related Species*, Williams and Wilkins, Baltimore, MD, 1996.
18. B. Lau, *Garlic for Health*, Lotus Light Publications, Wilmot, WI, 1988.
19. B. Mathew, *A Review of Allium sect. Allium*, Kew Royal Botanic Gardens, Kew, UK, 1996.
20. M. J. McGary (ed.), *Bulbs of North America*, Timber Press, Portland, OR, 2001.
21. T.J. Meredith, *The Complete Book of Garlic: A Guide for Gardeners, Growers, and Serious Cooks*, Timber Press, Portland, OR, 2008.
22. S. Moyers, *Garlic in Health, History, and World Cuisine*, Suncoast Press, St. Petersburg, FL, 1996.
23. E. S. Platt, *Garlic, Onion, & Other Alliums*, Stackpole Books, Mechanicsburg, PA, 2003.
24. H. D. Rabinowitch and L. Currah, *Allium Crop Science: Recent Advances*, CABI, Wallingford, UK, 2002.
25. H. D. Rabinowitch and J. L. Breswster (ed.), *Onions and Allied Crops*, CRC Press, Boca Raton, FL, 1990, vol. 1–3.
26. E. Regel, *Alliorum adhuc cognitorum monographia*, Petropolis, St. Petersburg, Russia, 1875 [in Latin].
27. E. Regel, *Allii species Asiae centralis in Asia media à Turcomania desertisque aralensibus et caspicis usque ad Mongoliam crescentes*, Petropolis, St. Petersburg, Russia, 1887 [in Latin].

28. M. Singer, *The Fanatic's Ecstatic Aromatic Guide to Onions, Garlic, Shallots, and Leeks*, Prentice Hall, Englewood Cliffs, NJ, 1981 [juvenile book].
29. L. Van Deven, *Onions and Garlic Forever*, Dover, Mineola, NY, 1992.
30. J. Wilen and L. Wilen, *Garlic: Nature's Super Healer*, Prentice Hall, Englewood Cliffs, NJ, 1997.
31. P. Woodward, *Garlic and Friends: The History, Growth and Use of Edible Alliums*, Hyland House, South Melbourne, Australia, 1996.

2 SELECTED REVIEW ARTICLES

1. M. Ali, M. Thomson and M. Afzal, Garlic and onions: their effect on eicosanoid metabolism and its clinical relevance, *Prostaglandins Leukotrienes Essential Fatty Acids*, 2000, **62**, 55–73.
2. E. Block, The chemistry of garlic and onions, *Sci. Am.*, 1985, **252**, 114–119.
3. E. Block, The organosulfur chemistry of the genus *Allium* – implications for organic sulfur chemistry, *Angew. Chem., Int. Edn. Engl.*, 1992, **31**, 1135–1178.
4. E. Block, E. M. Calvey, C. W. Gillies, J. Z. Gillies and P. Uden, Peeling the onion. Organosulfur and -selenium phytochemicals in genus *Allium* plants, *Recent Advances in Phytochemistry*, 1997, **31** (Functionality of Food Phytochemicals), 1–30.
5. E. Block, Garlic as a functional food: a status report, *ACS Symposium Series*, 1998, **702** (Functional Foods for Disease Prevention II: Medicinal Plants and Other Foods), 125–143.
6. W. Breu and W. Dorsch, *Allium cepa* L. (onion): Chemistry, analysis and pharmacology, *Economic and Medicinal Plant Research*, 1994, **6**, 115–147.
7. J. F. Carson, Chemistry and biological properties of onions and garlic, *Food Rev. Int.*, 1987, **3**, 71–103.
8. M. Corzo-Martínez, N. Corzoa and M. Villamiel, Biological properties of onions and garlic, *Trends Food Sci. Technol.*, 2007, **18**, 609–625.

9. G. R. Fenwick and A. B. Hanley, The genus *Allium* – Parts 1–3, *Crit. Rev. Food Sci. Nutr.*, 1985, **22**, 199–271; **22**, 273–377; **23**,1–73.
10. L. D. Lawson, Garlic: a review of its medicinal effects and indicated active compounds, *ACS Symposium Series*, 1998, **691** (Phytomedicines of Europe), 176–209.
11. M. S. Rahman, Allicin and other functional active components in garlic: health benefits and bioavailability, *Int. J. Food Prop.*, 2007, **10**, 245–268.
12. H. D. Reuter and A. Sendl, *Allium sativum* and *Allium ursinum*: Chemistry, pharmacology and medicinal applications, *Economic and Medicinal Plant Res.*, 1994, **6**, 55–113.
13. J. R. Whitaker, Development of flavor, odor, and pungency in onion and garlic, *Adv. Food Res.*, 1976, **22**, 73–133.

3 COOKBOOKS (SELECTED; SOME INCLUDE DETAILED INFORMATION ON ALLIUMS)

1. L. Bareham, *Onions Without Tears: Cooking with Onions & Shallots, Garlic, Leeks*, Michael Joseph, London, UK, 1995.
2. J. Bothwell, *The Onion Cookbook*, Dover Publications, New York, 1950.
3. C. A. Braida, *Glorious Garlic*, Storey Communications, Pownal, VT, 1986.
4. B. Cavage, *The Elegant Onion*, Storey Communications, Pownal, VT, 1987.
5. Gilroy Garlic Festival Association, *The Garlic Lovers' Cookbook*, Celestial Arts, Berkeley, CA, 1980.
6. Gilroy Garlic Festival Association, *The Garlic Lovers' Cookbook*, Volume II, Celestial Arts, Berkeley, CA, 1985.
7. Gilroy Garlic Festival Association, *Garlic Lovers' Greatest Hits*, Celestial Arts, Berkeley, CA, 1993.
8. L. and F. Griffith, *Garlic, Garlic, Garlic*, Houghton Mifflin, New York, 1998.
9. S. Kreitzman, *Garlic*, Harmony Books, New York, 1984.
10. J. Midgley, *The Goodness of Garlic*, Random House, New York 1992.
11. J. Norman, *Garlic & Onions*, Bantam Books, New York, 1992.

12. M. R. Shulman, *Garlic Cookery*, Butler & Tanner, London, UK, 1984.

4 INTERNET RESOURCES

1. http://www.kew.org/library/index.html Library catalogue for Kew Gardens (UK), one of the largest botanical libraries in the world.
2. http://www.mobot.org/ Website of the Missouri Botanical Gardens (U.S.A.) with a large collection of digitized images of Genus *Allium* plants.
3. http://www.nybg.org/edu/ Website of the New York Botanical Garden, with an extensive collection of alliums and rare books.

5 MOVIES ABOUT GARLIC

1. *Garlic is as Good as Ten Mothers*, Flower Films, Les Blank, 1980, 51 minutes, ISBN: 0-933621-16-7. http://www.lesblank.com/main.html
2. *The Gift of the Gods – The Vital History and Multiple Use of Garlic*, David C. Douglas Productions, Cremorne Australia, 1991, 70 minutes. http://www.afc.gov.au/filmsandawards/filmdbsearch.aspx?view=title&title=GIFTOF

References

L. A. Abayomi, L. A. Terry, S. F. White and P. J. Warner, Development of a disposable pyruvate biosensor to determine pungency in onions (*Allium cepa* L.), *Biosens. Bioelectron.*, 2006, **21**, 2176–2179.

ABC (Australian Broadcasting Corporation) News, Garlic-laced crops ward off insects, March 7, 2005; http://www.abc.net.au/news/newsitems/200503/s1318076.htm.

K. A. M. Abo-Elyousr and M. R. Asran, Antibacterial activity of certain plant extracts against bacterial wilt of tomato, *Arch. Phytopath. Plant Protect.*, 2009, **42**, 573–578.

G. A. Abrams and M. B. Fallon, Treatment of hepatopulmonary syndrome with *Allium sativum* L. (garlic): a pilot trial, *J. Clin. Gastroenterol.*, 1998, **27**, 232–235.

R. T. Ackermann, C. D. Mulrow, G. Ramirez, C. D. Gardner, L. Morbidoni and V. A. Lawrence, Garlic shows promise for improving some cardiovascular risk factors, *Arch. Intern. Med.*, 2001, **161**, 813–824.

M. Aguilar and F. Rincón, Improving knowledge of garlic paste greening through the design of an experimental strategy, *J. Agric. Food Chem.*, 2007, **55**, 10266–10274.

F. Akyüz, S. Kaymakoglu, K. Demir, N. Aksoy, I. Adalet and A. Okten, Is there any medical therapeutic option in hepatopulmonary syndrome? A case report, *Eur. J. Intern. Med.*, 2005, **16**, 126–128.

N. Ahmed, L. Laverick, J. Sammons, H. Zhang, D. J. Maslin and H. T. Hassan, Ajoene, a garlic-derived natural compound, enhances chemotherapy-induced apoptosis in human myeloid leukaemia CD34-positive resistant cells, *Anticancer Res.*, 2001, **21**, 3519–3523.

Garlic and Other Alliums: The Lore and the Science
By Eric Block

Published by the Royal Society of Chemistry, www.rsc.org

M. Ali, M. Afzal, R. A. Hassan, A. Farid and J. F. Burka, Comparative study of the *in vitro* synthesis of prostaglandins and thromboxanes in plants belonging to Liliaceae family, *Gen. Pharmacol.*, 1990, **21**, 273–276.

I. Alkorta and J. Elguero, Classical *versus* redox tautomerism: substituent effects on the keto/enol and sulfoxide/sulfenic acid equilibria, *Tetrahedron Lett.*, 2004, **45**, 4127–4129.

I. Alkorta, O. Picazo and J. Elguero, Chiral discrimination and isomerization processes in monomers, dimers and trimers of sulfoxides and thioperoxides, *Tetrahedron: Asymmetry*, 2004, **15**, 1391–1399.

D. E. Allen and G. Hatfield, *Medicinal Plants in Folk Tradition*, Timber Press, Portland, Oregon, 2004.

M. L. Allen, M. H. Mellow, M. G. Robinson and W. C. Orr, The effect of raw onions on acid reflux and reflux symptoms, *Amer. J. Gastroenterol.*, 1990, **85**, 377–380.

N. E. Allen, P. N. Appleby, A. W. Roddam, A. Tjønneland, N. F. Johnsen, K. Overvad, H. Boeing, S. Weikert, R. Kaaks, J. Linseisen, A. Trichopoulou, G. Misirli, D. Trichopoulos, C. Sacerdote, S. Grioni, D. Palli, R. Tumino, H. B. Bueno-de-Mesquita, L. A. Kiemeney, A. Barricarte, N. Larrañaga, M. J. Sánchez, A. Agudo, M. J. Tormo, L. Rodriguez, P. Stattin, G. Hallmans, S. Bingham, K. T. Khaw, N. Slimani, S. Rinaldi, P. Boffetta, E. Riboli and T. J. Key, Plasma selenium concentration and prostate cancer risk: results from the European Prospective Investigation into Cancer and Nutrition (EPIC), *Am. J. Clin. Nutr.*, 2008, **88**, 1567–1575.

S. A. Al-Nagdy, M. O. Abdel Rahman and H. I. Heiba, Extraction and identification of different prostaglandins in *Allium cepa*, *Comp. Biochem. Physiol. C.*, 1986, **85**, 163–166.

M. M. Al-Qattan, Garlic burns: case reports with an emphasis on associated and underlying pathology, *Burns*, 2009, **35**, 300–302.

M. S. Alvarez, S. Jacobs, S. B. Jiang, R. R. Brancaccio, N. A. Soter and D. E. Cohen, Photocontact allergy to diallyl disulfide, *Am. J. Contact Dermatol.*, 2003, **14**, 161–165.

L. Amarawardana, P. Bandara, V. Kumar, J. Pettersson, V. Ninkovic and R. Glinwood, Olfactory response of *Myzus persicae* (Homoptera: Aphididae) to volatiles from leek and chive: Potential for intercropping with sweet pepper, *Acta Agric. Scand., Sect. B – Plant Soil Sci.*, 2007, **57**, 87–91.

S. V. Amonkar and E. L. Reeves, Mosquito control with active principle of garlic, *Allium sativum, J. Econ. Entomol.*, 1970, **63**, 1172–1175.

S. V. Amonkar and A. Banerji, Isolation and characterization of larvicidal principle of garlic, *Science*, 1971, **174**, 1343–1344.

R. Amorati and G. F. Pedulli, Do garlic-derived allyl sulfides scavenge peroxyl radicals? *Org. Biomol. Chem.*, 2008, 1103–1107.

M. M. An, H. Shen, Y. B. Cao, J. D. Zhang, Y. Cai, R. Wang and Y. Y. Jiang, Allicin enhances the oxidative damage effect of amphotericin B against, *Candida albicans, Int. J. Antimicrobial Agents*, 2009, **33**, 258–263.

M. K. Ang-Lee, J. Moss and C. S. Yuan, Herbal medicines and perioperative care, *J. Am. Med. Assoc.*, 2001, **286**, 208–216.

B. Añibarro, J. L. Fontela and F. De La Hoz, Occupational asthma induced by garlic dust, *J. Allergy Clin. Immunol.*, 1997, **100**, 734–738.

S. Ankri, T. Miron, A. Rabinkov, M. Wilchek and D. Mirelman, Allicin from garlic strongly inhibits cysteine proteinases and cytopathic effects of *Entamoeba histolytica*, *Antimicrobial Agents Chemotherapy*, 1997, **41**, 2286–2288.

S. Ankri and D. Mirelman, Antimicrobial properties of allicin from garlic, *Microbes Infect.*, 1999, **1**, 125–129.

Anonymous, Eduard von Regel, *Nature*, 1892, **46**, 60–61 (obituary).

Anonymous, An old remedy revived, *Br. Med. J.*, July 13, 1912, **2689**, 90.

G. E. Anthon and D. M. Barrett, Modified method for the determination of pyruvic acid with dinitrophenylhydrazine in the assessment of onion pungency, *J. Sci. Food Agric.*, 2003, **83**, 1210–1213.

D. S. Antlsperger, V. M. Dirsch, D. Ferreira, J. L. Su, M. L. Kuo and A. M. Vollmar, Ajoene-induced cell death in human promyeloleukemic cells does not require JNK but is amplified by the inhibition of ERK, *Oncogene*, 2003, **22**, 582–589.

J. Antosiewicz, W. Ziolkowski, S. Kar, A. A. Powolny and S. V. Singh, Role of reactive oxygen intermediates in cellular responses to dietary cancer chemopreventive agents, *Planta Med.*, 2008, **74**, 1570–1579.

J. Antosiewicz, A. Herman-Antosiewicz, S. W. Marynowski and S. V. Singh, c-Jun NH_2-terminal kinase signaling axis regulates diallyl trisulfide-induced generation of reactive oxygen species and cell cycle arrest in human prostate cancer cells, *Cancer Res.*, 2006, **66**, 5379–5386.

R. Apitz-Castro, S. Cabrera, M. R. Cruz, E. Ledezma and M. K. Jain, Effects of garlic extract and of three pure components isolated from it on human platelet aggregation, arachidonate metabolism, release reaction and platelet ultrastructure, *Thromb. Res.*, 1983, **32**, 155–169.

K. Aquilano, G. Filomeni, S. Baldelli, S. Piccirillo, A. De Martino, G. Rotilio and M. R. Ciriolo, Neuronal nitric oxide synthase protects neuroblastoma cells from oxidative stress mediated by garlic derivatives, *J. Neurochem.*, 2007, **101**, 1327–1337.

F. D. Arditti, A. Rabinkov, T. Miron, Y. Reisner, A. Berrebi, M. Wilchek and D. Mirelman, Apoptotic killing of B-chronic lymphocytic leukemia tumor cells by allicin generated in situ using a rituximab–alliinase conjugate, *Mol. Cancer Therapy*, 2005, **4**, 325–331.

A. Arena, C. Cislaghi and P. Falagiani, Anaphylactic reaction to the ingestion of raw onion. A case report, *Allergol. Immunopathol. (Madr.)*, 2000, **28**, 287–289.

T. Ariga and T. Seki, Antithrombotic and anticancer effects of garlic-derived sulfur compounds: a review, *Biofactors*, 2006, **26**, 93–103.

T. Ariga, K. Tsuj, T. Seki, T. Moritomo and J. I. Yamamoto, Antithrombotic and antineoplastic effects of phyto-organosulfur compounds, *Biofactors*, 2000, **13**, 251–255.

A. Argento, E. Tiraferri and M. Marzaloni, Oral anticoagulants and medicinal plants. An emerging interaction, *Ann. Ital. Med. Int.*, 2000, **15**, 139–143.

H. Arimoto, S. Asano and D. Uemura, Total synthesis of allixin; an anti-tumor promoter from garlic, *Tetrahedron Lett.*, 1997, **38**, 7761–7762.

K. Ariyama, Y. Aoyama, A. Mochizuki, Y. Homura, M. Kadokura and A. Yasui, Determination of the geographic origin of onions between three main production areas in Japan and other countries by mineral composition, *J. Agric. Food Chem.*, 2007, **55**, 347–354.

K. Ariyama, H. Horita and A. Yasui, Application of inorganic element ratios to chemometrics for determination of the geographic origin of Welsh onions, *J. Agric. Food Chem.*, 2004, **52**, 5803–5809.

P. A. Ark and J. P. Thompson, Control of certain diseases of plants with antibiotics from garlic (*Allium sativum* L.), *Plant Disease Reporter*, 1959, **43**, 276–282.

R. Arnaud, P. Juvin and Y. Vallee, Density functional theory study of the dimerization of the sulfine H_2CSO, *J. Org. Chem.*, 1999, **64**, 8880–8886.

I. Arnault, N. Mondy, F. Cadoux and J. Auger, Possible interest of various sample transfer techniques for fast gas chromatography-mass spectrometric analysis of true onion volatiles, *J. Chromatogr. A*, 2000, **896**, 117–124.

I. Arnault, J. P. Christidès, N. Mandon, T. Haffner, R. Kahane and J. Auger, High-performance ion-pair chromatography method for simultaneous analysis of alliin, deoxyalliin, allicin and dipeptide precursors in garlic products using multiple mass spectrometry and UV detection, *J. Chromatogr. A.*, 2003, **991**, 69–75.

I. Arnault, N. Mondy, S. Diwo and J. Auger, Soil behaviour of sulfur natural fumigants used as methyl bromide substitutes, *Int. J. Environ. Anal. Chem.*, 2004, **84**, 75–82.

I. Arnault and J. Auger, Seleno-compounds in garlic and onions, *J. Chromatogr. A.*, 2006, **1112**, 23–30.

A. Arora, K. Seth and Y. Shukla, Reversal of P-glycoprotein-mediated multidrug resistance by diallyl sulfide in K562 leukemic cells and in mouse liver, *Carcinogenesis*, 2004, **25**, 941–949.

D. S. Arora and J. Kaur, Antimicrobial activity of spices, *Int. J. Antimicrobial Agents*, 1999, **12**, 257–262.

A. Arunkumar, M. R. Vijayababu, N. Gunadharini, G. Krishnamoorthy and J. Arunakaran, Induction of apoptosis and histone hyperacetylation by diallyl disulfide in prostate cancer cell line PC-3, *Cancer Lett.*, 2007, **251**, 59–67.

M. R. Aslani, M. Mohri and M. Chekani, Effects of garlic (*Allium sativum*) and its chief compound, allicin, on acute lethality of cyanide in rats, *Comp. Clin. Pathol.*, 2006, **15**, 211–213.

M. R. Aslani, M. Mohri and A. R. Movassaghi, Heinz body anaemia associated with onion (*Allium cepa*) toxicosis in a flock of sheep, *Comp. Clin. Pathol.*, 2005, **14**, 118–120.

J. Auger and E. Thibout, The action of volatile sulfur compounds from the leek (*Allium porrum*) on oviposition in *Acrolepiopsis assectella* (Lepidoptera: Hyponomeutoidea): the preponderance of thiosulfinates, *Can. J. Zool.*, 1979, **57**, 2223–2229.

J. Auger, C. Lecomte, J. Paris and E. Thibout, Identification of leek-moth and diamondback-moth frass volatiles that stimulate parasitoid, *Diadromus pulchellus. J. Chem. Ecol.*, 1989, **15**, 1391–1398.

J. Auger, C. Lecomte and E. Thibout, Leek odor analysis by gas chromatography and identification of the most active substance for the leek moth, *Acrolepiopsis assectella, J. Chem. Ecol.*, 1989, **15**, 1847–1854.

J. Auger, F. X. Lalau-Keraly and C. Belinsky, Thiosulfinates in vapor phase are stable and they can persist in the environment of *Allium*, *Chemosphere*, 1990, **21**, 837–843.

J. Auger, F. Cadoux and E. Thibout, Allium spp. thiosulfinates as substitute fumigants for methyl bromide, *Pestic. Sci.*, 1999, **55**, 200–202.

J. Auger, S. Dugravot, A. Naudin, A. Abo-Ghalia, D. Pierre and E. Thibout, Possible use of *Allium* allelochemicals in integrated control, *IOBC-WPRS Bulletin*, 2002, **25**(9), 295–306.

J. Auger, I. Arnault, S. Diwo-Allain, M. Ravier, F. Molia and M. Pettiti, Insecticidal and fungicidal potential of *Allium* substances as biofumigants, *Agroindustria*, 2004, **3**, 5–8.

K. T. Augusti and C. G. Sheela, Antiperoxide effect of *S*-allylcysteine sulfoxide, an insulin secretagogue, in diabetic rats, *Experientia*, 1996, **52**, 115–120.

M. J. Axley, A. Böck and T. C. Stadtman, Catalytic properties of an *Escherichia coli* formate dehydrogenase mutant in which sulfur replaces selenium, *Proc. Natl. Acad. Sci. U.S.A.*, 1991, **88**, 8450–8454.

A. Aydin, G. Ersoz, O. Tekesin, E. Akcicek, M. Tuncyurek and Y. Batur, Does garlic oil have a role in the treatment of *Helicobacter pylori* infection? *Turkish J. Gastroenterol.*, 1997, **8**, 181–184.

E. Ayrton and W. Loat, *Predynastic Cemetery at el-Mahasna*, Egypt Exploration Fund, London, 1911.

P. Badol, M. David-Dufilho, J. Auger, S. W. Whiteheart and F. Rendu, Thiosulfinates modulate platelet activation by reaction with surface free sulfhydryls and internal thiol-containing proteins, *Platelets*, 2007, **18**, 481–490.

R. D. Baechler, J. P. Hummel and K. Mislow, Reaction of allylic thioethers with elemental sulfur, *J. Am. Chem. Soc.*, 1973, **95**, 4442–4444.

S. Bagga, B. S. Thomas and M. Bhat, Garlic burn as self-inflicted mucosal injury – a case report and review of the literature, *Quintessence Int.*, 2008, **39**, 491–494.

H. Bahar, A. Islam, A. Mannan and J. Uddin, Effectiveness of some botanical extracts on bean aphids attacking yard-long beans, *J. Entomol.*, 2007, **4**, 136–142.

B. Bai, F. Chen, Z. Wang, X. Liao, G. Zhao and X. Hu, Mechanism of the greening color formation of "Laba" garlic, a traditional homemade Chinese food product, *J. Agric. Food Chem.*, 2005, **53**, 7103–7107.

J. S. Bajaj, R. Shaker and W. J. Hogan, Esophageal veggie spasms: a food-specific cause of chest distress, *Am. J. Gastroenterol.*, 2004, **99**, 1396–1398.

J. G. Baker, On the alliums of India, China and Japan, *J. Botany British Foreign*, 1874, **12**, 289–295.

M. Baker, Fur rubbing: Use of medicinal plants by capuchin monkeys (*Cebus capucinus*), *Am. J. Primatol.*, 1996, **38**, 263–270.

I. M. Bakri and C. W. Douglas, Inhibitory effect of garlic extract on oral bacteria, *Arch. Oral Biol.*, 2005, **50**, 645–651.

J. E. Baldwin, G. Hoefle and S. C. Choi, Rearrangement of strained dipolar species. I. Episulfoxides. Demonstration of the existence of thiosulfoxylates, *J. Am. Chem. Soc.*, 1971, **93**, 2810–2812.

J. O. Ban, D. Y. Yuk, K. S. Woo, T. M. Kim, U. S. Lee, H. S. Jeong, D. J. Kim, Y. B. Chung, B. Y. Hwang, K. W. Oh and J. T. Hong, Inhibition of cell growth and induction of apoptosis via inactivation of NF-kappaB by a sulfur compound isolated from garlic in human colon cancer cells, *J. Pharmacol. Sci.*, 2007, **104**, 374–383.

M. Bandell, L. J. Macpherson and A. Patapoutian, From chills to chilis: mechanisms for thermosensation and chemesthesis via thermoTRPs, *Current Opinion Neurobiol.*, 2007, **17**, 490–497.

S. K. Banerjee, P. K. Mukherjee and S. K. Maulik, Garlic as an antioxidant: the good, the bad and the ugly, *Phytotherapy Res.*, 2003, **17**, 97–106.

S. K. Banerjee, M. Maulik, S. C. Mancahanda, A. K. Dinda, S. K. Gupta and S. K. Maulik, Dose-dependent induction of endogenous antioxidants in rat heart by chronic administration of garlic, *Life Sci.*, 2002, **70**, 1509–1518.

S. K. Banerjee and S. K. Maulik, Effect of garlic on cardiovascular disorders: a review, *Nutrition J.*, 2002b, **1**, 4.

R. Banks, *Living in a Wild Garden*, St. Martin's Press, New York, 1980, p. 128.

L. Bareham, *Onions Without Tears: Cooking with Onions & Shallots, Garlic, Leeks*, Michael Joseph, London, 1995.

D. Barnard, The spontaneous decomposition of aryl thiolsulfinates, *J. Chem. Soc.*, 1957, 4675–4676.

D. Barnard, T. H. Houseman, M. Porter and B. K. Tidd, Thermal racemization and cis,*trans*-isomerization of allylically unsaturated di- and poly-sulphides: a mechanism involving branched sulphur chains, *Chem. Commun.*, 1969, 371–372.

P. M. Barnes, E. Powell-Griner, K. McFann and R. L. Nahin, Advance data: Complementary and Alternative Medicine Use Among Adults: United States, 2002, CDC (Centers for Disease Control and Prevention's National Center for Health Statistics (NCHS)), Number 343, May 24, 2004: http://www.cdc.gov/nchs/data/ad/ad343.pdf.

F. E. Barone and M. R. Tansey, Isolation, purification, identification, synthesis, and kinetics of activity of the anticandidal component of *Allium sativum*, and a hypothesis for its mode of action, *Mycologia*, 1977, **69**, 793–825.

S. A. Barrie, J. V. Wright and J. E. Pizzorno, Effects of garlic oil on platelet aggregation, serum lipids and blood pressure in humans, *J. Orthomol. Med.*, 1987, **2**, 15–21.

A. M. Baruchin, A. Sagi, B. Yoffe and M. Ronen, Garlic burns, *Burns*, 2001, **27**, 781–782.

S. I. Baskin, D. W. Porter, G. A. Rockwood, J. A. Romano Jr., H. C. Patel, R. C. Kiser, C. M. Cook and A. L. Ternay Jr., *In vitro* and *in vivo* comparison of sulfur donors as antidotes to acute cyanide intoxication, *J. Appl. Toxicol.*, 1999, **19**, 173–183.

B. Basnyat, High altitude cerebral and pulmonary edema, *Travel Med. Infect. Diseases*, 2005, **3**, 199–211.

K. Bassioukas, D. Orton and R. Cerio, Occupational airborne allergic contact dermatitis from garlic with concurrent Type 1 allergy, *Contact Dermatitis*, 2004, **50**, 39–50.

E. E. Battin, N. R. Perron and J. L. Brumaghim, The central role of metal coordination in selenium antioxidant activity, *Inorg. Chem.*, 2006, **45**, 499–501.

E. E. Battin and J. L. Brumaghim, Metal specificity in DNA damage prevention by sulfur antioxidants, *J. Inorg. Biochem.*, 2008, **47**, 6153–6161.

J.-B. Baudin, M.-G. Commenil, S. A. Julia and Y. Wang, A convergent synthesis of γ-unsaturated thioaldehyde- and thioketone *S*-oxides, *Bull. Soc. Chim. Fr.*, 1996, **133**, 515–529.

R. Bauer, W. Breu, H. Wagner and W. Weigand, Enantiomeric separation of racemic thiosulfinate esters by high-performance liquid chromatography, *J. Chromatog.*, 1991, **541**, 464–468.

D. M. Bautista, P. Movahed, A. Hinman, H. E. Axelsson, O. Sterner, E. D. Högestätt, D. Julius, S. E. Jordt and P. M. Zygmunt, Pungent products from garlic activate the sensory ion channel TRPA1, *Proc. Natl. Acad. Sci. U.S.A.*, 2005, **102**, 12248–12252.

D. M. Bautista, S. E. Jordt, T. Nikai, P. R. Tsuruda, A. J. Read, J. Poblete, E. N. Yamoah, A. I. Basbaum and D. Julius, TRPA1 mediates the inflammatory actions of environmental irritants and proalgesic agents, *Cell*, 2006, **124**, 1269–1282.

T. Bayer, H. Wagner, V. Wray and W. Dorsch, Inhibitors of cyclo-oxygenase and lipoxygenase in onions, *Lancet*, 1988a, **8616**, 906.

T. Bayer, H. Wagner and W. Dorsch, New biologically active sulfur-containing compounds from *Allium cepa, Planta Med.*, 1988b, **54**, 560.

T. Bayer, W. Breu, O. Seligmann, V. Wray and H. Wagner, Biologically active thiosulfinates and α-sulfinyldisulfides from *Allium cepa*, *Phytochemistry*, 1989a, **28**, 2373–2377.

T. Bayer, H. Wagner, E. Block, S. Grisoni, S.-H. Zhao and A. Neszmelyi, Zwiebelanes: novel biologically active 2,3-dimethyl-5,6-dithiabicyclo[2.1.1]-hexane 5-oxides from onion, *J. Am. Chem. Soc.*, 1989b, **111**, 3085–3086.

B. W. Beckert, M. J. Concannon, S. L. Henry, D. S. Smith and C. L. Puckett, The effect of herbal medicines on platelet function: an *in vivo* experiment and review of the literature, *Plastic Reconstructive Surgery*, 2007, **120**, 2044–2050.

S. Belman, Onion and garlic oils inhibit tumor promotion, *Carcinogenesis*, 1983, **4**, 1063–1065.

M. A. Belous and I. Ya. Postovskii, Pseudoallicin, *Zhur. Obshchei Khim.*, 1950, **20**, 1701–1710.

G. A. Benavides, G. L. Squadrito, R. W. Mills, H. D. Patel, T. S. Isbell, R. P. Patel, V. M. Darley-Usmar, J. E. Doeller and D. W. Kraus, Hydrogen sulfide mediates the vasoactivity of garlic, *Proc. Natl. Acad. Sci. U.S.A.*, 2007, **104**, 17977–17982.

R. Bentley and T. G. Chasteen, Microbial methylation of metalloids: arsenic, antimony, and bismuth, *Microbiol. Mol. Biol. Rev.*, 2002, **66**, 250–271.

R. Bentley, Role of sulfur chirality in the chemical processes of biology, *Chem. Soc. Rev.*, 2005, **34**, 609–624.

B. Berman, J. Patel, O. Perez and M. Viera, A prospective, randomized, investigator-blinded, placebo-controlled, comparative study evaluating the tolerability and efficacy of two topical medications versus placebo for the treatment of keloids and hypertrophic scars, *J. Am. Acad. Dermatol.*, 2008, **58**, AB45.

H. D. Betz, Ed., *The Greek Magical Papyri in Translation*, University of Chicago Press, IL, 1992.

J. M. Betz, Use of herbal medications before surgery, *J. Am. Med. Assoc.*, 2001, **286**, 2542.

J. Beuth, N. Hunzelmann, R. Van Leendert, R. Basten, M. Noehle and B. Schneider, Safety and efficacy of local administration of Contractubex to hypertrophic scars in comparison to corticosteroid treatment. Results of a multicenter, comparative epidemiological cohort study in Germany, *In vivo*, 2006, **20**, 277–283.

A. Bianchi, A. Zambonelli, A. -Z. D'Aulerio and F. Bellesia, Ultrastructural studies of the effects of *Allium sativum* on phytopathogenic fungi *in vitro*, *Plant Diseases*, 1997, **81**, 1241–1246.

J. Billings and P. W. Sherman, Antimicrobial functions of spices: Why some like it hot, *Q. Rev. Biol.*, 1998, **73**, 1–38.

G. P. Birrenkott, G. E. Brockenfelt, J. A. Greer and M. D. Owens, Topical application of garlic reduces northern fowl mite infestation in laying hens, *Poultry Sci.*, 2000, **79**, 1575–1577.

L. Blank with M. Gosling, *Garlic Is As Good As Ten Mothers*, 1980 (Flower Films, 10341 San Pablo Ave., El Cerrito, California 94530): http://www.lesblank.com/main.html.

M. A. Blankenhorn and C. E. Richards, Garlic breath odor, *J. Am. Med. Assoc.*, 1936, **107**, 409–410.

E. Block, Chemistry of alkyl thiosulfinate esters. II. Sulfenic acids from dialkyl thiolsulfinate esters, *J. Am. Chem. Soc.*, 1972, **94**, 642–644.

E. Block and J. O'Connor, Chemistry of alkyl thiolsulfinate esters. V. Novel synthesis of α-heteroatom substituted disulfides, *J. Am. Chem. Soc.*, 1973, **95**, 5048–5051.

E. Block and J. O'Connor, Chemistry of alkyl thiosulfinate esters. VI. Preparation and spectral studies, *J. Am. Chem. Soc.*, 1974a, **96**, 3921–3929.

E. Block and J. O'Connor, Chemistry of alkyl thiosulfinate esters. VII. Mechanistic studies and synthetic applications, *J. Am. Chem. Soc.*, 1974b, **96**, 3929–3944.

E. Block, α-Disulfide carbonium ions, *J. Org. Chem.*, 1974c, **39**, 734–736.

E. Block, R. E. Penn, R. J. Olsen and P. F. Sherwin, Sulfine, *J. Am. Chem. Soc.*, 1976, **98**, 1264–1265.

E. Block, *Reactions of Organosulfur Compounds*, Academic Press, New York, 1978.

E. Block, R. E. Penn and L. K. Revelle, Flash vacuum pyrolysis studies. 7. Structure and origin of the onion lachrymatory factor. A microwave study, *J. Am. Chem. Soc.*, 1979, **101**, 2200–2201.

E. Block, L. K. Revelle and A. A. Bazzi, The lachrymatory factor of the onion, *Tetrahedron Lett.*, 1980, **21**, 1277–1280.

E. Block, A. A. Bazzi and L. K. Revelle, The chemistry of sulfines. 6. Dimer of the onion lachrymatory factor: the first stable 1,2-dithietane derivative, *J. Am. Chem. Soc.*, 1980, **102**, 2490–2491.

E. Block, R. E. Penn, A. A. Bazzi and D. Cremer, The "syn-effect" in sulfines and carbonyl oxides: conformational preferences of CH_3CHSO and CH_3CHOO, *Tetrahedron Lett.*, 1981, **22**, 29–32.

E. Block, E. R. Corey, R. E. Penn, T. L. Renken, P. F. Sherwin, H. Bock, T. Hirabayshi, S. Mohmand and B. Solouki, Synthesis and thermal decomposition of 1,3-dithietane and its S-oxides, *J. Am. Chem. Soc.*, 1982, **104**, 3119–3130.

E. Block, S. Ahmad, M. K. Jain, R. W. Crecely, R. Apitz-Castro and M. R. Cruz, (*E*,*Z*)-Ajoene: a potent antithrombotic agent from garlic, *J. Am. Chem. Soc.*, 1984, **106**, 8295–8296.

E. Block, The chemistry of garlic and onions, *Sci. Am.*, 1985, **252**, 114–119.

E. Block, S. Ahmad, J. L. Catalfamo, M. K. Jain and R. Apitz-Castro, The chemistry of alkyl thiosulfinate esters. 9. Antithrombotic organosulfur compounds from garlic: structural, mechanistic, and synthetic studies, *J. Am. Chem. Soc.*, 1986, **108**, 7045–7055.

E. Block, A. J. Yencha, M. Aslam, V. Eswarakrishnan, J. Luo and A. Sano, Gas-phase determination of the geometric requirements of the silicon β-effect. Photoelectron and Penning ionization electron spectroscopic study of silylthiiranes and -oxiranes. Synthesis and chemistry of *trans*-2,3-bis(trimethylsilyl)thiirane, *J. Am. Chem. Soc.*, 1988a, **110**, 4748–4753.

E. Block, R. Iyer, S. Grisoni, C. Saha, S. Belman and F. P. Lossing, Lipoxygenase inhibitors from the essential oil of garlic. Markovnikov addition of the allyl-dithio radical to olefins, *J. Am. Chem. Soc.*, 1988b, **110**, 7813–7827.

E. Block and T. Bayer, (*Z*,*Z*)-d,l-2,3-Dimethyl-1,4-butanedithial *S*,*S*′-dioxide: a novel biologically active organosulfur compound from onion. Formation of vic-disulfoxides in onion extracts, *J. Am. Chem. Soc.*, 1990a, **112**, 4584–4585.

E. Block and S.-H. Zhao, Onion essential oil chemistry. *cis*- and *trans*-2-Mercapto-3,4-dimethyl-2,3-dihydrothiophene from pyrolysis of bis(1-propenyl) disulfide, *Tetrahedron Lett.*, 1990b, **31**, 4999–5002.

E. Block, The organosulfur chemistry of the genus *Allium* – implications for organic sulfur chemistry, *Angew. Chem., Int. Ed. Engl.*, 1992, **31**, 1135–1178.

E. Block and S.-H. Zhao, *Allium* chemistry: simple synthesis of antithrombotic cepaenes from onion and deoxycepaenes from oil of shallot by reaction 1-propenethiolate with sulfonyl halides, *J. Org. Chem.*, 1992, **57**, 5815–5817.

E. Block, S. Naganathan, D. Putman and S.-H. Zhao, *Allium* chemistry: HPLC quantitation of thiosulfinates from onion, garlic, wild garlic, leek, scallions, shallots, elephant (great-headed) garlic, chives and Chinese chives. Uniquely high allyl to methyl ratios in some garlic samples, *J. Agric. Food Chem.*, 1992b, **40**, 2418–2430.

E. Block, D. Putman and S.-H. Zhao, *Allium* chemistry: GC-MS analysis of thiosulfinates and related compounds from onion, leek, scallion, shallot, chive and Chinese chive, *J. Agric. Food Chem.*, 1992c, **40**, 2431–2438.

E. Block, P. F. Purcell and S. R. Yolen, Onions and heartburn, *Am. J. Gastroenterol.*, 1992, **87**, 679.

E. Block, Flavor artifacts, *J. Agric. Food Chem.*, 1993, **41**, 692.

E. Block and E. M. Calvey, Facts and Artifacts in *Allium* Chemistry, in *Sulfur Compounds in Food*, ed. C. J. Mussinan and M. E. Keelan, ACS Symposium Series 564, American Chemical Society, Washington DC, 1994a, pp. 63–79.

E. Block, C. Guo, M. Thiruvazhi and P. J. Toscano, Total synthesis of thiarubrine B [3-(3-buten-1- ynyl)–6-(1,3-pentadiynyl)- 1,2-dithiin], the antibiotic principle of giant ragweed (*Ambrosia trifida*), *J. Am. Chem. Soc.*, 1994b, **116**, 9403–9404.

E. Block, M. Thiruvazhi, P. J. Toscano, T. Bayer, S. Grisoni and S.-H. Zhao, *Allium* chemistry: structure, synthesis, natural occurrence in onion (*Allium cepa*), and reactions of 2,3-dimethyl-5,6-dithiabicyclo[2.1.1]hexane *S*-oxides, *J. Am. Chem. Soc.*, 1996a, **118**, 2790–2798.

E. Block, T. Bayer, S. Naganathan and S.-H. Zhao, *Allium* chemistry: synthesis and sigmatropic rearrangements of alk(en)yl 1-propenyl disulfide *S*-oxides from cut onion and garlic, *J. Am. Chem. Soc.*, 1996b, **118**, 2799–2810.

E. Block, J. Z. Gillies, C. W. Gillies, A. A. Bazzi, D. Putman, L. K. Revelle, D. Wang and X. Zhang, *Allium* chemistry: microwave spectroscopic identification, mechanism of formation, synthesis, and reactions of (*E*,*Z*)-propanethial *S*-oxide, the lachrymatory factor of the onion (*Allium cepa*), *J. Am. Chem. Soc.*, 1996c, **118**, 7492–7501.

E. Block, X.-J. Cai, P. C. Uden, X. Zhang, B. D. Quimby and J. J. Sullivan, *Allium* chemistry: natural abundance organoselenium compounds from garlic, onion, and related plants and in human garlic breath, *Pure Appl. Chem.*, 1996d, **68**, 937–944.

E. Block, H. Gulati, D. Putman, D. Sha, N. You and S.-H. Zhao, *Allium* chemistry: synthesis of 1-[alk(en)ylsulfinyl]propyl alk(en)yl disulfides (cepaenes), antithrombotic flavorants from homogenates of onion (*Allium cepa*), *J. Agric. Food Chem.*, 1997, **45**, 4414–4422.

E. Block, S. M. Bird, J. F. Tyson, P. C. Uden, X. Zhang and E. Denoyer, The search for anticarcinogenic organoselenium compounds from natural sources, *Phosphorus, Sulfur Silicon Related Elements*, 1998, **136–8**, 1–10.

E. Block, M. Birringer and C. He, 1,2-Dichalcogenins: simple syntheses of 1,2-diselenins, 1,2-dithiins, and 2-selenathiin, *Angew. Chem., Int. Ed. Engl.*, 1999, **38**, 1604–1607.

E. Block, M. Birringer, W. Jiang, T. Nakahodo, H. J. Thompson, P. J. Toscano, H. Uzar, X. Zhang and Z. Zhu, *Allium* chemistry: synthesis, natural occurrence, biological activity, and chemistry of *Se*-alk(en)ylselenocysteines and their γ-glutamyl derivatives and oxidation products, *J. Agric. Food Chem.*, 2001a, **49**, 458–470.

E. Block, Chemistry of analogous organoselenium and organosulfur compounds, *Phosphorus, Sulfur Silicon Related Elements*, 2001, **172**, 1–23.

E. Block and M. Groom, Garlic and onions: a source of environmentally benign nematicides? *Acta Horticulture*, 2009, in press.

E. Block, A. J. Dane, A. Vattekkatte, R. A. Musah, R. B. Cody and R. Kubec, unpublished results, 2009a.

E. Block, C. Wang, S.-Z. Zhang, R. Sheridan, unpublished results, 2009b.

K. I. Block, A. C. Koch, M. N. Mead, P. K. Tothy, R. A. Newman and C. Gyllenhaal, Impact of antioxidant supplementation on chemotherapeutic toxicity: a systematic review of the evidence from randomized controlled trials, *Int. J. Cancer*, 2008, **123**, 1227–1239.

S. A. Blum, R. G. Bergman and J. A. Ellman, Enantioselective oxidation of di-*tert*-butyl disulfide with a vanadium catalyst: progress toward mechanism elucidation, *J. Org. Chem.*, 2003, **68**, 150–155.

M. Blumenthal, A. Goldberg and J. Brinckman (ed.), Garlic, *Herbal Medicine: Expanded Commission E Monographs*, 2000, Lippincott, Williams and Wilkins, Newton, MA, pp. 139–148.

H. Bock, S. Mohmand, T. Hirabayashi and A. Semkow, Thioacrolein, *J. Am. Chem. Soc.*, 1982, **104**, 312–313.

M. Boelens, P. J. De Valois and H. J. Wobben, and A. Van der Gen, *Volatile flavor compounds from onion, J. Agric. Food Chem.*, 1971, **19**, 984–991.

D. Boivin, S. Lamy, S. Lord-Dufour, J. Jackson, E. Beaulieu, M. Cote, A. Moghrabi, S. Barrette, D. Gingras and R. Béliveau, Antiproliferative and antioxidant activities of common vegetables: a comparative study, *Food Chem.*, 2009, **112**, 374–380.

P. Bonaccorsi, C. Caristi, C. Gargiulli and U. Leuzzi, Flavonol glucosides in *Allium* species: a comparative study by means of HPLC–DAD–ESI–MS–MS, *Food Chem.*, 2008, **107**, 1668–1673.

I. Bonaduce, M. P. Colombini and S. J. Diring, Identification of garlic in old gildings by gas chromatography–mass spectrometry, *J. Chromatogr., A*, 2006, **1106**, 226–232.

F. Borrelli, R. Capasso and A. A. Izzo, Garlic (*Allium sativum* L.): adverse effects and drug interactions in humans, *Mol. Nutr. Food Res.*, 2007, **51**, 1386–1397.

V. Borelli, J. Lucioli, F. H. Furlan, P. G. Hoepers, J. F. Roveda, S. D. Traverso and A. Gava, Fatal onion (*Allium cepa*) toxicosis in water buffalo (*Bubalus bubalis*), *J. Vet. Diagn. Invest.*, 2009, **21**, 402–405.

J. Boscher, J. Auger, N. Mandon and S. Ferary, Qualitative and quantitative comparison of volatile sulphides and flavour precursors in different organs of some wild and cultivated garlics, *Biochem. Syst. Ecol.*, 1995, **23**, 787–791.

J. J. Boswell, ed., *English Botany*, George Bell and Sons, London, 1883, vol. **9**.

J. Bottéro, The Culinary Tablets at Yale, *J. Am. Oriental Soc.*, 1987, **107**, 11–19.

J. Bottéro, *The Oldest Cuisine in the World*, The University of Chicago Press, Chicago, IL, 2004.

L. D. Brace, Cardiovascular benefits of garlic (*Allium sativum* L), *J. Cardiovascular Nursing*, 2002, **16**(4), 33–49.

J. L. Brewster, *Onions and Other Vegetable Alliums*, 2nd edn., CABI, Wallingford, UK, 2008.

W. H. Briggs, J. D. Folts, H. E. Osman and I. L. Goldman, Administration of raw onion inhibits platelet-mediated thrombosis in dogs, *J. Nutr.*, 2001, **131**, 2619–2622.

W. H. Briggs, H. Xiao, K. L. Parkin, C. Shen and I. L. Goldman, Differential inhibition of human platelet aggregation by selected *Allium* thiosulfinates, *J. Agric. Food Chem.*, 2000, **48**, 5731–5735.

J. C. Brocklehurst, An assessment of chlorophyll as deodorant, *Br. Med. J.*, May 7, 1953, (4809), 541–544.

M. H. Brodnitz, C. L. Pollock and P. P. Vallon, Flavor components of onion oil, *J. Agric. Food Chem.*, 1969, **17**, 760–763.

M. H. Brodnitz and J. V. Pascale, Thiopropanal *S*-oxide. a lachrymatory factor in onions, *J. Agric. Food Chem.*, 1971a, **19**, 269–272.

M. H. Brodnitz, J. V. Pascale and L. V. Derslice, Flavor components of garlic extract, *J. Agric. Food Chem.*, 1971b, **19**, 273–275.

S. J. Brois, J. F. Pilot and H. W. Barnum, New synthetic concepts in organosulfur chemistry. I. New pathway to unsymmetrical disulfides. The thiol-induced fragmentation of sulfenyl thiocarbonates, *J. Am. Chem. Soc.*, 1970, **92**, 7629–7631.

B. Brône, P. J. Peeters, R. Marrannes, M. Mercken, R. Nuydens, T. Meert and H. J. Gijsen, Tear gasses CN, CR, and CS are potent activators of the human TRPA1 receptor, *Toxicol. Appl. Pharmacol.*, 2008, **231**, 150–156.

D. N. Brooks, An onion in your ear, *J. Laryngol. Otol.*, 1986, **100**, 1043–1046.

M. Brvar, T. Ploj, G. Kozelj, M. Mozina, M. Noc and M. Bunc, Case report: fatal poisoning with Colchicum *autumnale*, *Critical Care*, 2004, **8**, R56–R59.

E. Buiatti, D. Palli, A. Decarli, D. Amadori, C. Avellini, S. Bianchi, R. Biserni, F. Cipriani, P. Cocco, A. Giacosa, E. Marubini, R. Puntoni, C. Vindigni, J. Fraumeni Jr. and W. Blot, A case-control study of gastric cancer and diet in Italy, *Int. J. Cancer*, 1989, **44**, 611–616.

E. Bulska, I. A. Wysocka, M. H. Wierzbicka, K. Proost, K. Janssens and G. Falkenberg, *In vivo* investigation of the distribution and the local speciation of selenium in *Allium cepa* L. by means of microscopic X-ray absorption near-edge structure spectroscopy and confocal microscopic X-ray fluorescence analysis, *Anal. Chem.*, 2006, **78**, 7616–7624.

J. M. Burke, A. Wells, P. Casey and J. E. Miller, Garlic and papaya lack control over gastrointestinal nematodes in goats and lambs, *Veterinary Parasitol.*, 2009, **159**, 171–174.

B. E. Burnham, Garlic as a possible risk for postoperative bleeding, *Plastic Reconstructive Surgery*, 1995, **95**, 213.

R. C. Cahn, C. K. Ingold and V. Prelog, Specification of molecular chirality, *Angew. Chem., Int. Edn. Engl.*, 1966, **5**, 385–415.

M. A. Cahours, *Leçons de Chimie Générale'lémentaire*, Mallet-Bachelier, Paris, 1856.

M. A. Cahours and A. W. Hofmann, Note on a new class of alcohols, *Proc. R. Soc. London*, 1856–1857, **8**, 33–40.

X.-J Cai, P. C. Uden, J. J. Sullivan, B. D. Quimby and E. Block, Headspace–gas chromatography with atomic emission and mass selective detection for the determination of organoselenium compounds in elephant garlic, *Anal. Proc.*, 1994a, **31**, 325–327.

X.-J. Cai, P. C. Uden, E. Block, X. Zhang, B. D. Quimby and J. J. Sullivan, *Allium* chemistry: identification of natural abundance organoselenium volatiles from garlic, elephant garlic, onion, and Chinese chive using headspace gas chromatography with atomic emission detection, *J. Agric. Food Chem.*, 1994b, **42**, 2081–2084.

X.-J. Cai, E. Block, P. C. Uden, B. D. Quimby and J. J. Sullivan, *Allium* chemistry: identification of natural abundance organoselenium compounds in human breath after ingestion of garlic using gas chromatography with atomic emission detection, *J. Agric. Food Chem.*, 1995, **43**, 1751–1753.

X.-J. Cai, E. Block, P. C. Uden, X. Zhang, B. D. Quimby and J. J. Sullivan, *Allium* chemistry: identification of selenoamino acids in ordinary and selenium-enriched garlic, onion, and broccoli using gas chromatography with atomic emission detection, *J. Agric. Food Chem.*, 1995, **43**, 1754–1757.

S. H. Caldwell, L. J. Jeffers, O. S. Narula, E. A. Lang, K. R. Reddy and E. R. Schiff, Ancient remedies revisited: does *Allium sativum* (garlic) palliate the hepatopulmonary syndrome? *J. Clin. Gastroenterol.*, 1992, **15**, 248–250.

E. M. Calvey, J. A. G. Roach and E. Block, Supercritical fluid chromatography of garlic (*Allium sativum*) extracts with mass spectrometric identification of allicin., *J. Chromatogr. Sci.*, 1994, **32**, 93–96.

E. M. Calvey, J. E. Matusik, K. D. White, R. DeOrazio, D. Sha and E. Block, *Allium* chemistry: supercritical fluid extraction and LC–APCI–MS of thiosulfinates and related compounds from homogenates of garlic, onion and ramp. Identification in garlic and ramp and synthesis of 1-propanesulfinothioic acid *S*-allyl ester, *J. Agric. Food Chem.*, 1997, **45**, 4406–4413.

E. M. Calvey, K. D. White, J. E. Matusik, D. Sha and E. Block, *Allium* chemistry: identification of organosulfur compounds in ramp (*Allium tricoccum*) homogenates, *Phytochemistry*, 1998, **49**, 359–364.

V. Canduela, I. Mongil, M. Carrascosa, S. Docio and P. Cagigas, Garlic: always good for the health? *Br. J. Dermatol.*, 1995, **132**, 161–162.

J. Cannon and M. Cannon, *Dye Plants and Dyeing*, Timber Press, Portland, Oregon, 2003.

G. Cao, E. Sofic and R. L. Prior, Antioxidant capacity of tea and common vegetables, *J. Agric. Food Chem.*, 1996, **44**, 3426–3431.

M. Capraz, M. Dilek and T. Akpolat, Garlic, hypertension and patent education, *Int. J. Cardiology*, 2007, **121**, 130–131.

J. T. Carbery, A case of onion poisoning in a cow, *N. Z. Veterinary J.*, 1999, **47**, 184.

A. Cardelle-Cobas, F. J. Moreno, N. Corzo, A. Olano and M. Villamiel, Assessment of initial stages of Maillard reaction in dehydrated onion and garlic samples, *J. Agric. Food Chem.*, 2005, **53**, 9078–9082.

J. F. Carson and F. P. Wong, The volatile flavor components of onions, *J. Agric. Food Chem.*, 1961a, **9**, 140–143.

J. F. Carson and F. F. Wong, Isolation of (+)-*S*-methyl-L-cysteine sulfoxide and of (+)-*S*-N-propyl-L-cysteine sulfoxide from onions as their *N*-2,4-dinitrophenyl derivatives, *J. Org. Chem.*, 1961b, **26**, 4997–5000.

J. F. Carson and F. F. Wong, Synthesis of *cis*-*S*-(prop-1-enyl)-l-cysteine, *Chem. Ind.*, 1963, 1764–1765.

J. F. Carson, R. E. Lundin and T. M. Lukes, The configuration of (+)-*S*-(1-propenyl)-L-cysteine *S*-oxide from *Allium cepa*, *J. Org. Chem.*, 1966, **31**, 1634–1635.

M. J. Caterina, Chemical biology: sticky spices, *Nature*, 2007, **445**, 491–492.

B. Cavage, *The Elegant Onion*, Garden Way, Pownal, VT, 1987.

C. J. Cavallito and J. H. Bailey, Allicin, the antibacterial principle of *Allium sativum*. I. Isolation, physical properties and antibacterial action, *J. Am. Chem. Soc.*, 1944a, **66**, 1950–1951.

C. J. Cavallito, J. S. Buck and C. M. Suter, Allicin, the antibacterial principle of *Allium sativum*. II. Determination of the chemical structure, *J. Am. Chem. Soc.*, 1944b, **66**, 1952–1954.

C. J. Cavallito and J. H. Bailey, Preliminary note on the inactivation of antibiotics, *Science*, 1944c, **66**, 390.

C. J. Cavallito, J. H. Bailey and J. S. Buck, The Antibacterial principle of *Allium sativum*. III. Its precursor and "essential oil of garlic", *J. Am. Chem. Soc.*, 1945, **67**, 1032–1033.

C. J. Cavallito, Relationship of thiol structures to reaction with antibiotics, *J. Biol. Chem.*, 1946, **164**, 29–34.

C. J. Cavallito and L. V. D. Small, Hydrocarbon esters of hydrocarbonylthiolsulfinic acids and their process of preparation, *U.S. Pat.*, US2508745, 1950.

C. J. Cavallito, Extraction of garlic, *U.S. Pat.*, US2554088, 1951.

C. Cerella, C. Scherer, S. Cristofanon, E. Henry, A. Anwar, C. Busch, M. Montenarh, M. Dicato, C. Jacob and M. Diederich, Cell cycle arrest in early mitosis and induction of caspase-dependent apoptosis in U937 cells by diallyltetrasulfide (Al_2S_4), *Apoptosis*, 2009, **14**, 641–654.

CFSAN (Center for Food Safety and Applied Nutrition of the U.S. Food and Drug Administration), July 2007: http://www.cfsan.fda.gov/~dms/hclmgui5.html.

F. Challenger and D. Greenwood, Sulphur compounds of the genus *Allium*. Detection of *n*-propylthiol in the onion. The fission and methylation of diallyl disulphide in cultures of *Scopulariopsis brevicaulis*, *Biochem. J.*, 1949, **44**, 87–91.

C. C. Chan, H. C. Wu, C. H. Wu and C. Y. Hsu, Hepatopulmonary syndrome in liver cirrhosis: report of a case, *J. Formosan Med. Assoc.*, 1995, **94**, 185–188.

H.-P. Chang and Y.-H. Chen, Differential effects of organosulfur compounds from garlic oil on nitric oxide and prostaglandin E2 in stimulated macrophages, *Nutrition*, 2005, **21**, 530–536.

T. G. Chasteen, M. Wiggli and R. Bentley, Historical review. Of garlic, mice and Gmelin: the odor of trimethylarsine, *Appl. Organomet. Chem.*, 2002, **16**, 281–286.

H. Chen, A. Wortmann, W. Zhang and R. Zenobi, Rapid *in vivo* fingerprinting of nonvolatile compounds in breath by extractive electrospray ionization quadrupole time-of-flight mass spectrometry, *Angew. Chem., Int. Ed.*, 2007, **46**, 580–583.

W. Y. Chiam, Y. Huang, S. X. Chen and S. H. Ho, Toxic and antifeedant effects of allyl disulfide on *Tribolium castaneum* (Coleoptera: Tenebrionidae) and *Sitophilus zeamais* (Coleoptera: Curculionidae), *J. Econ. Entomol.*, 1999, **92**, 239–245.

J. Child, L. Bertholle and S. Beck, *Mastering the Art of French Cooking*, Knopf, New York, 1966.

H.-W. Chin and R. C. Lindsay, Mechanisms for formation of volatile sulfur compounds following the action of cysteine sulfoxide lyases, *J. Agric. Food Chem.*, 1994, **42**, 1529–1536.

J. Cho, E. J. Lee, K. S. Yoo, S. K. Lee and B. S. Patil, Identification of candidate amino acids involved in the formation of blue pigments in crushed garlic cloves (*Allium sativum* L.), *J. Food Sci.*, 2009, **74**, C11–16.

I.-H. Choi, S.-C. Shin and I.-K. Park, Nematicidal activity of onion (*Allium cepa*) oil and its components against the pine wood nematode (*Bursaphelenchus xylophilus*), *Nematology*, 2007, **9**, 231–235.

J. H. Choi and K. H. Kyung, Allyl alcohol is the sole antiyeast compound in heated garlic extract, *J. Food Sci.*, 2005, **70**, M305–M309.

R. Chowdhury, A. Dutta, S. R. Chaudhuri, N. Sharma, A. K. Giri and K. Chaudhuri, *In vitro* and *in vivo* reduction of sodium arsenite induced toxicity by aqueous garlic extract, *Food Chem, Toxicol.*, 2008, **46**, 740–751.

Q. Chu, M. T. Ling, H. Feng, H. W. Cheung, S. W. Tsao, X. Wang and Y. C. Wong, A novel anticancer effect of garlic derivatives: inhibition of cancer cell invasion through restoration of E-cadherin expression, *Carcinogenesis*, 2006, **27**, 2180–2189.

Q. Chu, D. T. Lee, S. W. Tsao, X. Wang and Y. C. Wong, *S*-allylcysteine, a water-soluble garlic derivative, suppresses the growth of a human androgen-independent prostate cancer xenograft, CWR22R, under *in vivo* conditions, *Br. J. Urol. Int.*, 2007, **99**, 925–932.

S. C. Chuah, P. K. Moore and Y. Z. Zhu, *S*-Allylcysteine mediates cardioprotection in an acute myocardial infarction rat model via a hydrogen sulfide-mediated pathway, *Am. J. Physiol. Heart Circulatory Physiol.*, 2007, **293**, H2693–H2701.

I. Chung, S. H. Kwon, S. T. Shim and K. H. Kyung, Synergistic antiyeast activity of garlic oil and allyl alcohol derived from alliin in garlic, *J. Food Sci.*, 2007, **72**, M437–440.

V. Q. Chung, L. Kelley, D. Marra and S. B. Jiang, Onion extract gel *versus* petrolatum emollient on new surgical scars: prospective double-blinded study, *Dermatol. Surgery*, 2006, **32**, 193–197.

L. C. Clark, G. F. Combs Jr., B. W. Turnbull, E. H. Slate, D. K. Chalker, J. Chow, L. S. Davis, R. Glover, G. F. Graham, E. G. Gross, A. Kongract, J. L. Lesher, H. K. Vark, B. B. Sanders Jr., C. L. Smith and J. R. Taylor, Effects of selenium supplementation for cancer prevention in patients with carcinoma of the skin. A randomized controlled trial. Nutritional Prevention of Cancer Study Group, *J. Am. Med. Assoc.*, 1996, **276**, 1957–1963.

L. F. Clarke, B. Baker, C. Trahan, L. Meyers and S. E. Metzinger, A prospective double-blinded study of Mederma skin care *vs.* placebo for post-traumatic scar reduction, *Cosmetic Dermatol.*, 1999, **12**, 19–26.

R. B. Cody, J. A. Laramee and H. D. Durst, Versatile new ion source for the analysis of materials in open air under ambient conditions, *Anal. Chem.*, 2005, **77**, 2297–2302.

D. A. Cogan, G. Liu, K. Kim, B. J. Backes and J. A. Ellman, Catalytic asymmetric oxidation of *tert*-butyl disulfide. Synthesis of *tert*-butanesulfinamides, *tert*-butyl sulfoxides, and *tert*-butanesulfinimines, *J. Am. Chem. Soc.*, 1998, **120**, 8011–8019.

J. R. Coley-Smith and D. Parfitt, Some effects of diallyl bisulphide on sclerotia of *Sclerotium cepivorum*: possible novel control method for white rot disease of onions, *Pest. Sci.*, 1986, **17**, 587–594.

S. Colonna, V. Pironti, J. Drabowicz, F. Brebion, L. Fensterbank and M. Malacria, Enantioselective synthesis of thiosulfinates and of acyclic alkylidenemethylene sulfide sulfoxides, *Eur. J. Org. Chem.*, 2005, 1727–1730.

R. G. Cooks, Z. Ouyang, Z. Takats and J. M. Wiseman, Ambient mass spectrometry, *Science*, 2006, **311**, 1566–1570.

R. B. Cope, *Allium* species poisoning in dogs and cats, *Veterinary Med.*, 2005, 562–566.

A. Coppi, M. Cabinian, D. Mirelman and P. Sinnis, Antimalarial activity of allicin, a biologically active compound from garlic cloves, *Antimicrob. Agents Chemotherapy*, 2006, **50**, 1731–1737.

M. Corzo-Martínez, N. Corzo and M. Villamiel, Biological properties of onions and garlic, *Trends Food Sci. Technol.*, 2007, **18**, 609–625.

D. Crawford, Garlic-growing and agricultural specialization in Graeco-Roman Egypt, *Chron. d'Égypte*, 1973, **48**, 350–363.

F. Crowe, Garlic and mosquitoes, *Science*, 1995, **269**, 1804–1805.

J. W. Cubbage, Y. Guo, R. D. McCulla and W. S. Jenks, Thermolysis of alkyl sulfoxides and derivatives: a comparison of experiment and theory, *J. Org. Chem.*, 2001, **66**, 8722–8736.

H. Curtis, U. Noll, J. Stoermann and A. J. Slusarenko, Broad-spectrum activity of the volatile phytoanticipin allicin in extracts of garlic (*Allium sativum* L.) against plant pathogenic bacteria, fungi and oomycetes, *Physiol. Mol. Plant Pathol.*, 2004, **65**, 79–89.

R. Dadd, The discovery and introduction of *Allium giganteum*, *Curtis's Botanical Magazine*, 1987, **4**, 91–96.

E. Daebritz and A. I. Virtanen, *S*-Vinylcysteine *S*-oxide; precursor of a new lachrymator, ethenesulfenic acid, *Acta Chem. Scand.*, 1964, **18**, 837–838.

S. Daiches and I. W. Slotki, (trans.) Kethuboth 75a, *The Babylonian Talmud*, ed. Isidore Epstein, Soncino Press, London, 1936, Vol. 4:470 (*Seder Nashim*).

A. Das, N. L. Banik and S. K. Ray, Garlic compounds generate reactive oxygen species leading to activation of stress kinases and cysteine proteases for apoptosis in human glioblastoma T98G and U87MG cells, *Cancer*, 2007, **110**, 1083–1095.

A. Davidson, *The Oxford Companion to Food*, Oxford University Press, Oxford, 1999.

D. Davies, *Alliums: the Ornamental Onions*, Timber Press, Portland, OR, 1992.

L. E. Davis, J. K. Shen and Y. Cai, Antifungal activity in human cerebrospinal fluid and plasma after intravenous administration of *Allium sativum*, *Antimicrob. Agents Chemotherapy*, 1990, **34**, 651–653.

L. E. Davis, J. Shen and R. E. Royer, *In vitro* synergism of concentrated *Allium sativum* extract and amphotericin B against *Cryptococcus neoformans*, *Planta Med.*, 1994, **60**, 546–549.

S. R. Davis, An overview of the antifungal properties of allicin and its breakdown products – the possibility of a safe and effective antifungal prophylactic, *Mycoses*, 2005, **48**, 95–100.

S. R. Davis, R. Perrie and R. Apitz-Castro, The *in vitro* susceptibility of *Scedosporium prolificans* to ajoene, allitridium and a raw extract of garlic (*Allium sativum*), *J. Antimicrob. Chemotherapy*, 2003, **51**, 593–597.

U. C. Davis, 2008: Postharvest handling systems: underground vegetables (roots, tubers and bulbs), University of California Cooperative Extension Vegetable Research and Information Center: http://vric.ucdavis.edu/selectnewtopic.undergnd.htm.

J. C. Day, New nitrogen bases with severe steric hindrance due to flanking *tert*-butyl groups. *cis*-2,6-Di-*tert*-butylpiperidine. Possible steric blocking of olfaction, *J. Org. Chem.*, 1978, **43**, 3646–3649.

B. M. de Rooij, P. J. Boogaard and D. A. Rijksen, J. N. Commandeur SNF N. P. Vermeulen, Urinary excretion of *N*-acetyl-*S*-allyl-L-cysteine upon garlic consumption by human volunteers, *Arch Toxicol.*, 1996, **70**, 635–639.

P. M. De Wet, H. Rode, D. Sidler and A. J. Lastovica, Allicin: a possible answer to antibiotic resistant campylobacter diarrhoeal infection? *Arch. Diseases Childhood*, 1999, **81**, 278.

W. Debin, G. Jiande and L. Guangshu, General situation of *Allium* crops in China, *Acta Horticulture*, 2005, **688**, 327–332.

M. S. Defelice, Wild garlic, *Allium vineale* L. – little to crow about, *Weed Technol.*, 2003, **17**, 890–895.

T. A. Delaney and A. M. Donnelly, Garlic dermatitis, *Australas. J. Dermatol.*, 1996, **37**, 109–110.

C. M. Dentinger, W. A. Bower, O. V. Nainan, S. M. Cotter, G. Myers, L. M. Dubusky, S. Fowler, E. D. Salehi and B. P. Bell, An outbreak of hepatitis A associated with green onions, *J. Infectious Diseases*, 2001, **183**, 1273–1276.

R. G. Deshpande, M. B. Khan, D. A. Bhat and R. G. Navalkar, Inhibition of *Mycobacterium avium* complex isolates from AIDS patients by garlic (*Allium sativum*), *J. Antimicrob. Chemotherapy*, 1993, **32**, 623–626.

E. Devrim and I. Durak, Is garlic a promising food for benign prostatic hyperplasia and prostate cancer? *Mol. Nutr. Food Res.*, 2007, **51**, 1319–1323.

M. Diemling, As the Jews like to eat Garlick, in *Food and Judaism. Studies in Jewish Civilization*, ed. L. J. Greenspoon, R. A. Simkins and G. Shapiro, Creighton University Press, Omaha, NE, 2002, vol. 15.

D. M Dietz, J. R. Varcelotti and K. R. Stahlfeld, Garlic burns: a not-so-rare complication of a naturopathic remedy? *Burns*, 2004, **30**, 612–613.

M. Dillon, *Girls and Women in Classical Greek Religion*, Routledge, London, UK, 2003.

H. Diner, *Hungering for America: Italian, Irish, and Jewish Foodways in the Age of Migration*, Harvard University Press, Cambridge, MA, 2001.

I. Dini, G. C. Tenore and A. Dini, *S*-Alkenyl cysteine sulfoxide and its antioxidant properties from *Allium cepa* var. *tropeana* (red onion) seed, *J. Nat. Prod.*, 2008, **71**, 2036–2037.

V. M. Dirsch, A. L. Gerbes and A. M. Vollmar, Ajoene, a compound of garlic, induces apoptosis in human promyeloleukemic cells, accompanied by generation of reactive oxygen species and activation of nuclear factor kappaB, *Mol. Pharmacol.*, 1998a, **53**, 402–407.

V. M. Dirsch, A. K. Kiemer, H. Wagner and A. M. Vollmar, Effect of allicin and ajoene, two compounds of garlic, on inducible nitric oxide synthase, *Atherosclerosis*, 1998b, **139**, 333–339.

V. M. Dirsch, D. S. Antlsperger, H. Hentze and A. M. Vollmar, Ajoene, an experimental anti-leukemic drug: mechanism of cell death, *Leukemia*, 2002, **16**, 74–83.

B. Dittmann, B. Zimmermann, C. Engelen, G. Jany and S. Nitz, Use of the MS-sensor to discriminate between different dosages of garlic flavoring in tomato sauce, *J. Agric. Food Chem.*, 2000, **48**, 2887–2892.

W. E. Dixon, The specific action of drugs in tuberculosis, *Br. Med. J.*, May 2, 1925, **3357**, 813–815.

L. Djurdjevic, A. Dinic, P. Pavlovic, M. Mitrovic, B. Karadzic and V. Tesevic, Allelopathic potential of *Allium ursinum* L., *Biochem. Syst. Ecol.*, 2004, **32**, 533–544.

Y. Dong, D. Lisk, E. Block and C. Ip, Characterization of the biological activity of γ-glutamyl-*Se*-methylselenocysteine: a novel, naturally occurring anticancer agent from garlic, *Cancer Res.*, 2001, **61**, 2923–2928.

J. A. Doran, J. S. O'Donnell, L. L. Lairson, M. R. McDonald, A. L. Schwan and B. Grodzinski, *S*-Alk(en)yl-L-cysteine sulfoxides and relative pungency measurements of photosynthetic and nonphotosynthetic tissues of *Allium porrum*, *J. Agric. Food Chem.*, 2007, **55**, 8243–8250.

W. Dorsch, H. Wagner, T. Bayer, B. Fessler, G. Hein, J. Ring, P. Scheftner, W. Sieber, T. Strasser and E. Weiss, Anti-asthmatic effects of onions. Alk(e-n)ylsulfinothioic acid alk(en)yl-esters inhibit histamine release, leukotriene and thromboxane biosynthesis *in vitro* and counteract PAF and allergen-induced bronchial obstruction *in vivo*, *Biochem. Pharmacol.*, 1988, **37**, 4479–4486.

W. Dorsch, E. Schneider, T. Bayer, W. Breu and H. Wagner, Anti-inflammatory effects of onions: inhibition of chemotaxis of human polymorphonuclear leukocytes by thiosulfinates and cepaenes, *Int. Arch. Allergy Appl. Immunol.*, 1990, **92**, 39–42.

J. Dostrovsky, as quoted in a University of Toronto news release, 10/3/2003: http://www.news.utoronto.ca/bios/askus39.htm.

Z. D. Draelos, The ability of onion extract gel to improve the cosmetic appearance of postsurgical scars, *J. Cosmetic Dermatol.*, 2008, **7**, 101–104.

G. Duda, J. Suliburska and D. Pupek-Musialik, Effects of short-term garlic supplementation on lipid metabolism and antioxidant status in hypertensive adults, *Pharmacol. Reports*, 2008, **60**, 163–170.

S. Dugravot, E. Thibout, A. Abo-Ghalia and J. Huignard, How a specialist and a non-specialist insect cope with dimethyl disulfide produced by *Allium porrum*, *Entomol. Exp. Appl.*, 2004, **113**, 173–179.

S. Dugravot, N. Mondy, N. Mandon and E. Thibout, Increased sulfur precursors and volatiles production by the leek *Allium porrum* in response to specialist insect attack, *J. Chem. Ecol.*, 2005, **31**, 1299–1314.

S. Dugravot and E. Thibout, Consequences for a specialist insect and its parasitoid of the response of *Allium porrum* to conspecific herbivore attack, *Physiol. Entomol.*, 2006, **31**, 73–79.

E. Dumont, Y. Ogra, F. Vanhaecke, K. T. Suzuki and R. Cornelis, Liquid chromatography–mass spectrometry (LC–MS): a powerful combination for selenium speciation in garlic (*Allium sativum*), *Anal. Bioanal. Chem.*, 2006, **384**, 1196–1206.

I. Durak, E. Yilmaz, E. Devrim, H. Perk and M. Kacmaz, Consumption of aqueous garlic extract leads to a significant improvement in patients with benign prostate hyperplasia and prostate cancer, *Nutr. Res.*, 2003, **23**, 199–204.

R. D. Durbin and T. F. Uchytil, Role of allicin in the resistance of garlic to Penicillium species, *Phytopathologia Mediterranea*, 1971, **10**, 227–230.

M. Durenkamp and L. J. De Kok, Impact of pedospheric and atmospheric sulphur nutrition on sulphur metabolism of *Allium cepa* L., a species with a potential sink capacity for secondary sulphur compounds, *J. Exp. Botany*, 2004, **55**, 1821–1830.

M. Durenkamp, L. J. De Kok and S. Kopriva, Adenosine 59-phosphosulphate reductase is regulated differently in *Allium cepa* L. and *Brassica oleracea* L. upon exposure to H_2S, *J. Exp. Botany*, 2007, **58**, 1571–1579.

C. C. Eady, T. Kamoi, M. Kato, N. G. Porter, S. Davis, M. Shaw, A. Kamoi and S. Imai, Silencing onion lachrymatory factor synthase causes a significant change in the sulfur secondary metabolite profile, *Plant Physiol.*, 2008, **147**, 2096–2106.

C. Eberhardie, Nutritional supplements and the EU: is anyone happy? *Proc. Nutr. Soc.*, 2007, **66**, 508–511.

A. I. M. Ebid, A. G. Hassan and S. A. Mohammed, Aspirin, garlic, and morbidity in patients with cardiovascular disorders: a prospective study, *Egyptian J. Biomed. Sci.*, 2006, **22**, 219–236.

A. J. Edelstein, Dermatitis caused by garlic, *Arch. Dermatol.*, 1950, **61**, 111.

D. M. Eisenberg, R. B. Davis, S. L. Ettner, S. Appel, S. Wilkey, M. Van Rompay and R. C. Kessler, Trends in alternative medicine use in the United States, 1990–1997: results of a follow-up national survey, *J. Am. Med. Assoc.*, 1998, **280**, 1569–1575.

A. L. Ekeowa-Anderson, B. Shergill and P. Goldsmith, Allergic contact cheilitis to garlic, *Contact Dermatitis*, 2007, **56**, 174–175.

K. El-Bayoumy, R. Sinha, J. T. Pinto and R. S. Rivlin, Cancer chemoprevention by garlic and garlic-containing sulfur and selenium compounds, *J. Nutr*, 2006, **136** (3 Suppl), 864S–869S.

G. S. Ellmore and R. S. Feldberg, Alliin lyase localization in bundle sheaths of the garlic clove (*Allium sativum*), *Am. J. Botany*, 1994, **81**, 89–94.

J. W. Elrod, J. W. Calvert, J. Morrison, J. E. Doeller, D. W. Kraus, L. Tao, X. Jiao, R. Scalia, L. Kiss, C. Szabo, H. Kimura, C. W. Chow and D. J. Lefer, Hydrogen sulfide attenuates myocardial ischemia-reperfusion injury by preservation of mitochondrial function, *Proc. Natl. Acad. Sci. U.S.A.*, 2007, **104**, 15560–15565.

F. El-Sabban and H. Abouazra, Effect of garlic on atherosclerosis and its factors, *East Mediterr. Health J.*, 2008, **14**, 195–205.

K. A. A. El-Shaikh and M. A. Bekheet, Effect of intercropping faba bean and garlic on sugar beet in the newly reclaimed soils, *Assiut J. Agric. Sci.*, 2004, **35**, 187–204.

E. Ernst, Garlic, *Mol. Nutr. Food Res.*, 2007, **51**, 1317.

S. Espirito Santo, H. -P. Keiss, K. Meyer, R. Buytenhek, Th. Roos, V. Dirsch, G. Buniatian, C. Ende, J. Günther, K. Heise, D. Kellert, K. Lerche, S. Pavlica, F. Struck, E. Usbeck, J. Voigt, S. Zellmer, J. M. G. Princen, A. M. Vollmar and R. Gebhardt, Garlic and cardiovascular diseases, *Medicinal Aromatic Plant Sci. Biotech.*, 2007, **1**, 31–36.

J. A. Esté and E. De Clercq, Ajoene [(*E*,*Z*)-4,5,9-trithiadodeca-1,6,11-triene 9-oxide] does not exhibit antiviral activity at subtoxic concentrations, *Biomed. Pharmacotherapy*, 1998, **52**, 236–238.

T. Etoh and P. W. Simon, Diversity, fertility and seed production in garlic, in *Allium Crop Science: Recent Advances*, ed. H. D. Rabinowitch and L. Currah, CABI, Wallingford, UK, 2002, ch. 5.

T. Ettala and A. I. Virtanen, Labeling of sulfur-containing amino acids and γ-glutamylpeptides after injection of labeled sulfate into onion (*Allium cepa*), *Acta Chem. Scand.*, 1962, **16**, 2061–2063.

European Court of Justice, Garlic extract powder capsules are not medicinal products, 11/15/2007, Case C–319/05.

European Scientific Cooperative on Phytotherapy, *Allii Sativi Bulbus* – Garlic, *ESCOP Monographs*, ESCOP–Georg Thieme Verlag, Exeter, UK, 2nd edn, 2003, pp. 14–25.

V. Evans, Herbs and the brain: friend or foe? The effects of ginkgo and garlic on warfarin use, *J. Neurosci. Nursing*, 2000, **32**, 229–232.

S. A. Everett, L. K. Folkes, P. Wardman and K. D. Asmus, Free-radical repair by a novel perthiol: reversible hydrogen transfer and perthiyl radical formation, *Free Radical Res.*, 1994, **20**, 387–400.

S. A. Everett and P. Wardman, Perthiols as antioxidants: radical-scavenging and prooxidative mechanisms, *Methods Enzymol.*, 1995, **251** (Biothiols, Part A), 55–69.

A. E. Falleroni, C. R. Zeiss and D. Levitz, Occupational asthma secondary to inhalation of garlic dust, *J. Allergy Clin. Immunol.*, 1981, **68**, 156–160.

M. B. Fallon, G. A. Abrams, T. T. Abdel-Razek, J. Dai, S.-J. Chen, Y.-F. Chen, B. Luo, S. Oparil and D. D. Ku, Garlic prevents hypoxic pulmonary hypertension in rats, *Am. J. Physiol. Lung Cell Mol. Physiol.*, 1998, **275**, L283–287.

FAO (Food and Agricultural Organization of the United Nations), 2005 statistics: http://faostat.fao.org/site/336/DesktopDefault.aspx.

P. Farkas, P. Hradsky and M. Kovac, Novel flavor components identified in the steam distillate of onion (*Allium cepa*), *Z. Lebensm.-Unters. -Forsch.*, 1992, **195**, 459–462.

A. M. Farrell and R. C. D. Staughton, Garlic burns mimicking herpes zoster, *Lancet*, 1996, **347** (9009), 1195.

K. T. Farrell, *Spices, Condiments and Seasonings*, Van Nostrand Reinhold, New York, 2nd edn, 1990.

FDA, 2006, ephedra ban: http://www.fda.gov/bbs/topics/NEWS/2006/NEW01434.html.

FDA, 2007, CGMPs: http://www.fda.gov/consumer/updates/dietarysupps062207.html.

FDA, 2009, information on DSHEA and DSNDCPA: http://www.cfsan.fda.gov/~dms/supplmnt.html; http://frwebgate.access.gpo.gov/cgi-bin/getdoc.cgi?dbname=109_ cong_public_laws&docid=f:publ462.109.pdf.

S. Ferary, E. Thibout and J. Auger, Direct analysis of odors emitted by freshly cut *Allium* using combined high-performance liquid chromatography and mass spectrometry, *Rapid Commun. Mass Spectrom.*, 1996a, **10**, 1327–1332.

S. Ferary and J. Auger, What is the true odor of cut *Allium*? Complementarity of various hyphenated methods: gas chromatography–mass spectrometry and high-performance liquid chromatography–mass spectrometry with particle beam and atmospheric pressure ionization interfaces in sulfenic acids rearrangement components discrimination, *J. Chromatogr., A*, 1996b, **750**, 63–74.

S. Ferary, J. Keller, J. Boscher and J. Auger, Fast narrow-bore HPLC-DAD analysis of biologically active thiosulfinates obtained without solvent from wild *Allium* species, *Biomed. Chromatog.*, 1998, **12**, 104–106.

N. Ferri, K. Yokoyama, M. Sadilek, R. Paoletti, R. Apitz-Castro, M. H. Gelb and A. Corsini, Ajoene, a garlic compound, inhibits protein prenylation and arterial smooth muscle cell proliferation, *Brit. J. Pharmacol.*, 2003, **138**, 811–818.

B. Field, F. Jordán and A. Osbourn, First encounters – deployment of defence-related natural products by plants, *New Phytol.*, 2006, **172**, 193–207.

G. Filomeni, G. Rotilio and M. R. Ciriolo, Molecular transduction mechanisms of the redox network underlying the antiproliferative effects of allyl compounds from garlic, *J. Nutr.*, 2008, **138**, 2053–2057.

H. M. Flint, N. J. Parks, J. E. Holmes, J. A. Jones and C. M. Higuera, Tests of garlic oil for the control of the silverleaf whitefly, *Bemisia argentifolia* Bellows and Perring (Homoptera: Aleyrodidae) in cotton, *Southwestern Entomologist*, 1995, **20**, 137–150.

S. J. Flora, A. Mehta and R. Gupta, Prevention of arsenic-induced hepatic apoptosis by concomitant administration of garlic extracts in mice, *Chem.-Biol. Interact.*, 2009, **177**, 227–233.

T. Fossen and O. M. Andersen, Malonated anthocyanins of garlic *Allium sativum* L., *Food Chem.*, 1997, **58**, 215–217.

T. E. Fox, C. Atherton, J. R. Dainty, D. J. Lewis, N. J. Langford, M. J. Baxter, H. M. Crews and S. J. Fairweather-Tait, Absorption of selenium from wheat, garlic, and cod intrinsically labeled with Se-77 and Se-82 stable isotopes, *Int. J. Vitamin Nutr. Res.*, 2005, **75**, 179–186.

F. Freeman, *vic*-Disulfoxides and OS-sulfenyl sulfinates, *Chem. Rev.*, 1984, **84**, 117–135.

T. Friedman, A. Shalom and M. Westreich, Self-inflicted garlic burns: our experience and literature review, *Int. J. Dermatol.*, 2006, **45**, 1161–1163.

R. M. Fritsch and M. Keusgen, Occurrence and taxonomic significance of cysteine sulphoxides in the genus *Allium L.* (Alliaceae), *Phytochemistry*, 2006, **67**, 1127–1135.

R. A. Fromtling and G. S. Bulmer, *In vitro* effect of aqueous extract of garlic (*Allium sativum*) on the growth and viability of *Cryptococcus neoformans*, *Mycologia*, 1978, **70**, 397–405.

H. Fujisawa, K. Suma, K. Origuchi, H. Kumagai, T. Seki and T. Ariga, Biological and chemical and stability of garlic-derived allicin, *J. Agric. Food Chem.*, 2008a, **56**, 4229–4235.

H. Fujisawa, K. Suma, K. Origuchi, T. Seki and T. Ariga, Thermostability of allicin determined by chemical and biological assays, *Biosci. Biotechnol. Biochem.*, 2008, **72**, 2877–2883.

M. Fujiwara, M. Yoshimura and S. Tsuno, Allithiamine, a newly found derivative of vitamin B_1. III. Allicin homologs in *Allium* plants, *J. Biochem. (Tokyo)*, 1955, **42**, 591–601 (*Chem. Abstr.*, 1956, **50**, 5213).

M. Fujiwara, M. Yoshimura, S. Tsuno and F. Murakami, Allithiamine, a newly found derivative of vitamin B_1. IV. The alliin homologs in vegetables, *J. Biochem.* (Tokyo), 1958, **45**, 141–149 (*Chem. Abstr.*, 1958, **52**, 73016).

W. Gaffield, F. F. Wong and J. F. Carson, Configurational relationships among sulfinyl amino acids, *J. Org. Chem.*, 1965, **30**, 951–952.

M. H. Gail, R. M. Pfeiffer, L. M. Brown, L. Zhang, J. L. Ma, K. F. Pan, W. D. Liu and W. C. You, Garlic, vitamin, and antibiotic treatment for *Helicobacter pylori*: a randomized factorial controlled trial, *Helicobacter*, 2007, **12**, 575–578.

C. Galeone, C. Pelucchi, F. Levi, E. Negri, S. Franceschi, R. Talamini, A. Giacosa and C. La Vecchia, Onion and garlic use and human cancer, *Am. J. Clin. Nutr.*, 2006, **84**, 1027–1032.

C. Galeone, C. Pelucchi, L. Dal Maso, E. Negri, M. Montella, A. Zucchetto, R. Talamini and C. La Vecchia, *Allium* vegetables intake and endometrial cancer risk, *Public Health Nutr.*, 2009, **12**, 1576–1579.

C. Galeone, A. Tavani, C. Pelucchi, E. Negri and C. La Vecchia, *Allium* vegetable intake and risk of acute myocardial infarction in Italy, *Eur. J. Nutr.*, 2009, **48**, 120–123.

M. R. Gamboa-León, I. Aranda-González, M. Mut-Martín, M. R. García-Miss and E. Dumonteil, *In vivo* and *in vitro* control of *Leishmania mexicana* due to garlic-induced NO production, *Scand. J. Immunol.*, 2007, **66**, 508–514.

P. H. Gann, Randomized trials of antioxidant supplementation for cancer prevention, *J. Am. Med. Assoc.*, 2009, **301**, 102–103.

L. A. Gapter, O. Z. Yuin and K. Y. Ng, *S*-Allylcysteine reduces breast tumor cell adhesion and invasion, *Biochem. Biophys. Res. Commun.*, 2008, **367**, 446–451.

J. García-Añoveros and K. Nagata, TRPA1, *Handb. Exp. Pharmacol.*, 2007, **179**, 347–362.

C. D. Gardner, L. D. Lawson, E. Block, L. M. Chatterjee, A. Kiazand, R. R. Balise and H. C. Kraemer, The effect of raw garlic *vs.* garlic supplements on plasma lipids concentrations in adults with moderate hypercholesterolemia: a randomized clinical trial, *Arch. Int. Med.*, 2007, **167**, 346–353.

C. D. Gardner, L. D. Lawson and E. Block, Effects of Garlic on Cholesterol: not down but not out either, *Arch. Int. Med.*, 2008, **168**, 111–112.

B.-Z. Garty, Garlic burns, *Pediatrics*, 1993, **91**, 658–659.

H. Gautier, J. Auger, C. Legros and B. Lapied, Calcium-activated potassium channels in insect pacemaker neurons as unexpected target site for the novel fumigant dimethyl disulfide, *J. Pharmacol. Exp. Ther.*, 2008, **324**, 149–159.

R. Gebhardt and H. Beck, Differential inhibitory effects of garlic-derived organosulfur compounds on cholesterol biosynthesis in primary rat hepatocyte cultures, *Lipids*, 1996, **31**, 1269–1276.

K. C. George, S. V. Amonkar and J. Eapen, Effect of garlic oil on incorporation of amino acids into proteins of *Culex pipiens quinquefasciatus* Say larvae, *Chem.-Biol. Interac.*, 1973, **6**, 169–75.

E. Germain, J. Auger, C. Ginies, M.-H. Siess and C. Teyssier, *In vivo* metabolism of diallyl disulphide in the rat: identification of two new metabolites, *Xenobiotica*, 2002, **32**, 1127–1138.

E. Germain, J. Chevalier, M. H. Siess and C. Teyssier, Hepatic metabolism of diallyl disulphide in rat and man, *Xenobiotica*, 2003, **33**, 1185–99.

E. Germain, E. Semon, M. H. Siess and C. Teyssier, Disposition and metabolism of dipropyl disulphide *in vivo* in rat, *Xenobiotica*, 2008, **38**, 87–97.

M. R. Gholami and M. Izadyar, A joint experimental and computational study on the kinetic and mechanism of diallyl disulfide pyrolysis in the gas phase, *Chem. Phys.*, 2004, **301**, 45–51.

J. Gill, *Essential Gaudí*, Paragon, Bath, UK, 2001.

J. Z. Gillies, E. A. Cotter, C. W. Gillies, H. E. Warner and E. Block, The rotational spectra, molecular structure and electric dipole moment of propanethial *S*-oxide, *J. Phys. Chem.*, 1999, **103**, 4948–4954.

R. Gmelin, H.-H. Huxa, K. Roth and G. Höfle, Dipeptide precursor of garlic odour in Marasmius species, *Phytochemistry*, 1976, **15**, 1717–1721.

A. Gonen, D. Harats, A. Rabinkov, T. Miron, D. Mirelman, M. Wilchek, L. Weiner, E. Ulman, H. Levkovitz, D. Ben-Shushan and A. Shaish, The antiatherogenic effect of allicin: possible mode of action, *Pathobiology*, 2005, **72**, 325–334.

C. A. Gonzalez and E. Riboli, Diet and cancer: where we are, where we are going, *Nutr. Cancer*, 2006, **56**, 225–231.

E. A. Goreshnik, D. Shollmeier and V. V. Oliinik, Copper(I) tetrafluoroborate with diallyl sulfide: the synthesis and crystal structure of the π-complex of equimolar composition, *Russ. J. Coord. Chem.*, 1997, **23**, 725–728.

E. A. Goreshnik and M. G. Mys'kiv, Synthesis and crystal structure of the copper(I) chloride π-complex with diallyl sulfide, $2CuCl.(C_3H_5)_2S$, *Russ. J. Coord. Chem.*, 1999, **25**, 137–140.

S. Gorinstein, Z. Jastrzebski, J. Namiesnik, H. Leontowicz, M. Leontowicz and S. Trak, The atherosclerotic heart disease and protecting properties of garlic: contemporary data, *Mol. Nutr. Food Res.*, 2007, **51**, 1365–1381.

R. Gorton, Garlic song: http://www.mudcat.org/@displaysong.cfm?SongID=2178.

K. Goto, M. Holler and R. Okazaki, Synthesis, structure, and reactions of a sulfenic acid bearing a novel bowl-type substituent: the first synthesis of a stable sulfenic acid by direct oxidation of a thiol, *J. Am. Chem. Soc.*, 1997, **119**, 1460–1461.

E. Gowers, *The Loaded Table. Representations of Food in Roman Literature*, Clarendon, Oxford, 1993.

D. Y. Graham, S.-Y. Anderson and T. Lang, Garlic or jalapeno peppers for the treatment of *Helicobacter pylori* infection, *Am. J. Gastroenterol.*, 1994, **94**, 1200–1202.

B. Granroth, Separation of *Allium* sulfur amino acids and peptides by thin-layer electrophoresis and thin-layer chromatography, *Acta Chem. Scand.*, 1968, **22**, 3333–3335.

B. Granroth, Biosynthesis and decomposition of cysteine derivatives in onion and other *Allium* species, *Ann. Acad. Sci. Fenn. Ser. A2*, 1970, **154**, 1–71.

M. Granvogl, M. Christlbauer and P. Schieberle, Quantitation of the intense aroma compound 3-mercapto-2-methylpentan-1-ol in raw and processed onions (*Allium cepa*) of different origins and in other *Allium* varieties using a stable isotope dilution assay, *J. Agric. Food Chem.*, 2004, **52**, 2797–2802.

M. Graubard, *Man's Food, its Rhyme and Reason*, Macmillan, New York, 1943.

D. D. Gregor and W. S. Jenks, Computational investigation of vicinal disulfoxides and other sulfinyl radical dimers, *J. Phys. Chem. A*, 2003, **107**, 3414–3423.

M. Gregory, R. M. Fritsch, N. W. Friesen, F. O. Khassanov and D. W. McNeal, *Nomenclator Alliorum*, Royal Botanic Gardens, Kew, UK, 1998.

R. M. Gries, G. G. Gries, G. Khaskin, N. Avelino and C. Cambell, Compounds, compositions and methods of repelling blood-feeding arthropods and deterring their landing and feeding, *U.S. Pat., Appl.* US2009/0069407 A1, 2009.

M. R. Groom, ECOspray products as a nematicide: a report submitted to the UK Pesticide Safety Directorate, 2004.

M. Groom and D. Sadler-Bridge, A pesticide and repellant. *PCT Int. Appl.*, WO 2006109028 A1 20061019, 2006.

F. C. Groppo, J. C. Ramacciato, R. P. Simoes, F. M. Florio and A. Sartoratto, Antimicrobial activity of garlic, tea tree oil, and chlorhexidine against oral microorganisms, *Int. Dentistry J.*, 2002, **52**, 433–437.

F. C. Groppo, J. C. Ramacciato, R. H. Motta, P. M. Ferraresi and A. Sartoratto, Antimicrobial activity of garlic against oral streptococci, *Int. J. Dental Hygiene*, 2007, **5**, 109–15.

R. T. Gunther, *The Greek Herbal of Dioscorides*, Hafner Publishing Company, 1959.

N. L. Guo, D. P. Lu, G. L. Woods, E. Reed, G. Z. Zhou, L. B. Zhang and R. H. Waldman, Demonstration of the anti-viral activity of garlic extract against human cytomegalovirus *in vitro*, *Chin. Med. J. (Engl.)*, 1993, **106**, 93–96.

R. Gupta and N. K. Sharma, A study of the nematicidal activity of allicin – an active principle in garlic, *Allium sativum* L., against root-knot nematode, *Meloidogyne incognita* (Kofoid and White, 1919) Chitwood, 1949, *Int. J. Pest Management*, 1993, **39**, 390–392.

B. J. Gurley, S. F. Gardner, M. A. Hubbard, D. K. Williams, W. B. Gentry, Y. Cui and C. Y. Ang, Clinical assessment of effects of botanical supplementation on cytochrome P450 phenotypes in the elderly: St John's wort, garlic oil, *Panax ginseng* and *Ginkgo biloba*, *Drugs Aging*, 2005, **22**, 525–539.

G. Gurusubramanian and S. S. Krishna, The effects of exposing eggs of four cotton insect pests to volatiles of *Allium sativum* (Liliaceae), *Bull. Entomol. Res.*, 1996, **86**, 29–31.

I. Guvenc and E. Yildirim, Increasing productivity with intercropping systems in cabbage production, *J. Sustainable Agric.*, 2006, **28**, 29–44.

H. O. Hall, The onion (*Allium cepa*) and garlic (*Allium sativa*) as a remedy for pneumonia and pulmonary tuberculosis, *Am. Med.*, 1913, **19**, 26–34.

C. H. Halsted, Dietary supplements and functional foods: 2 sides of a coin? *Am. J. Clin. Nutr.*, 2003, **77** (4 Suppl), 1001S–1007S.

C. M. Harris, S. C. Mitchell, R. H. Waring and G. L. Hendry, The case of the black-speckled dolls: an occupational hazard of unusual sulphur metabolism, *Lancet*, 1986, **1**, 492–493.

C. M. Harris, Curiosity, *J. R. Soc. Med.*, 1986, **79**, 319–322.

J. C. Harris, S. Plummer, M. P. Turner and D. Lloyd, The microaerophilic flagellate *Giardia intestinalis: Allium sativum* (garlic) is an effective antigiardial, *Microbiology*, 2000, **146**, 3119–3127.

L. Harris, *The Book of Garlic*, Addision-Wesley, Reading, MA, 1995.

W. L. Hasler, Garlic breath explained: why brushing your teeth won't help, *Gastroenterology*, 1999, **117**, 1248–1250.

H. T. Hassan, Ajoene (natural garlic compound): a new anti-leukaemia agent for AML therapy, *Leukemia Res.*, 2004, **28**, 667–671.

J. Hasserodt, H. Pritzkow and W. Sundermeyer, Partially and perfluorinated thioketones and thioaldehydes: chemical storage, in situ generation and surprising reactivity towards bis(trimethylstannyl)diazomethane and C,N-bis (triisopropylsilyl)nitrilimine, *Liebigs Ann.*, 1995, 95–104.

D. L. Hatfield, B. A. Carlson, X.-M. Xu, H. Mix and V. N. Gladyshev, *Prog. Nucleic Acid Res. Mol. Biol.*, 2006, **81**, 97.

M. J. Havey, Advances in new alliums, in *Perspectives on New Crops and New Uses*, ed. J. Janick, ASHS Press, Alexandria, VA, 1999.

Health Canada, Natural Health Products, Monograph on Garlic (May 2008): http://www.hc-sc.gc.ca/dhp-mps/prodnatur/applications/licen-prod/monograph/mono_garlic-ail-eng.php.

A. M. Heck, B. A. DeWitt and A. L. Lukes, Potential interactions between alternative therapies and warfarin, *Am. J. Health-System Pharmacy*, 2000, **57**, 1221–1227.

D. W. Held, P. Gonsiska and D. A. Potter, Evaluating companion planting and non-host masking odors for protecting roses from the Japanese beetle (*Coleoptera: Scarabaeidae*), *J. Economic Entomol.*, 2003, **96**, 81–87.

A. Herman-Antosiewicz and S. V. Singh, Signal transduction pathways leading to cell cycle arrest and apoptosis induction in cancer cells by *Allium* vegetable-derived organosulfur compounds: a review, *Mutation Res./Fundamental. Mol. Mechanisms Mutagenesis*, 2004, **555**, 121–131.

A. Herman-Antosiewicz, A. A. Powolny and S. V. Singh, Molecular targets of cancer chemoprevention by garlic-derived organosulfides, *Acta Pharmacol. Sinica*, 2007, **28**, 1355–1364.

O. Higuchi, K. Tateshita and H. Nishimura, Antioxidative activity of sulfur-containing compounds in *Allium* species for human low-density lipoprotein (LDL) oxidation *in vitro*, *J. Agric. Food Chem.*, 2003, **51**, 7208–7214.

A. G. Hile, Z. Shan, S.-Z. Zhang and E. Block, Aversion of European starlings (*Sturnus vulgaris*) to garlic oil treated granules: garlic oil as an avian repellent, *J. Agric. Food Chem.*, 2004, **52**, 2192–2196.

M. A. Hilliard, C. Bergamasco, S. Arbucci, R. H. Plasterk and P. Bazzicalupo, Worms taste bitter: ASH neurons, QUI-1, GPA-3 and ODR-3 mediate quinine avoidance in *Caenorhabditis elegans*, *EMBO J.*, 2004, **23**, 1101–1011.

R. Hine, The crystal structure and molecular configuration of (+)-*S*-methyl-L-cysteine sulfoxide, *Acta Crystallogr.*, 1962, **15**, 635–642.

A. Hinman, H. H. Chuang, D. M. Bautista and D. Julius, TRP channel activation by reversible covalent modification, *Proc. Natl. Acad. Sci. U.S.A.*, 2006, **103**, 19564–19568.

K. Hirsch, M. Danilenko, J. Giat, T. Miron, A. Rabinkov, M. Wilchek, D. Mirelman, J. Levy and Y. Sharoni, Effect of purified allicin, the major ingredient of freshly crushed garlic, on cancer cell proliferation, *Nutr. Cancer*, 2000, **38**, 245–254.

S. D. Hiscock, N. S. Isaacs, M. D. King, R. E. Sue, R. H. White and D. J. Young, Desulfination of allylic sulfinic acids: characterization of a retro-ene transition state, *J. Org. Chem.*, 1995, **60**, 7166–7169.

B. Hiyasat, D. Sabha, K. Grötzinger, J. Kempfert, J. W. Rauwald, F. W. Mohr and S. Dhein, Antiplatelet activity of *Allium ursinum* and *Allium sativum*, *Pharmacology*, 2009, **83**, 197–204.

W. S. Ho, S. Y. Ying, P. C. Chan and H. H. Chan, Use of onion extract, heparin, allantoin gel in prevention of scarring in chinese patients having laser removal of tattoos: a prospective randomized controlled trial, *Dermatol. Surgery*, 2006, **32**, 891–896.

G. Hodge, S. Davis, M. Rice, H. Tapp, B. Saxon and T. Revesz, Garlic compounds selectively kill childhood pre-B acute lymphoblastic leukemia cells *in vitro* without reducing T-cell function: potential therapeutic use in the treatment of ALL, *J. Biol.: Targets Therapy*, 2008, **1**, 143–149.

G. Hoefle and J. E. Baldwin, Thiosulfoxides. Intermediates in rearrangement and reduction of allylic disulfides, *J. Am. Chem. Soc.*, 1971, **93**, 6307–6308.

J. H. Hofenk de Graaff, *The Colourful Past. Origins, Chemistry, and Identification of Natural Dyestuffs*, Archetype Publications Ltd., London, 2004.

A. W. Hofmann, Remarks on a new class of alcohols (second note.)[abstract]. *Proc. R. Soc. London*, 1856–1857, **8**, 511–515.

A. W. Hofmann and A. Cahours, Researches on a new class of alcohols, *Philos. Trans. R. Soc. London*, 1857, **147**, 555–574.

K. Holder and G. Duff, *A Clove of Garlic*, Reader's Digest Association, London, UK, 1996.

Y. S. Hong, Y. A. Ham, J. H. Choi and J. Kim, Effects of allyl sulfur compounds and garlic extract on the expression of Bcl-2, Bax, and p53 in non small cell lung cancer cell lines, *Exp. Mol. Med.*, 2000, **32**, 127–134.

H. Horie and K. Yamashita, Non-derivatized analysis of methiin and alliin in vegetables by capillary electrophoresis, *J. Chromatogr., A*, 2006, **1132**, 337–339.

R. K. Hosamani, R. Gulati, S. K. Sharma and R. Kumar, Efficacy of some botanicals against ectoparasitic mite, *Tropilaelaps clareae* (Acari: Laelapidae) in *Apis mellifera* colonies, *Systematic Appl. Acarol.*, 2007, **12**, 99–108.

M. Hosnuter, C. Payasli, A. Isikdemir and B. Tekerekoglu, The effects of onion extract on hypertrophic and keloid scars, *J. Wound Care*, 2007, **16**, 251–254.

T. Hosono, T. Hosono-Fukao, K. Inada, R. Tanaka, H. Yamada, Y. Iitsuka, T. Seki, I. Hasegawa and T. Ariga, Alkenyl group is responsible for the disruption of microtubule network formation in human colon cancer cell line HT-29 cells, *Carcinogenesis*, 2008, **29**, 1400–1406.

T. Hosono, T. Fukao, J. Ogihara, Y. Ito, H. Shiba, T. Seki and T. Ariga, Diallyl trisulfide suppresses the proliferation and induces apoptosis of human colon cancer cells through oxidative modification of beta-tubulin, *J. Biol. Chem.*, 2005, **280**, 41487–41493.

Y. Hou, I. A. Abu-Yousef, Y. Doung and D. N. Harpp, Sulfur-atom insertion into the S-S bond – formation of symmetric trisulfides, *Tetrahedron Lett.*, 2001, **42**, 8607–8610.

Y. Hou, I. A. Abu-Yousef, A. Imad and D. N. Harpp, Three sulfur atom insertion into the S-S bond, pentasulfide preparation, *Tetrahedron Lett.*, 2000, **41**, 7809–7812.

T. M. Hovell, Garlic in whooping cough, *Br. Med. J.*, July 1, 1916, **2896**, 15.

M. H. Y. Hovius and M. R. McDonald, Management of *Allium* white rot [*Sclerotium cepivorum*] in onions on organic soil with soil-applied diallyl disulfide and di-*n*-propyl disulfide, *Can. J. Plant Pathol.*, 2002, **24**, 281–286.

E. W. Howard, M. T. Ling, C. W. Chua, H. W. Cheung, X. Wang and Y. C. Wong, Garlic-derived *S*-allylmercaptocysteine is a novel *in vivo* antimetastatic agent for androgen-independent prostate cancer, *Clin. Cancer Res.*, 2007, **13**, 1847–1856.

E. W. Howard, D. T. Lee, Y. T. Chiu, C. W. Chua, X. Wang and Y. C. Wong, Evidence of a novel docetaxel sensitizer, garlic-derived *S*-allylmercaptocysteine, as a treatment option for hormone refractory prostate cancer, *Int. J. Cancer*, 2008, **122**, 1941–1948.

Q. Hu, Q. Yang, O. Yamato, M. Yamasaki, Y. Maede, Y. Teruhiko and T. Yoshihara, Isolation and identification of organosulfur compounds oxidizing canine erythrocytes from garlic (*Allium sativum*), *J. Agric. Food Chem.*, 2002, **50**, 1059–1062.

D. Huang, B. Ou and R. L. Prior, The chemistry behind antioxidant capacity assays, *J. Agric. Food Chem.*, 2005, **53**, 1841–1856.

T.-H. Huang, R. C. Muehlbauer, C.-H. Tang, H.-I. Chen, G.-L. Chang, Y.-W. Huang, Y.-T. Lai, H.-S. Lin, W.-T. Yang and R.-S. Yang, Onion decreases the ovariectomy-induced osteopenia in young adult rats, *Bone*, 2008, **42**, 1154–1163.

Y. Huang, S. X. Chen and S. H. Ho, Bioactivities of methyl allyl disulfide and diallyl trisulfide from essential oil of garlic to two species of stored-product pests, *Sitophilus zeamais* (Coleoptera: Curculionidae) and *Tribolium castaneum* (Coleoptera: Tenebrionidae), *J. Economic Entomol.*, 2000, **93**, 537–543.

V. G. Hubbard and P. Goldsmith, Garlic-fingered chefs, *Contact Dermatitis*, 2005, **52**, 165–166.

J. Hughes, A. Tregova, A. B. Tomsett, M. G. Jones, R. Cosstick and H. A. Collin, Synthesis of the flavour precursor, alliin, in garlic tissue cultures, *Phytochemistry*, 2005, **66**, 187–194.

I. G. Hwang, K. S. Woo, D. J. Kim, J. T. Hong, B. Y. Hwang, Y. R. Lee and H. S. Jeong, Isolation and identification of an antioxidant substance from heated garlic (*Allium sativum* L.), *Food Sci. Biotechnol.*, 2007, **16**, 963–966.

J. R. Hwu and D. A. Anderson, Zwitterion-accelerated [3,3]-sigmatropic rearrangement of allyl vinyl sulfoxides to sulfines. A specific class of charge-accelerated rearrangement, *Tetrahedron Lett.*, 1986, **27**, 4965–4968.

J. R. Hwu and D. A. Anderson, Zwitterion-accelerated [3,3]-sigmatropic rearrangements and [2,3]-sigmatropic rearrangements of sulfoxides and amine oxides, *J. Chem. Soc., Perkin Trans. 1*, 1991, 3199–3206.

B. Iberl, G. Winkler, B. Muller and K. Knoblauch, Products of allicin transformation: ajoenes and dithiins, characterization and their determination by HPLC, *Planta Med.*, 1990, **56**, 202–211.

B. Iberl, G. Winkler, B. Muller and K. Knoblauch, Quantitative determination of allicin and alliin from garlic by HPLC, *Planta Med.*, 1990, **56**, 320–326.

M. Ichikawa, K. Ryu, J. Yoshida, N. Ide, Y. Kodera, T. Sasaoka and R. T. Rosen, Identification of six phenylpropanoids from garlic skin as major antioxidants, *J. Agric. Food Chem.*, 2003, **51**, 7313–7317.

M. Ichikawa, N. Ide, J. Yoshida, Y. Hiroyuki and K. Ono, Determination of seven organosulfur compounds in garlic by high-performance liquid chromatography, *J. Agric. Food Chem.*, 2006, **54**, 1535–1540.

M. Iciek, A. Bilska, L. Ksiazek, Z. Srebro and L. Wlodek, Allyl disulfide as donor and cyanide as acceptor of sulfane sulfur in the mouse tissues, *Pharmacol. Reports*, 2005, **57**, 212–218.

M. Iciek, I. Kwiecien and L. Wlodek, Biological properties of garlic and garlic-derived organosulfur compounds, *Environ. Mol. Mutagenesis*, 2009, **50**, 247–265.

S. Imai, N. Tsuge, M. Tomotake, Y. Nagatome, H. Sawada, T. Nagata and H. Kumagai, An onion enzyme that makes the eyes water, *Nature*, 2002, **419**, 685.

S. Imai, K. Akita, M. Tomotake and H. Sawada, Identification of two novel pigment precursors and a reddish-purple pigment involved in the blue-green discoloration of onion and garlic, *J. Agric. Food Chem.*, 2006, **54**, 843–847.

S. Imai, K. Akita, M. Tomotake and H. Sawada, Model studies on precursor system generating blue pigment in onion and garlic, *J. Agric. Food Chem.*, 2006, **54**, 848–852.

E. Ingals, Bites of serpents, *J. Am. Med. Assoc.*, 1883, **1**, 249–250.

C. Ip and D. J. Lisk, Bioavailability of selenium from selenium-enriched garlic, *Nutr. Cancer*, 1993, **20**, 129–137.

C. Ip, D. J. Lisk and G. S. Stoewsand, Mammary cancer prevention by regular garlic and selenium-enriched garlic, *Nutr. Cancer*, 1992, **17**, 279–286.

C. Ip, M. Birringer, E. Block, M. Kotrebai, J. F. Tyson, P. C. Uden and D. J. Lisk, Chemical speciation influences comparative activity of selenium-enriched garlic and yeast in mammary cancer prevention, *J. Agric. Food Chem.*, 2000, **48**, 2062–2070.

M. Ipek, A. Ipek, S. G. Almquist and P. W. Simon, Demonstration of linkage and development of the first low-density genetic map of garlic, based on AFLP markers, *Theoretical Appl. Genetics*, 2005, **110**, 228–236.

M. Ipek, A. Ipek and P. W. Simon, Genetic characterization of *Allium tuncelianum*: An endemic edible *Allium* species with garlic odor, *Sci. Horticulture*, 2008, **115**, 409–415.

A. Ishii, K. Komiya and J. Nakayama, Synthesis of a stable sulfenic acid by oxidation of a sterically hindered thiol (thiophenetriptycene-8-thiol) and its characterization, *J. Am. Chem. Soc.*, 1996, **118**, 12836–12837.

A. Ishii, M. Ohishi, K. Matsumoto and T. Takayanagi, Synthesis and properties of a dithiirane *trans*-1,2-dioxide, a three-membered *vic*-disulfoxide, *Org. Lett.*, 2006, **8**, 91–94.

A. Jabbari, H. Argani, A. Ghorbanihaghjo and R. Mahdavi, Comparison between swallowing and chewing of garlic on levels of serum lipids, cyclosporine, creatinine and lipid peroxidation in renal transplant recipients, *Lipids Health Disease*, 2005, **4**, 11.

B. A. Jackson and A. J. Shelton, Pilot study evaluating topical onion extract as treatment for postsurgical scars, *Dermatol. Surgery*, 1999, **25**, 267–269.

C. Jacob, A. Anwar and T. Burkholz, Perspective on recent developments on sulfur-containing agents and hydrogen sulfide signaling, *Planta Med.*, 2008a, **74**, 1580–1592.

C. Jacob and A. Anwar, The chemistry behind redox regulation with a focus on sulphur redox systems, *Physiol. Plant*, 2008b, **133**, 469–480.

H. Jansen, B. Müller and K. Knobloch, Allicin characterization and its determination by HPLC, *Planta Med.*, 1987, **53**, 559–562.

M. S. Jarial, Toxic effect of garlic extracts on the eggs of *Aedes aegypti* (Diptera: Culicidae): a scanning electron microscopic study, *J. Med. Entomol.*, 2001, **38**, 446–450.

U. Jappe, B. Bonnekoh, B. M. Hausen and H. Gollnick, Garlic-related dermatoses: case report and review of the literature, *Am. J. Contact Dermatol*, 1999, **10**, 37–39.

Z. Jastrzebski, H. Leontowicz, M. Leontowicz, J. Namiesnik, Z. Zachwieja, H. Barton, E. Pawelzik, P. Arancibia-Avila, F. Toledo and S. Gorinstein, The bioactivity of processed garlic (*Allium sativum* L.) as shown *in vitro* and *in vivo* studies on rats, *Food Chem. Toxicol.*, 2007, **45**, 1626–1633.

J. Jedelská, A. Vogt, U. M. Reinscheid and M. Keusgen, Isolation and identification of a red pigment from *Allium* subgenus *Melanocrommyum*, *J. Agric. Food Chem.*, 2008, **56**, 1465–1470.

G. A. Jelodar, M. Maleki, M. H. Motadayen and S. Sirus, Effect of fenugreek, onion and garlic on blood glucose and histopathology of pancreas of alloxan-induced diabetic rats, *Indian J. Med. Sci.*, 2005, **59**, 64–69.

X. W. Jiang, J. Hu and F. I. Mian, A new therapeutic candidate for oral aphthous ulcer: allicin, *Med. Hypotheses*, 2008, **71**, 897–899.

L. Jirovetz, W. Jaeger, H. P. Koch and G. Remberg, Investigations of the volatile constituents of the essential oil of Egyptian garlic by GC–MS and GC–FTIR, *Z. Lebensm. Unters Forsch.*, 1992, **194**, 363–365.

D. R. Johnson and F. X. Powell, Microwave detection of thioformaldehyde, *Science*, 1970, **169**, 679–680.

H. A. Jones and L. K. Mann, *Onions and their Allies*, Interscience, New York, 1963.

M. G. Jones, J. Hughes, A. Tregova, J. Milne, A. B. Tomsett and H. A. Collin, Biosynthesis of the flavour precursors of onion and garlic, *J. Exp. Botany*, 2004, **55**, 1903–1918.

M. G. Jones, H. A. Collin, A. Tregova, L. Trueman, L. Brown, R. Cosstick, J. Hughes, J. Milne, M. C. Wilkinson, A. B. Tomsett and B. Thomas, The biochemical and physiological genesis of alliin in garlic, *Med. Aromatic Plant Sci. Biotechnol.*, 2007, **1**, 21–24.

P. Josling, Preventing the common cold with a garlic supplement: a double-blind, placebo-controlled survey, *Adv. Therapies*, 2001, **18**, 189–193.

M. A. Joslyn and R. G. Peterson, Food discoloration, reddening of white onion bulb purees, *J. Agric. Food Chem.*, 1958, **6**, 754–765.

R. Kamenetsky and R. M. Fritsch, Ornamental alliums, in *Allium Crop Science: Recent Advances*, ed. H. D. Rabinowitch and L. Currah, CABI, Wallingford, UK, 2002, ch. 19.

R. Kamenetsky, I. L. Shafir, F. Khassanov, C. Kik, A. W. Van Heusden, M. Vrielink-Van Ginkel, K. Burger-Meijer, J. Auger, I. Arnault and H. D. Rabinowitch, Diversity in fertility potential and organo-sulphur compounds among garlics from Central Asia, *Biodiversity Conservation*, 2005, **14**, 281–295.

R. Kamenetsky, F. Khassanov, H. D. Rabinowitch, J. Auger and C. Kik, Garlic biodiversity and genetic resources, *Med. Aromatic Plant Sci. Biotechnol.*, 2007a, **1**, 1–5.

R. Kamenetsky, Garlic: botany and horticulture, *Horticultural Rev.*, 2007b, **33**, 123–172.

S. H. Kao, C. H. Hsu, S. N. Su, W. T. Hor, T. W. H. Chang and L. P. Chow, Identification and immunologic characterization of an allergen, alliin lyase, from garlic (*Allium sativum*), *J. Allergy Clin. Immunol.*, 2004, **113**, 161–168.

M. M. Kaplan, S. Cerutti, J. A. Salonia, J. A. Gásquez and L. D. Martinez, Preconcentration and determination of tellurium in garlic samples by hydride generation atomic absorption spectrometry, *J. AOAC Int.*, 2005, **88**, 1242–1246.

E. Kápolna, M. Shah, J. A. Caruso and P. Fodor, Selenium speciation studies in Se-enriched chives (*Allium schoenoprasum*) by HPLC–ICP–MS, *Food Chem.*, 2007, **101**, 1398–1406.

D. N. Karma, N. Agarwal and L. C. Chaudhary, Inhibition of rumen methanogenesis by tropical plants containing secondary metabolites, *Int. Cong. Ser.*, 2006, **1293**, 156–163.

S. Kawakishi and Y. Morimitsu, New inhibitor of platelet aggregation in onion oil, *Lancet*, 1988, **8606**, 330.

H. P. Keiss, V. M. Dirsch, T. Hartung, T. Haffner, L. Trueman, J. Auger, R. Kahane and A. M. Vollmar, Garlic (*Allium sativum* L.) modulates cytokine expression in lipopolysaccharide-activated human blood thereby inhibiting NF-kappaB activity, *J Nutr.*, 2003, **33**, 2171–2175.

D. A. Kessler, Cancer and herbs, *N. Engl. J. Med.*, 2000, **324**, 1742–1743.

M. Keusgen, H. Schultz, J. Glodek, I. Krest, H. Krüger, N. Herchert and J. Keller, Characterization of some *Allium* hybrids by aroma precursors, aroma profile and alliinase activity, *J. Agric. Food Chem.*, 2002, **50**, 2884–2890.

M. Keusgen, M. Jünger, I. Krest and M. J. Schöning, Development of a biosensor specific for cysteine sulfoxides, *Biosens. Bioelectron.*, 2003, **18**, 805–812.

M. Keusgen, R. M. Fritsch, H. Hisoriev, P. A. Kurbonova and F. O. Khassanov, Wild *Allium* species (Alliaceae) used in folk medicine of Tajikistan and Uzbekistan, *J. Ethnobiol. Ethnomed.*, 2006, **2**(18).

M. Keusgen, R. M. Fritsch, Onions of the *Allium* subgenus *Melanocrommyum* – the better garlic? *Planta Med.*, 2006, **72**, Meeting abstracts: 54th Annual Congress on Medicinal Plant Research, Book of Abstracts.

M. Keusgen, Unusual cystine lyase activity of the enzyme alliinase: direct formation of polysulphides, *Planta Med.*, 2008, **74**, 73–79.

F. A. Khalid, N. M. Abdalla, H. E. O. Mohomed, A. M. Toum, M. M. A. Magzoub and M. S. Ali, *In vitro* assessment of anti-cutaneous leishmaniasis activity of some Sudanese Plants, *Acta Parasitologica Turcica*, 2005, **29**, 3–6.

F. Khanum, K. R. Anilakumar and K. R. Viswanathan, Anticarcinogenic properties of garlic: a review, *Crit. Rev. Food. Sci. Nutr.*, 2004, **44**, 479–488.

Y. S. Khoo and Z. Aziz, Garlic supplementation and serum cholesterol: a meta-analysis, *J. Clin. Pharm. Therapy*, 2009, **34**, 133–145.

J. L. Kice, Mechanisms and reactivity in reactions of organic oxyacids of sulfur and their anhydrides, *Adv. Phys. Org. Chem.*, 1980, **17**, 65–181.

C. Kik, *Garlic & Health, 2000–2004*, European Union project.

A. Kim, J. Y. Jung, M. Son, S. H. Lee, J. S. Lim and A. S. Chung, Long exposure of non-cytotoxic concentrations of methylselenol suppresses the invasive potential of B16F10 melanoma, *Oncol. Reports*, 2008b, **20**, 557–565.

H. K. Kim, J. S. Kim, Y. S. Cho, Y. W. Park, H. S. Son, S. S. Kim and H. S. Chae, Endoscopic removal of an unusual foreign body: a garlic-induced acute esophageal injury, *Gastrointestinal Endoscopy*, 2008c, **68**, 565–566.

J. W. Kim, J. H. Choi, Y. -S. Kim and K. H. Kyung, Antiyeast potency of heated garlic in relation to the content of allyl alcohol thermally generated from alliin, *J. Food Sci.*, 2006, **71**, M185–M189.

J. Y. Kim, Y. S. Kim and K. H. Kyung, Inhibitory activity of essential oils of garlic and onion against bacteria and yeasts, *J. Food Protection*, 2004, **67**, 499–504.

J. Y. Kim and O. Kwon, Garlic intake and cancer risk: an analysis using the Food and Drug Administration's evidence-based review system for the scientific evaluation of health claims, *Am. J. Clin. Nutr.*, 2009, **89**, 265–272.

K. M. Kim, S. B. Chun, M. S. Koo, W. J. Choi, T. W. Kim, Y. G. Kwon, H. T. Chung, T. R. Billiar and Y. M. Kim, Differential regulation of NO availability from macrophages and endothelial cells by the garlic component *S*-allyl cysteine, *Free Radical Biol. Med.*, 2001, **30**, 747–756.

S. M. Kim, C. M. Wu, K. Kubota and A. Kobayashi, Effect of soybean oil on garlic volatile compounds isolated by distillation, *J. Agric. Food Chem.*, 1995, **43**, 449–452.

S.-Y. Kim, K. W. Park, J. Y. Kim, I. Y. Jeong, M. W. Byun, J. E. Park, S. T. Yee, K. H. Kim, J. S. Rhim, K. Yamada and K. I. Seo, Thiosulfinates from *Allium tuberosum* L. induce apoptosis via caspase-dependent and -independent pathways in PC-3 human prostate cancer cells, *Bioorg. Med. Chem. Lett.*, 2008a, **18**, 199–204.

S.-Y. Kim, K. W. Park, J. Y. Kim, I. Y. Shon, M. Y. Yee, K. H. Kim, J. S. Rhim, K. Yamada and K. I. Seo, Induction of apoptosis by thiosulfinates in primary human prostate cancer cells, *Int. J. Oncol.*, 2008b, **32**, 869–875.

Y. A. Kim, D. Xiao, H. Xiao, A. A. Powolny, K. L. Lew, M. L. Reilly, Y. Zeng, Z. Wang and S. V. Singh, Mitochondria-mediated apoptosis by diallyl trisulfide in human prostate cancer cells is associated with generation of reactive oxygen species and regulated by Bax/Bak, *Mol. Cancer Therapy*, 2007, **6**, 1599–1609.

A. C. Kimbaris, E. Kioulos, G. Koliopoulos, M. G. Polissiou and A. Michaelakis, Coactivity of sulfide ingredients: a new perspective of the larvicidal activity of garlic essential oil against mosquitoes, *Pest Management Sci.*, 2009, **65**, 249–254.

E. A. Kimmel, *Onions and Garlic: an Old Tale*, Holiday House, New York, 1996.

J. F. King and T. Durst, Geometrical isomerism about a carbon-sulfur double bond, *J. Am. Chem. Soc.*, 1963, **85**, 2676–2677.

K. F. Kiple and K.C. Ornelas, (Ed.), *The Cambridge World History of Food*, Cambridge University Press, Cambridge, UK, 2000.

P. Knekt, R. Jarvinen, A. Reunanen and J. Maatela, Flavonoid intake and coronary mortality in Finland: a cohort study, *Br. Med. J.*, 1996, **312**, 478–481.

A. P. Knight, D. Lassen, T. McBride, D. Marsh, C. Kimberling, M. G. Delgado and D. Gould, Adaptation of pregnant ewes to an exclusive onion diet, *J. Vet. Hum. Toxicol.*, 2000, **42**, 1–4.

H. P. Koch and L. D. Lawson, (ed.), *Garlic The Science and Therapeutic Applications of Allium sativum L. and Related Species*, Williams and Wilkins, Baltimore, MD, 1996.

Y. Kodera, H. Matsuura, S. Yoshida, T. Sumida, Y. Itakura, T. Fuwa and H. Nishino, Allixin, a stress compound from garlic, *Chem. Pharm. Bull.*, 1989, **37**, 1656–1658.

Y. Kodera, M. Ayabe, K. Ogasawara, S. Yoshida, N. Hayashi and K. Ono, Allixin accumulation with long-term storage of garlic, *Chem. Pharm. Bull.*, 2002a, **50**, 405–407.

Y. Kodera, M. Ichikawa, J. Yoshida, N. Kashimoto, N. Uda, I. Sumioka, N. Ide and K. Ono, Pharmacokinetic study of allixin, a phytoalexin produced by garlic, *Chem. Pharm. Bull.*, 2002b, **50**, 354–363.

P. Koelewijn and H. Berger, Mechanism of the antioxidant action of dialkyl sulfoxides, *Rec. Trav. Chim. Pays-Bas*, 1972, **91**, 1275–1286.

E. F. Kohman, The chemical components of onion vapor responsible for wound-healing qualities, *Science*, 1947, **106**, 625–627.

K. Koizumi, Y. Iwasaki, M. Narukawa, H. Izuka, T. Fukao, T. Seki, T. Ariga and T. Watanabe, Diallyl sulfides in garlic activate both TRPA1 and TRPV1, *Biochem. Biophys. Res. Commun.*, 2009, **382**, 545–548.

G. König, M. Brunda, H. Puxbaum, C. N. Hewitt, C. S. Duckham and J. Rudolph, Relative contribution of oxygenated hydrocarbons to the total biogenic VOC emissions of selected mid-European agricultural and natural plant species, *Atmos. Environ.*, 1995, **29**, 861–874.

M. Kotrebai, M. Birringer, J. F. Tyson, E. Block and P. C. Uden, Identification of the principal selenium compounds in selenium-enriched natural sample extracts by ion-pair liquid chromatography with inductively coupled plasma- and electrospray ionization–mass spectrometric detection, *Anal. Commun.*, 1999, **36**, 249–252.

M. Kotrebai, M. Birringer, J. F. Tyson, E. Block and P. C. Uden, Characterization of selenium compounds in selenium-enriched and natural sample extracts using perfluorinated carboxylic acids as HPLC ion-pairing agents with ICP- and electrospray ionization–mass spectrometric detection, *Analyst*, 2000, **125**, 71–78.

S. Kreitzman, *Garlic*, Harmony Books, New York, 1984.

I. Krest and M. Keusgen, Quality of herbal remedies from *Allium sativum*. Differences between alliinase from garlic powder and fresh garlic, *Planta Med.*, 1999, **65**, 139–143.

I. Krest, J. Glodek and M. Keusgen, Cysteine sulfoxides and alliinase activity of some *Allium* species, *J. Agric. Food Chem.*, 2000, **48**, 3753–3760.

B. H. Kroes, European perspective on garlic and its regulation, *J. Nutr.*, 2006, **136**, 732S–735S.

R. Kubec, M. Svobodová and J. Velísek, Gas chromatographic determination of *S*-alk(en)ylcysteine sulfoxides, *J. Chromatogr., A*, 1999, **862**, 85–94.

R. Kubec, M. Svobodova and J. Velísek, Distribution of *S*-alk(en)ylcysteine sulfoxides in some *Allium* species. Identification of a new flavor precursor: *S*-ethylcysteine sulfoxide (ethiin), *J. Agric. Food Chem.*, 2000, **48**, 428–433.

R. Kubec and R. A. Musah, Cysteine sulphoxide derivatives in *Petiveria alliacea, Phytochemistry*, 2001, **58**, 981–985.

R. Kubec and R. A. Musah, γ-Glutamyl dipeptides in *Petiveria alliacea, Phytochemistry*, 2005, **66**, 2494–2497.

R. Kubec, S. Kim, D. M. McKeon and R. A. Musah, Isolation of *S*-n-butyl-cysteine sulfoxide and six n-butyl-containing thiosulfinates from *Allium siculum*, *J. Nat. Prod.*, 2002, **65**, 960–964.

R. Kubec, S. Kim, J. Velísek and R. A. Musah, The amino acid precursors and odor formation in society garlic (*Tulbaghia violacea* Harv.), *Phytochemistry*, 2002a, **60**, 21–25.

R. Kubec and J. Velísek, Allium discoloration: The color-forming potential of individual thiosulfinates and amino acids: Structual requirements for the color-developing precursors, *J. Agric. Food Chem.*, 2007, **55**, 3491–3497.

R. Kubec and R. Dadáková, Quantitative determination of *S*-alk(en)ylcysteine-*S*-oxides by micellar electrokinetic capillary chromatography, *J. Chromatogr., A*, 2008, **1212**, 154–157.

R. Kubec, personal communication, 2009.

K. Kubota, H. Hirayama, Y. Sato, A. Kobayashi and F. Sugawara, Amino acid precursors of the garlic-like odour in *Scorodocarpus borneensis, Phytochemistry*, 1998, **49**, 99–102.

E. B. Kuettner, R. Hilgenfeld and M. S. Weiss, The active principle of garlic at atomic resolution, *J. Biol. Chem.*, 2002, **277**, 46402–46407.

M. C. Kuo, M. Chien and C. T. Ho, Novel polysulfides identified in the volatile components from Welsh onions (*Allium fistulosum* L. var. maichuon) and scallions (*Allium fistulosum* L. var. caespitosum), *J. Agric. Food Chem.*, 1990, **38**, 1378–1381.

M. C. Kuo and C. T. Ho, Volatile constituents of the distilled oils of Welsh onions (*Allium fistulosum* L. variety maichuon) and scallions (*Allium fistulosum* L. variety caespitosum), *J. Agric. Food Chem.*, 1992, **40**, 111–117.

M. C. Kuo and C. T. Ho, Volatile constituents of the solvent extracts of Welsh onions (*Allium fistulosum* L. variety maichuon) and scallions (*A. fistulosum* L. variety caespitosum), *J. Agric. Food Chem.*, 1992, **40**, 1906–1910.

I. Laakso, T. Seppanen-Laakso, R. Hiltunen, B. Mueller, H. Jansen and K. Knobloch, Volatile garlic odor components: gas phases and adsorbed exhaled air analyzed by headspace gas chromatography-mass spectrometry, *Planta Med.*, 1989, **55**, 257–261.

S. Lacombe, M. Loudet, E. Banchereau, M. Simon and G. Pfister-Guillouzo, Sulfenic acids in the gas phase: a photoelectron study, *J. Am. Chem. Soc.*, 1996, **118**, 1131–1138.

S. Lacombe, M. Loudet, A. Dargelos and E. Robert-Banchereau, Oxysulfur compounds derived from dimethyl disulfide: an *ab initio* study, *J. Org. Chem.*, 1998, **63**, 2281–2291.

L. Laisheng, H. Yang, X. Chen and L. Xu, Qualitative and quantitative analysis of diallyl sulfides in garlic powder with Ag (I) induced ionization by liquid chromatography-mass spectrometry, *Fenxi Huaxue*, 2006, 34, 1183–1186 (*Chem. Abstr.*, 2006, **146**, 1216171).

J. E. Lancaster and H. A. Collin, Presence of alliinase in isolated vacuoles and of alkyl cysteine sulphoxides in the cytoplasm of bulbs of onion (*Allium cepa*), *Plant Sci. Lett.*, 1981, **22**, 169–176.

J. E. Lancaster, M. L. Shaw, M. D. P. Joyce, J. A. McCallum and M. T. McManus, A novel alliinase from onion roots. Biochemical characterization and cDNA cloning, *Plant Physiol.*, 2000a, **122**, 1269–1279.

J. E. Lancaster, M. L. Shaw and E. F. Walton, *S*-Alk(en)yl-L-cysteine sulfoxides, alliinase and aroma in *Leucocoryne*, *Phytochemistry*, 2000, **55**, 127–130.

J.-F. Landry, Taxonomic review of the leek moth genus Acrolepiopsis (Lepidoptera *Acroolepiidate*) in North America, *The Canadian Entomologist*, 2007, **139**, 319–353.

A. Lang, M. Lahav, E. Sakhnini, I. Barshack, H. H. Fidder, B. Avidan, E. Bardan, R. Hershkoviz, S. Bar-Meir and Y. Chowers, Allicin inhibits spontaneous and TNF-alpha induced secretion of proinflammatory cytokines and chemokines from intestinal epithelial cells, *Clin. Nutr.*, 2004, **23**, 1199–1208.

V. Lanzotti, The analysis of onion and garlic, *J. Chromatog., A*, 2006, **1112**, 3–22.

M. Laska, R. M. R. Bautista, D. Höfelmann, V. Sterlemann and L. T. H. Salazar, Olfactory sensitivity for putrefaction-associated thiols and indoles in three species of non-muman primate, *J. Exp. Biol.*, 2007, **210**, 4169–4178.

B. D. Lawenda, K. M. Kelly, E. J. Ladas, S. M. Sagar, A. Vickers and J. B. Blumberg, Should supplemental antioxidant administration be avoided during chemotherapy and radiation therapy? *J. Natl. Cancer Inst.*, 2008, **100**, 773–783.

L. D. Lawson, S. G. Wood and B. G. Hughes, HPLC analysis of allicin and other thiosulfinates in garlic clove homogenates, *Planta Med.*, 1991a, **57**, 263–270.

L. D. Lawson, Z. Y. J. Wang and B. G. Hughes, Identification and HPLC quantitation of the sulfides and dialk(en)yl thiosulfinates in commercial garlic products, *Planta Med.*, 1991b, **57**, 363–370.

L. D. Lawson, Z. J. Wang and B. G. Hughes, γ-Glutamyl-*S*-alkylcysteines in garlic and other *Allium* spp.: precursors of age-dependent *trans*-1-propenyl thiosulfinates, *J. Nat. Prod.*, 1991c, **54**, 436–444.

L. D. Lawson and B. G. Hughes, Characterization of the formation of allicin and other thiosulfinates from garlic, *Planta Med.*, 1992, **58**, 345–350.

L. D. Lawson, in *Garlic: The Science and Therapeutic Applications of Allium sativum L. and Related Species*, ed. H.P. Koch and L.D. Lawson, Williams and Wilkins, Baltimore, MD, 1996.

L. D. Lawson, Garlic: a review of its medicinal effects and indicated active compounds, in *Phytomedicines of Europe: Chemistry and Biological Activity*, ed. L. D. Lawson, R. Bauer, ACS Symposium Series 691, American Chemical Society, Washington, DC, 1998, pp. 176–209.

L. D. Lawson and Z. J. Wang, Allicin release from garlic supplements: a major problem due to the sensitivities of alliinase activity, *J. Agric. Food Chem.*, 2001a, **49**, 2592–2599.

L. D. Lawson, Z. J. Wang and D. Papadimitriou, Allicin release under simulated gastrointestinal conditions from garlic powder tablets employed in clinical trials on serum cholesterol, *Planta Med.*, 2001b, **67**, 13–18.

L. D. Lawson and Z. J. Wang, Allicin and allicin-derived garlic compounds increase breath acetone through allyl methyl sulfide: use in measuring allicin bioavailability, *J. Agric. Food Chem.*, 2005a, **53**, 1974–1983.

L. D. Lawson and C. D. Gardner, Composition, stability, and bioavailability of garlic products used in a clinical trial, *J. Agric. Food. Chem.*, 2005b, **53**, 6254–6261.

L. D. Lawson, Effect of consuming raw or boiled garlic on agonist-induced ex vivo platelet aggregation: an open trial with healthy volunteers, *FASEB J.*, 2007, **21**, 864–867.

E. Ledezma, L. De Sousa, A. Jorquera, A. Lander, J. Sanchez, E. Rodriguez, M. K. Jain and R. Apitz-Castro, Effectiveness of ajoene, an organosulfur derived from garlic, in the short-term therapy of *Tinea pedis* in humans, *Mycoses*, 1996, **39**, 393–395.

E. Ledezma, K. Marcano, A. Jorquera, L. De Sousa, M. Padilla, M. Pulgar and R. Apitz-Castro, Efficacy of ajoene in the treatment of tinea pedis: a double-blind and comparative study with terbinafine, *J. Am. Acad. Dermatol.*, 2000, **43**, 829–832.

J. H. Lee, H. S. Yang, K. W. Park, J. Y. Kim, M. K. Lee, I. Y. Jeong, K. H. Shim, Y. S. Kim, K. Yamada and K. I. Seo, Mechanisms of thiosulfinates from *Allium tuberosum* L.-induced apoptosis in HT-29 human colon cancer cells, *Toxicol. Lett.*, 2009, **188**, 142–147.

K. W. Lee, O. Yamato, M. Tajima, M. Kuraoka, S. Omae and Y. Maede, Hematologic changes associated with the appearance of eccentrocytes after intragastric administration of garlic extract to dogs, *Am. J. Veterinary Res.*, 2000, **61**, 1446–1450.

S. O. Lee, J. Yeon Chun, N. Nadiminty, D. L. Trump, C. Ip, Y. Dong and A. C. Gao, Monomethylated selenium inhibits growth of LNCaP human prostate cancer xenograft accompanied by a decrease in the expression of androgen receptor and prostate-specific antigen (PSA), *Prostate*, 2006, **66**, 1070–1075.

S. U. Lee, J. H. Lee, S. H. Choi, J. S. Lee, M. Ohnisi-Kameyama, N. Kozukue, C. E. Levin and M. Friedman, Flavonoid content in fresh, home-processed, and light-exposed onions and in dehydrated commercial onion products, *J. Agric. Food Chem.*, 2008, **56**, 8541–8548.

T. Y. Lee and T. H. Lam, Contact dermatitis due to topical treatment with garlic in Hong Kong, *Contact Dermatitis*, 1991, **24**, 193–196.

Y. Lee, Induction of apoptosis by *S*-allylmercapto-L-cysteine, a biotransformed garlic derivative, on a human gastric cancer cell line, *Int. J. Mol. Med.*, 2008, **21**, 765–770.

D. J. Lefer, A new gaseous signaling molecule emerges: cardioprotective role of hydrogen sulfide, *Proc. Natl. Acad. Sci. U.S.A.*, 2007, **104**, 17907–17908.

K. M. Lemar, M. A. Aon, S. Cortassa, B. O'Rourke, C. T. Müller and D. Lloyd, Diallyl disulfide depletes glutathione in *Candida albicans*: oxidative stress-mediated cell death studied by two-photon microscopy, *Microbiology*, 2005, **151**, 3257–3265.

K. M. Lemar, O. Passa, M. A. Aon, S. Cortassa, C. T. Müller, S. Plummer, B. O'Rourke and D. Lloyd, Allyl alcohol and garlic (*Allium sativum*) extract produce oxidative stress in *Candida albicans*, *Yeast*, 2007, **24**, 695–706.

T. Leustek and K. Saito, Sulfate transport and assimilation in plants, *Plant Physiol.*, 1999, **120**, 637–643.

A. Leuthner von Grundt, *Gründtliche Darstellung der fünff Seüllen, wie solche von...Vitruuio, Scamozzio und andern...Baumeistren...uerfasset worden...Mit schönen Grundtrissen, Lusthaussern, Capellen, Klöstern, Schlössern...zusamben gebracht, gerissen auf hundert und mehr Kupffer radirt undt vorgestellet durch Abraham Leüthner von Grundt, Burger... der...newen Stadt Prag*. . .Prague, 1677. [Avery Architectural Library, Columbia University].

W. H. Lewis and M. P. F. Elvin-Lewis, *Medical Botany: Plants Affecting Man's Health*, John Wiley & Sons, New York, 1977.

I. J. Levy and R. L. Zumwalt, *Ritual Medical lore of Sephardic Women: Sweetening the Spirits, Healing the Sick*, University of Illinois Press, Urbana, IL, 2002.

H. Li, H.-Q. Li, Y. Wang, H. Xu, W. Fan, M. Wang, P. Sun and X. Xie, An intervention study to prevent gastric cancer by micro-selenium and large dose of allitridum, *Chin. Med. J.*, 2004, **117**, 1155–1160.

L. Li, D. Hu, Y. Jiang, F. Chen, X. Hu and G. Zhao, Relationship between γ-glutamyl transpeptidase activity and garlic greening, as controlled by temperature, *J. Agric. Food Chem.*, 2008, **56**, 941–945.

M. Li, J. R. Ciu, Y. Ye, J. M. Min, L. H. Zhang, K. Wang, M. Gares, J. Cros, M. Wright and J. Leung-Tack, Antitumor activity of Z-ajoene, a natural compound purified from garlic: antimitotic and microtubule-interaction properties, *Carcinogenesis*, 2002, **23**, 573–579.

N. Li, R. Guo, W. Li, J. Shao, S. Li, K. Zhao, X. Chen, N. Xu, S. Liu and Y. Lu, A proteomic investigation into a human gastric cancer cell line BGC823 treated with diallyl trisulfide, *Carcinogenesis*, 2006, **7**, 1222–1231.

J. G. Lin, G. W. Chen, C. C. Su, C. F. Hung, C. C. Yang, J. H. Lee and J. G. Chung, Effects of garlic components diallyl sulfide and diallyl disulfide on arylamine *N*-acetyltransferase activity and 2-aminofluorene–DNA adducts in human promyelocytic leukemia cells, *Am. J. Chin. Med.*, 2002, **30**, 315–325.

K. Linde, G. ter Riet, M. Hondras, A. Vickers, R. Saller and D. Melchart, Systematic reviews of complementary therapies – an annotated bibliography. Part 2: herbal medicine, *BioMed Central Complementary Alternative Med.*, 2001, **1**, 5.

S. A. Lippman, P. J. Goodman, E. A. Klein, H. L. Parnes, I. M. Thompson Jr., A. R. Kristal, R. M. Santella, J. L. Probstfield, C. M. Moinpour, D. Albanes, P. R. Taylor, L. M. Minasian, A. Hoque, S. M. Thomas, J. J. Crowley, J. M. Gaziano, J. L. Stanford, E. D. Cook, N. E. Fleshner, M. M. Lieber, P. J. Walther, F. R. Khuri, D. D. Karp, G. G. Schwartz, L. G. Ford and C. A. Coltman, Jr., Designing the Selenium and Vitamin E Cancer Prevention Trial (SELECT), *J. Natl. Cancer Inst.*, 2005, **97**, 94–102.

S. A. Lippman, E. A. Klein, P. J. Goodman, M. S. Lucia, I. M Thompson, L. G Ford, H. L. Parnes, L. M. Minasian, J. M Gaziano, J. A. Hartline, J. K. Parsons, J. D. Bearden 3rd, E. D. Crawford, G. E. Goodman, J. Claudio, E. Winquist, E. D. Cook, D. D. Karp, P. Walther, M. M. Lieber, A. R. Kristal, A. K. Darke, K. B. Arnold, P. A. Ganz, R. M. Santella, D. Albanes, P. R. Taylor, J. L. Probstfield, T. J. Jagpal, J. J. Crowley, F. L. Meyskens. Jr., L. H. Baker, C. A. Coltman, Jr., Effect of selenium and vitamin E on risk of prostate cancer and other cancers: The Selenium and Vitamin E Cancer Prevention Trial (SELECT), *J. Am. Med. Assoc.*, 2009, **301**, 39–51.

E. Lissiman, A. L. Bhasale and M. Cohen, Garlic for the common cold, *Cochrane Database Syst. Rev.* 2009, July 8 (3), CD006206.

C.-T. Liu, L.-Y. Sheen and C.-K. Lii, Does garlic have a role as an antidiabetic agent? *Mol. Nutr. Food Res.*, 2007, **51**, 1353–1364.

T. X. Liu and P. A. Stansley, Toxicity and repellency of some biorational insecticides to *Bemisia argentifolia* on tomato plants, *Entomol. Exper. Appl.*, 1995, **74**, 137–143.

J. Longrigg, *Greek Medicine*, Routledge, New York, 1998.

J. N. Losso and S. Nakai, Molecular size of garlic fructooligosaccharides and fructopolysaccharides by matrix-assisted laser desorption ionization mass spectrometry, *J. Agric. Food Chem.*, 1997, **45**, 4342–4346.

Mrs. [J.W.] Loudon, *Ladies' Flower-Garden of Ornamental Bulbous Plants*, William Smith, London, 1841.

C. F. Low, P. P. Chong, P. V. Yong, C. S. Lim, Z. Ahmad and F. Othman, Inhibition of hyphae formation and SIR2 expression in *Candida albicans* treated with fresh *Allium sativum* (garlic) extract, *J. Appl. Microbiol.*, 2008, **105**, 2169–2177.

G. Lucier, Garlic: flavor of the ages, *Agric. Outlook*, 2000, June/July, 7–10.

T. M. Lukes, Factors governing the greening of garlic puree, *J. Food Sci.*, 1986, **51**, 1577.

B. Lundegårdh, P. Botek, V. Schulzov, J. Hajslov, A. Strömberg and H. C. Andersson, Impact of different green manures on the content of *S*-alk(en)yl-L-cysteine sulfoxides and L-ascorbic acid in leek (*Allium porrum*), *J. Agric. Food Chem.*, 2008, **56**, 2102–2111.

Z. R. Lun, C. Burri, M. Menzinger and R. Kaminsky, Antiparasitic activity of diallyl trisulfide (Dasuansu) on human and animal pathogenic protozoa (*Trypanosoma* sp., *Entamoeba histolytica* and *Giardia lamblia*) *in vitro*, *Ann. Soc. Belg. Med. Tropicale*, 1994, **74**, 51–59.

D. Q. Luo, J. H. Guo, F. J. Wang, Z. X. Jin, X. L. Cheng, J. C. Zhu, C. Q. Peng and C. Zhang, Anti-fungal efficacy of polybutylcyanoacrylate nanoparticles of allicin and comparison with pure allicin, *J. Biomater. Sci. Polym. Ed.*, 2009, **20**, 21–31.

J. A. Lybarger, J. S. Gallagher, D. W. Pulver, A. Litwin, S. Brooks and I. L. Bernstein, Occupational asthma induced by inhalation and ingestion of garlic, *J. Allergy Clin. Immunol.*, 1982, **69**, 448–454.

K. L. MacDonald, R. F. Spengler, C. L. Hatheway, N. T. Hargrett and M. L. Cohen, Type A botulism from sauteed onions. Clinical and epidemiologic observations, *J. Am. Med. Assoc.*, 1985, **253**, 1275–1278.

M. B. MacDonald and M. Jacob, Inhalation the chief factor in onion or garlic contamination of milk, *Science*, 1928, **68**, 568–569.

L. J. Macpherson, B. H. Geierstanger, V. Viswanath, M. Bandell, S. R. Eid, S. Hwang and A. Patapoutian, The pungency of garlic: activation of TRPA1 and TRPV1 in response to allicin, *Curr. Biol.*, 2005, **15**, 929–934.

E. Malézieux, Y. Crozat, C. Dupraz, M. Laurans, D. Makowski, H. Ozier-Lafontaine, B. Rapidel, S. de Tourdonnet and M. Valantin-Morison, Mixing plant species in cropping systems: concepts, tools and models. A review, *Agron. Sustainable Development*, 2008, **29**, 43–62.

S. B. Mallek, T. S. Prather and J. J. Stapleton, Interaction effects of *Allium* spp. residues, concentrations and soil temperature on seed germination of four weedy plant species, *Appl. Soil Ecol.*, 2007, **37**, 233–239.

T. Manabe, A. Hasumi, M. Sugiyama, M. Yamazaki and K. Saito, Alliinase [*S*-alk(en)yl-L-cysteine sulfoxide lyase] from *Allium tuberosum* (Chinese chive) – purification, localization, cDNA cloning and heterologous functional expression, *Eur. J. Biochem.*, 1998, **257**, 21–30.

S. Mandal and D. K. Choudhuri, Cholesterol metabolism in *Lohita grandis* Gray (Hemiptera: Pyrrhocoridae: Insecta). Effect of corpora allatectomy and garlic extract, *Curr. Sci.*, 1982, **51**, 367–369.

L. Manniche, *An Ancient Egyptian Herbal*, British Museum Press, London, UK, 2006.

P. Marcos, M. P. Lue-Meru, R. Ricardo, G. Maximo, V. Maribel, B. J. Luis and B. Marcela, Pungency evaluation of onion cultivars from the Venezuelan West-Center region by flow injection analysis-UV-visible spectroscopy pyruvate determination, *Talanta*, 2004, **64**, 1299–1303.

D. M. Marcus and A. P. Grollman, Botanical medicines – the need for new regulations, *N. Engl. J. Med.*, 2002, **347**, 2073–2076.

K. W. Martin and E. Ernst, Herbal medicines for treatment of bacterial infections: a review of controlled clinical trials, *J. Antimicrob. Chemotherapy*, 2003, **51**, 241–246.

D. Maslin, Effects of garlic on cholesterol: not down but not out either, *Arch. Int. Med.*, 2008, **168**, 111.

J. R. Mason and G. Linz, Repellency of garlic extract to European starlings, *Crop Protection*, 1997, **16**, 107–108.

E. M. Matheson, A. G. Mainous 3rd and M. A. Carnemolla, The association between onion consumption and bone density in perimenopausal and postmenopausal non-Hispanic white women 50 years and older, *Menopause*, 2009, **16**, 756–759.

B. Mathews, *A Review of Allium section Allium*, Richmond, Royal Botanic Gardens, Kew, UK, 1996.

T. Matsukawa, S. Yurugi and T. Matsuoka, *Science*, 1953, **118**, 325–327.

Y. Matsumura, K. Shirai, T. Maki, Y. Itakura and Y. Kodera, Facile synthesis of allixin and its related compounds, *Tetrahedron Lett.*, 1998, **39**, 2339–2340.

V. Maurer, E. Perler and F. Heckendorn, *In vitro* efficacies of oils, silicas and plant preparations against the poultry red mite *Dermanyssus gallinae*, *Exp. Appl. Acarol.*, 2009, **48**, 31–41.

R. McCann, 2009, cited in *Macarthur Chronicle*, Wollondilly, Australia, January 13, 2009, p. 3: http://macarthur-chronicle-wollondilly.whereilive.com.au/news/story/oh-my-father-that-S-one-for-the-cooks/.

H. McGee, *On Food and Cooking. The Science and Lore of the Kitchen*, Macmillan, NewYork, 1984.

H. McGee, *In victu veritas*, *Nature*, 1998, **392**, 649–650.

T. McLean, *Medieval English Gardens*, Viking Press, New York, 1980.

C. A. McNulty, M. P. Wilson, W. Havinga, B. Johnston, E. A. O'Gara and D. J. Maslin, A pilot study to determine the effectiveness of garlic oil capsules in the treatment of dyspeptic patients with *Helicobacter pylori*, *Helicobacter*, 2001, **6**, 249–253.

S. McSheehy, W. Yang, F. Pannier, J. Szpunar, R. Lobinski, J. Auger and M. Potin-Gautier, Speciation analysis of selenium in garlic by two-dimensional high-performance liquid chromatography with parallel inductively coupled plasma mass spectrometric and electrospray tandem mass spectrometric detection, *Anal. Chim. Acta*, 2000, **421**, 147–153.

S. Meher and L. Duley, Garlic for preventing pre-eclampsia and its complications, *Cochrane Database Syst. Rev.*, 2006, **3**, CD006065.

J. A. Mennella, A. Johnson and G. K. Beauchamp, Garlic ingestion by pregnant women alters the odor of amniotic fluid, *Chem. Senses*, 1995, **20**, 207–209.

J. A. Mennella and G. K. Beauchamp, The effects of repeated exposure to garlic-flavored milk on the nursling's behavior, *Pediatrics Res.*, 1993, **34**, 805–808.

J. A. Mennella and G. K. Beauchamp, Maternal diet alters the sensory qualities of human milk and the nursling's behavior, *Pediatrics*, 1991, **88**, 737–744.

F. Merhi, J. Auger, F. Rendu and B. Bauvois, *Allium* compounds: dipropyl and dimethyl thiosulfinates as antiproliferative and differentiating agents of human acute myeloid leukemia cell lines, *Biol.: Targets Therapy*, 2008, **2**, 885–895.

H. Meunier, O. Petit and J. L. Deneubourg, Social facilitation of fur rubbing behavior in white-faced capuchins, *Am. J. Primatol.*, 2008, **70**, 161–168.

H. Meunier, O. Petit and J. L. Deneubourg, Resource influence on the form of fur rubbing behaviour in white-faced capuchins, *Behavioural Processes.*, 2008, **77**, 320–326.

M. A. Meyers, *Happy Accidents. Serendipity in Modern Medical Breakthroughs*, Arcade Publishing, New York, 2007.

M. R. Mezzabotta, What was "ulpicum?", *The Classical Quarterly*, 2000, **50**, 230–237.

H. Miething, HPLC-Analysis of the volatile oil of garlic bulbs, *Phytotherapy Res.*, 1988, **2**, 149–151.

A. E. Millen, A. F. Subar, B. I. Graubard, U. Peters, R. B. Hayes, J. L. Weissfeld, L. A. Yokochi and R. G. Ziegler, Fruit and vegetable intake and prevalence of colorectal adenoma in a cancer screening trial, *Am. J. Clin. Nutr.*, 2007, **86**, 1754–1764.

H. Miller, Identity takeout: how American Jews made Chinese food their ethnic cuisine, *J. Popular Culture*, 2006, **39**, 430–465.

J. A. Milner, Preclinical perspectives on garlic and cancer, *J. Nutr.*, 2006, **136**, 827S–831S.

W. C. Minchin, *The Treatment, Prevention and Cure of Tuberculosis and Lupus with Allyl Sulphide*, Bailliere, Tindall and Cox, London, UK, 1912.

W. C. Minchin, *A Study in Tubercle Virus, Polymorphism and the Treatment of Tuberculosis and Lupus with Oleum Alii*, Bailliere, Tindall & Cox, London, UK, 1927.

D. Mirelman, D. Monheit and S. Varon, Inhibition of growth of *Entamoeba histolytica* by allicin, the active principle of garlic extract (*Allium sativum*), *J. Infectious Diseases*, 1987, **156**, 243–244.

T. Miron, A. Rabinkov, D. Mirelman, M. Wilchek and L. Weiner, The mode of action of allicin: its ready permeability through phospholipid membranes may contribute to its biological activity, *Biochim Biophys Acta*, 2000, **1463**, 20–30.

T. Miron, I. Shin, G. Feigenblat, L. Weiner, D. Mirelman, M. Wilchek and A. Rabinkov, A spectrophotometric assay for allicin, alliin, and alliinase (alliin lyase) with a chromogenic thiol: reaction of 4-mercaptopyridine with thiosulfinates, *Anal. Biochem.*, 2002, **307**, 76–83.

T. Miron, M. Mironchik, D. Mirelman, M. Wilchek and A. Rabinkov, Inhibition of tumor growth by a novel approach: *in situ* allicin generation using targeted alliinase delivery, *Mol. Cancer Therapy*, 2003, **2**, 1295–1301.

T. Miron, H. SivaRaman, A. Rabinkov, D. Mirelman and M. Wilchek, A method for continuous production of allicin using immobilized alliinase, *Anal. Biochem.*, 2006, **351**, 152–154.

T. Miron, M. Wilchek, A. Sharp, Y. Nakagawa, M. Naoi, Y. Nozawa and Y. Akao, Allicin inhibits cell growth and induces apoptosis through the mitochondrial pathway in HL60 and U937 cells, *J. Nutr. Biochem.*, 2008, **19**, 524–535.

M. I. Mogahed, Influence of intercropping on population dynamics of major insect-pests of potato (*Solanum tuberosum*), *Indian J. Agric. Sci.*, 2003, **73**, 546–549.

N. Mondy, D. Duplat, J. P. Christides, I. Arnault and J. Auger, Aroma analysis of fresh and preserved onions and leek by dual solid-phase microextraction-liquid extraction and gas chromatography-mass spectrometry, *J Chromatogr., A.*, 2002, **963**, 89–93.

G. S. Moore and R. D. Atkins, The fungicidal and fungistatic effects of an aqueous garlic extract on medically important yeast-like fungi, *Mycologia*, 1977, **69**, 341–348.

T. L. Moore and D. E. O'Connor, The reaction of methanesulfenyl chloride with alkoxides and alcohols. Preparation of aliphatic sulfenate and sulfinate esters, *J. Org. Chem.*, 1966, **31**, 3587–3592.

V. Morelli and R. J. Zoorob, Alternative therapies: Part I. Depression, diabetes, obesity, *Am. Family Physician*, 2000, **62**, 1051–1060.

Y. Morimitsu and S. Kawakishi, Inhibitors of platelet aggregation from onion, *Phytochemistry*, 1990, **29**, 3435–3439.

Y. Morimitsu and S. Kawakishi, Optical resolution of 1-(methylsulfinyl)propyl alk(en)yl disulfides, inhibitors of platelet aggregation isolated from onion, *Agric. Biol. Chem.*, 1991, **55**, 889–890.

Y. Morimitsu, Y. Morioka and S. Kawakishi, Inhibitors of platelet aggregation generated from mixtures of Allium species and/or *S*-alk(en)nyl-L-cysteine sulfoxides, *J. Agric. Food Chem.*, 1992, **40**, 368–372.

C. A. Morris and J. Avorn, Internet marketing of herbal products, *J. Am. Med. Assoc.*, 2003, **290**, 1505–1509.

C. J. Morris and J. F. Thompson, The identification of (+)-*S*-methyl-l-cysteine sulfoxide in plants, *J. Am. Chem. Soc.*, 1956, **78**, 1605–1608.

V. C. Morris, Selenium content of foods, *J. Nutr.*, 1970, **100**, 1385–1386.

D. L. Morse, L. K. Pickard, J. J. Guzewich, B. D. Devine and M. Shayegani, Garlic-in-oil associated botulism: episode leads to product modification, *Am. J. Public Health*, 1990, **80**, 1372–1373.

A. W. Mott and G. Barany, A new method for the synthesis of unsymmetrical trisulfanes, *Synthesis*, 1984, 658–660.

S. Moyers, *Garlic in Health, History, and World Cuisine*, Suncoast Press, St. Petersburg, Florida, 1996.

M. Moyle, K. Frowen and R. Nixon, Use of gloves in protection from diallyl disulphide allergy, *Australasian J. Dermatol.*, 2004, **45**, 223–225.

L. Mskhiladze, J. Legault, S. Lavoie, V. Mshvildadze, J. Kuchukhidze, R. Elias and A. Pichette, Cytotoxic steroidal saponins from the flowers of *Allium leucanthum*, *Molecules*, 2008, **13**, 2925–2934.

A. L. Mueller and A. I. Virtanen, Synthesis of *S*-(buten-1-yl)-L-cysteine sulfoxide and its enzymic cleavage to 1-butenylsulfenic acid, *Acta Chem. Scand.*, 1966, **20**, 1163–1165.

S. Mukherjee, I. Lekli, S. Goswami and D. K. Das, Freshly crushed garlic is a superior cardioprotective agent than processed garlic, *J. Agric. Food Chem.*, 2009, **57**, 7137–7144.

C. Mulrow, V. Lawrence, R. Ackermann, G. Gilbert Ramirez, L. Morbidoni, C. Aguilar, J. Arterburn, E. Block, E. Chiquette, C. Gardener, M. Harris, P. Heidenreich, D. Mullins, M. Richardson, N. Russell, A. Vickers and V. Young, Garlic: effects on cardiovascular risks and disease, protective effects against cancer, and clinical adverse effects, *AHRQ Evid. Rep. Technol. Assess.* (*Summ.*), 2000, No. 20, 1–4. [AHRQ Publication No. 01–E023: http://www.ahrq.gov/clinic/garlicsum.htm].

R. Munday, J. S. Munday and C. M. Munday, Comparative effects of mono-, di-, tri-, and tetrasulfides derived from plants of the *Allium* family: redox cycling *in vitro* and hemolytic activity and Phase 2 enzyme induction *in vivo*, *Free Radical Biol. Med.*, 2003, **34**, 1200–1211.

A. Murakami, H. Ashida and J. Terao, Multitargeted cancer prevention by quercetin, *Cancer Lett.*, 2008, **269**, 315–325.

F. Murakami, Studies on the nutritional value of *Allium* plants. XXXVII. Decomposition of alliin homologues by acetone-powdered enzyme preparation of *Bacillus subtilis*, *Bitamin (Kyoto)*, 1960, **20**, 131–135.

R. W. Murray, R. D. Smetana and E. Block, Oxidation of disulfides with triphenyl phosphite ozonide, *Tetrahedron Lett.*, 1971, 299–302.

N. B. K. Murthy and S. V. Amonkar, Effect of a natural insecticide from garlic (*Allium sativum* L.) and its synthetic form (diallyl disulphide) on plant pathogenic fungi, *Indian J. Exp. Biol.*, 1974, **12**, 208–209.

L. J. Musselman, Is *Allium kurrat* the leek of the Bible? *Economic Botany*, 2002, **56**, 399–400.

M. Mütsch-Eckner, O. Sticher and B. Meier, Reversed-phase high-performance liquid chromatography of *S*-alk(en)yl-L-cysteine derivatives in *Allium sativum* including the determination of (+)-*S*-allyl-L-cysteine sulfoxide, γ-L-glutamyl-*S*-allyl-l-cysteine and γ-L-glutamyl-*S*-(*trans*-1-propenyl)-L-cysteine, *J. Chromatogr.*, 1992, **625**, 183–190.

M. Mütsch-Eckner, C. A. Erdelmeier, O. Sticher and H. D. Reuter, A novel amino acid glycoside and three amino acids from *Allium sativum*, *J. Nat. Prod.*, 1993, **56**, 864–869.

R. Naganawa, N. Iwata, K. Ishikawa, H. Fukuda, T. Fujino and A. Suzuki, Inhibition of microbial growth by ajoene, a sulfur-containing compound derived from garlic, *Appl. Environ. Microbiol.*, 1996, **62**, 4238–4242.

S. Nagini, Cancer chemoprevention by garlic and its organosulfur compounds-panacea or promise? *Anticancer Agents Med. Chem.*, 2008, **8**, 313–321.

V. Najar-Nezhad, M. R. Aslani and M. Balali-Mood, Evaluation of allicin for the treatment of experimentally induced subacute lead poisoning in sheep, *Biol. Trace Elem. Res.*, 2008, **126**, 141–147.

Y. K. Nakamura, T. Matsuo, K. Shimoi, Y. Nakamura and I. Tomita, *S*-Methyl methanethiosulfonate, bio-antimutagen in homogenates of *Cruciferae* and *Liliaceae* vegetables, *Biosci. Biotechnol. Biochem.*, 1996, **60**, 1439–1443.

J. C. Namyslo and C. Stanitzek, A palladium-catalyzed synthesis of isoalliin, the main cysteine sulfoxide in onions (*Allium cepa*), *Synthesis*, 2006, 3367–3369.

A. Nath, N. K. Sharma, S. Bhardwaj and C. D. Thapa, Nematicidal properties of garlic, *Nematologica*, 1982, **28**, 253–255.

National Cancer Institute, 2004 ("antioxidants"): http://www.cancer.gov/.

National Center for Complementary and Alternative Medicine (NCCAM: NIH, U.SA.) Publication No. D274, March 2008: http://nccam.nih.gov/health/garlic/index.htm.

National Public Radio (NPR), 2006: Does a bit of steel get rid of that garlic smell? http://www.npr.org/templates/story/story.php?storyId=6473350.

A. Nault and D. Gagnon, Ramet demography of *Allium tricoccum*, a spring ephemeral, perennial forest herb, *J. Ecol.*, 1993, **81**, 101–119.

M. T. Naznin, M. Akagawa, K. Okukawa, T. Maeda and N. Morita, Characterization of E- and Z-ajoene obtained from different varieties of garlics, *Food Chem.*, 2008, **106**, 1113–1119.

M. Negbi, E. E. Goldschmidt and N. Serikoff, Classical and Hebrew sages on cultivated biennial plants. Part II, *Scr. Classica Israelica*, 2004, **23**, 81–94.

O. Negishi, Y. Negishi and T. Ozawa, Effects of food materials on removal of *Allium*-specific volatile sulfur compounds, *J. Agric. Food Chem.*, 2002, **50**, 3856–3861.

H. A. Neil, C. A. Silagy, T. Lancaster, J. Hodgeman, K. Vos, J. W. Moore, L. Jones, J. Cahill and G. H. Fowler, Garlic powder in the treatment of moderate hyperlipidaemia: a controlled trial and meta-analysis, *J. R. Coll. Physicians Lond.*, 1996, **30**, 329–334.

New York State Task Force on Life and the Law, *Dietary Supplements: Balancing Consumer Choice & Safety*, 2005: http://www.health.state.ny.us/regulations/task_force/docs/dietary_supplement_safety.pdf.

New York Times Editorial: The 1993 snake oil protection act, October 5, 1993.

S. N. Ngo, D. B. Williams, L. Cobiac and R. J. Head, Does garlic reduce risk of colorectal cancer? A systematic review, *J. Nutr.*, 2007, **137**, 2264–2269.

D. Nicodemo and R. H. Nogueira-Couto, Use of repellents for honeybees (*Apis mellifera* L.) *in vitro* in the yellow passion-fruit (Passiflora edulis Deg) crop and in confined beef cattle feeders, *J. Venomous Animals Toxins Including Tropical Diseases*, 2004, **10**, 77–85.

P. T. Nicholson and I. Shaw, (ed.), *Ancient Egyptian Materials and Technology*, Cambridge University Press, Cambridge, UK, 2000.

W. D. Niegisch and W. H. Stahl, The onion: gaseous emanation products, *Food Res.*, 1956, **21**, 657–665.

G. S. Nielsen and L. Poll, Determination of odor active aroma compounds in freshly cut leek (*Allium ampeloprasum* Var. Bulga) and in long-term stored frozen unblanched and blanched leek slices by gas chromatography olfactometry analysis, *J. Agric. Food Chem.*, 2004, **52**, 1642–1646.

M. Nishida, T. Hada, K. Kuramochi, H. Yoshida, Y. Yonezawa, I. Kuriyama, F. Sugawara, H. Yoshida and Y. Mizushina, Diallyl sulfides: selective inhibitors of family X DNA polymerases from garlic (*Allium sativum* L.), *Food Chem.*, 2008, **108**, 551–560.

T. Nishikawa, N. Yamada, A. Hattori, H. Fukuda and T. Fujino, Inhibition by ajoene of skin-tumor promotion in mice, *Biosci. Biotechnol. Biochem.*, 2002, **66**, 2221–2223.

H. Nishimura, O. Higuchi, K. Tateshita, K. Tomobe, Y. Okuma and Y. Nomura, Antioxidative activity and ameliorative effects of memory impairment of sulfur-containing compounds in *Allium* species, *Biofactors*, 2006, **26**, 135–146.

H. Nishino, A. Nishino, J. Takayasu, A. Iwashima, Y. Itakura, Y. Kodera, H. Matsuura and T. Fuwa, Antitumor-promoting activity of allixin, a stress compound produced by garlic, *Cancer J.*, 1990, **3**, 20–21.

D. M. Oaks, H. Hartmann and K. P Dimick, Analysis of S compounds with electron capture/H flame dual channel gas chromatography, *Anal. Chem.*, 1964, **36**, 1560–1565.

J. Obagwu, Control of brown blotch of bambara groundnut with garlic extract and benomyl, *Phytoparasitica*, 2003a, **31**, 207–209.

J. Obagwu and L. Korsten, Control of citrus green and blue molds with garlic extracts, *Eur. J. Plant Pathol.*, 2003b, **109**, 221–225.

G. O'Donnell and S. Gibbons, Antibacterial activity of two canthin-6-one alkaloids from *Allium* neapolitanum, *Phytotherapy Res.*, 2007, **21**, 653–657.

G. O'Donnell, R. Poeschl, O. Zimhony, M. Gunaratnam, J. B. Moreira, S. Neidle, D. Evangelopoulos, S. Bhakta, J. P. Malkinson, H. I. Boshoff, A. Lenaerts and S. Gibbons, Bioactive pyridine-*N*-oxide disulfides from *Allium stipitatum*, *J. Nat. Prod.*, 2009, **72**, 360–365.

E. A. O'Gara, D. J. Hill and D. J. Maslin, Activities of garlic oil, garlic powder, and their diallyl constituents against *Helicobacter pylori*, *Appl. Environ. Microbiol.*, 2000, **66**, 2269–2273.

E. A. O'Gara, D. J. Maslin, A. M. Nevill and D. J. Hill, The effect of simulated gastric environments on the anti-Helicobacter activity of garlic oil, *J. Appl. Microbiol.*, 2008, **104**, 1324–1331.

R. Ohta, N. Yamada, H. Kaneko, K. Ishikawa, H. Fukuda, T. Fujino and A. Suzuki, *In vitro* inhibition of the growth of *Helicobacter pylori* by oil-macerated garlic constituents, *Antimicrob. Agents Chemotherapy*, 1999, **43**, 1811–1812.

Y. Okada, K. Tanaka, I. Fujita, E. Sato and H. Okajima, Antioxidant activity of thiosulfinates derived from garlic, *Redox Report*, 2005, **10**, 96–102.

Y. Okada, K. Tanaka, E. Sato and H. Okajima, Kinetic and mechanistic studies of allicin as an antioxidant, *Org. Biomol. Chem.*, 2006, **4**, 4113–4117.

Y. Okada, K. Tanaka, E. Sato and H. Okajima, Antioxidant activity of the new thiosulfinate derivative, *S*-benzyl phenylmethanethiosulfinate, from *Petiveria alliacea* L., *Org. Biomol. Chem.*, 2008, **6**, 1097–1102.

T. Okuyama, K. Miyake, T. Fueno, T. Yoshimura, S. Soga and E. Tsukurimichi, Equilibrium and kinetic studies of reactions of 2-methyl-2-propanesulfenic acid, *Heteroatom Chem.*, 1992, **3**, 577–583.

V. V. Oliinik, E. A. Goreshnik, Z. Zhonchinska and T. Glovyak, The π-complex of copper(I) bromide with diallyl sulfide 5CuBr · 2DAS: synthesis and crystal structure, *Russ. J. Coord. Chem.*, 1997, **23**, 595–598.

V. V. Oliinik, E. A. Goreshnik, V. N. Davydov and M. G. Mys'kiv, π-Complex of silver(I) perchlorate with diallyl sulfide: Synthesis and crystal structure of [Ag(DAS)ClO_4], *Russ. J. Coord. Chem.*, 1998, **24**, 512–514.

S. H. Omar, A. Hasan, N. Hunjul, J. Ali and M. Aqil, Historical, chemical and cardiovascular perspectives on garlic: a review, *Pharmacog. Rev.*, 2007, **1**, 80–87.

H. T. Ong and J. S. Cheah, Statin alternatives or just placebo: an objective review of omega-3, red yeast rice and garlic in cardiovascular therapeutics, *Chin. Med. J.*, 2008, **121**, 1588–1594.

Onion goggles: http://www.broadwaypanhandler.com/broadway/shopper lookup.asp.

S. Oommen, R. J. Anto, G. Srinivas and D. Karunagaran, Allicin (from garlic) induces caspase-mediated apoptosis in cancer cells, *Eur. J. Pharmacol.*, 2004, **485**, 97–103.

A. M. Oparaeke, M. C. Dike and C. I. Amatobi, Effect of application of different concentrations and appropriate schedules of aqueous garlic (*Allium sativum* L.) bulb extracts against *Maruca vitrata* and *Clavigralla tomentosicollis* on cowpea, *Vigna unguiculata* (L.) Walp, *Arch. Phytopath. Plant Protection*, 2007, **40**, 246–251.

Oxford University, Electronic Text Corpus of Sumerian Literature (ETCSL) project, 2006: http://www-etcsl.orient.ox.ac.uk/.

C. Papageorgiou, J. P. Corbet, F. Menezes-Brandao, M. Pecegueiro and C. Benezra, Allergic contact dermatitis to garlic (*Allium sativum*). Identification of the allergens: the role of mono-, di-, and trisulfides present in garlic, *Arch. Dermatol. Res.*, 1983, **275**, 229–234.

L. V. Papp, J. Lu, A. Holmgren and K. K. Khanna, From selenium to selenoproteins: synthesis, identity, and their role in human health, *Antioxidant Redox Signaling*, 2007, **9**, 775–806.

K. Parejko, Pliny the Elder's Silphium: first recorded species extinction, *Conservation Biol.*, 2003, **17**, 925–927.

L. Pari, P. Murugavel, S. L. Sitasawad and K. S. Kumar, Cytoprotective and antioxidant role of diallyl tetrasulfide on cadmium induced renal injury: an *in vivo* and *in vitro* study, *Life Sci.*, 2007, **80**, 650–658.

R. A. Parish, S. McIntire and D. M. Heimbach, Garlic burns: a naturopathic remedy gone awry, *Pediatric Emergency Care*, 1987, **3**, 258–260.

I.-K. Park, J.-Y. Park and S.-C. Shin, Fumigant activity of plant essential oils and components from garlic (*Allium sativum*) and clove bud (*Eugenia caryophyllata*) oils against the Japanese termite (*Reticulitermes speratus* Kolbe), *J. Agric. Food Chem.*, 2005, **53**, 4388–4392.

I.-K. Park, J.-Y. Park, K.-H. Kim, K. Sik Choi, I.-H. Choi, C.-S. Kim and S.-C. Shin, Nematicidal activity of plant essential oils and components from garlic (*Allium sativum*) and cinnamon (*Cinnamomum verum*) oils against the pine wood nematode (*Bursaphelenchus xylophilus*), *Nematology*, 2005, **7**, 767–774.

K. W. Park, S. Y. Kim, I. Y. Jeong, M. W. Byun, K. H. Park, K. Yamada and K. I Seo, Cytotoxic and antitumor activities of thiosulfinates from *Allium tuberosum* L, *J. Agric. Food Chem.*, 2007, **55**, 7957–7961.

S. Y. Park, S. J. Cho, H. C. Kwon, K. R. Lee, D. K. Rhee and S. Pyo, Caspase-independent cell death by allicin in human epithelial carcinoma cells: involvement of PKA, *Cancer Lett.*, 2005, **224**, 123–132.

J. Parkinson, *Paradisi in Sole*, Richard Thrale, London, 1656.

R. J. Parry and G. R. Sood, Investigations of the biosynthesis of *trans*-(+)-*S*-1-propenyl-L-cysteine sulfoxide in onions (*Allium cepa*), *J. Am. Chem. Soc.*, 1989, **111**, 4514–4515.

R. J. Parry and F. L. Lii, Investigations of the biosynthesis of *trans*-(+)-*S*-1-propenyl-L-cysteine sulfoxide. Elucidation of the stereochemistry of the oxidative decarboxylation process, *J. Am. Chem. Soc.*, 1991, **113**, 4704–4706.

R. Parsons, *How to Read a French Fry*, Houghton Mifflin, New York, 2001.

K. Parton, Onion toxicity in farmed animals, *N. Z. Veterinary J.*, 2000, **48**, 89.

L. Pasteur, Mémoire sur la fermentation appelée lactique, *Ann. Chim. Phys., S3*, 1858, **52**, 404–418.

A. K. Patra, D. N. Karma and N. Agarwal, Effect of spices on rumen fermentation, methanogenesis and protozoa counts in *in vitro* gas production test, *Int. Congr. Ser.*, 2006, **1293**, 176–179.

A. Paukner and S. J. Suomi, The effects of fur rubbing on the social behavior in tufted capuchin monkeys, *Am. J. Primatol.*, 2008, **70**, 1007–1012.

PDR (Physicians' Desk Reference) for Herbal Medicines, Thompson Healthcare, Montvale, NJ, 4th edn, 2007.

W. Pearson, H. J. Boermans, W. J. Bettger, B. W. McBride and M. I. Lindinger, Association of maximum voluntary dietary intake of freeze-dried garlic with Heinz body anemia in horses, *Am. J. Veterinary Res.*, 2005, **66**, 457–465.

J. Pechey, *The English Herbal of Physical Plants*, 1694 (reprinted by Medical Publications Ltd., London, UK, 1951).

J. Pedraza-Chaverrí, M. Gil-Ortiz, G. Albarrán, L. Barbachano-Esparza, M. Menjívar and O. N. Medina-Campos, Garlic's ability to prevent *in vitro* Cu^{2+}-induced lipoprotein oxidation in human serum is preserved in heated garlic: effect unrelated to Cu^{2+}-chelation, *Nutr. J.*, 2004, **3**, 10.

R. M. Peek, *Helicobacter pylori* infection and disease: from humans to animal models, *Disease Models Mechanisms*, 2008, **1**, 50–55.

R. E. Penn, E. Block and L. K. Revelle, Flash vacuum pyrolysis studies. 5. Methanesulfenic acid, *J. Am. Chem. Soc.*, 1978, **100**, 3622–3623.

R. Pérez-Calderón, M. A. Gonzalo-Garijo and R. Fernández de Soria, Exercise-induced anaphylaxis to onion, *Allergy*, 2002, **57**, 752–753.

A. J. Perez-Piniento, I. Moneo, M. Santaolalla, S. de Paz, B. Fernandez-Parra and A. R. Dominguez-Lazaro, Anaphylactic reaction to young garlic, *Allergy*, 1999, **54**, 626–629.

E. B. Peffley, Genome complexity of *Allium*, *The Plant Genome*, 2006, **1**, 111–130.

A. G. Perkin and J. J. Hummel, Occurrence of quercetin in the outer skins of the bulb of the onion (*Allium cepa*), *J. Chem. Soc., Trans.*, 1896, **69**, 1295–1298.

H. Perrin, *British Flowering Plants*, Bernard Quaritch, London, UK, 1914.

N. R. Perron, J. N. Hodges, M. Jenkins and J. L. Brumaghim, Predicting how polyphenol antioxidants prevent DNA damage by binding to iron, *Inorg. Chem.*, 2008, **47**, 6153–6161.

C. C. Perry, M. Weatherly, T. Beale and A. Randriamahefa, Atomic force microscopy study of the antimicrobial activity of aqueous garlic *versus* ampicillin against *Escherichia coli* and *Staphylococcus aureus*, *J. Sci. Food Agric.*, 2009, **89**, 958–964.

J. Peterson, The *Allium* species (onions, garlic, leeks, chives, and shallots, in *The Cambridge World History of Food*, Cambridge University Press, Cambridge, 2000, vol. 1, pp. 249–271.

Pharmaceutical Society of Great Britain, *The British Pharmaceutical Codex*, The Pharmaceutical Press, London, UK, 1934.

E. Pichersky and D. R. Gang, Genetics and biochemistry of secondary metabolites in plants: an evolutionary perspective, *Trends Plant Sci.*, 2000, **5**, 439–445.

I. J. Pickering, E. Y. Sneeden, R. C. Prince, E. Block, H. H. Harris, G. Hirsch and G. N. George, Localizing the chemical forms of sulfur in vivo using X-ray fluorescence spectroscopic imaging: application to onion (*Allium cepa*) tissues, *Biochemistry*, 2009, **48**, 6846–6853.

G. Pires, E. Pargana, V. Loureiro, M. M. Almeida and J. R. Pinto, Allergy to garlic, *Allergy*, 2002, **57**, 957–958.

S. C. Piscitelli, A. H. Burstein, N. Welden, K. D. Gallicano and J. Falloon, The effect of garlic supplements on the pharmacokinetics of saquinavir, *Clin. Infectectious Diseases*, 2002, **34**, 234–238.

M. H. Pittler and E. Ernst, Clinical effectiveness of garlic (*Allium sativum*), *Mol. Nutr. Food Res.*, 2007, **51**, 1382–1385.

E. S. Platt, *Garlic, Onion, & Other Alliums*, Stackpole Books, Mechanicsburg, PA, 2003.

Z. A. Polat, A. Vural, F. Ozan, B. Tepe, S. Oezcelik and A. Cetin, *In vitro* evaluation of the amoebicidal activity of garlic (*Allium sativum*) extract on *Acanthamoeba castellanii* and its cytotoxic potential on corneal cells, *J. Ocular Pharmacol. Therapeutics*, 2008, **24**, 8–14.

D. Portz, E. Koch and A. J. Slusarenko, Effects of garlic (*Allium sativum*) juice containing allicin on *Phytophthora infestans* and downy mildew of cucumber caused by *Pseudoperonospora cubensis*, *Eur. J. Plant Pathol.*, 2008, **122**, 197–206.

M. J. Potts and N. Gunadi, The influence of intercropping with *Allium* on some insect populations in potato (*Solatium tuberosum*), *Ann. Appl. Biol.*, 1991, **119**, 207–213.

A. A. Powolny and S. V. Singh, Multitargeted prevention and therapy of cancer by diallyl trisulfide and related *Allium* vegetable-derived organosulfur compounds, *Cancer Lett.*, 2008, **269**, 305–314.

M. Prager-Khoutorsky, I. Goncharov, A. Rabinkov, D. Mirelman, B. Geiger and A. D. Bershadsky, Allicin inhibits cell polarization, migration and division *via* its direct effect on microtubules, *Cell Motility Cytoskeleton*, 2007, **64**, 321–337.

E. D. Pribitkin and G. Boger, Herbal therapy: what every plastic surgeon must know, *Arch. Facial Plastic Surgery*, 2001, **3**, 127–132.

K. R. Price and M. J. C. Rhodes, Analysis of the major flavonol glycosides present in four varieties of onion (*Allium cepa*) and changes in composition resulting from autolysis, *J. Sci. Food Agric.*, 1997, **74**, 331–339.

C. L. Prince, M. L. Shuler and Y. Yamada, Altering flavor profiles in onion (*Allium cepa* L.) root cultures through directed biosynthesis, *Biotechnol. Prog.*, 1997, **13**, 506–510.

R. L. Prior, X. Wu and K. Schaich, Standardized methods for the determination of antioxidant capacity and phenolics in foods and dietary supplements, *J. Agric. Food Chem.*, 2005, **53**, 4290–4302.

G. M. Prowse, T. S. Galloway and A. Foggo, Insecticidal activity of garlic juice in two dipteran pests, *Agric. Forest Entomol.*, 2006, **8**, 1–6.

H. Puxbaum and G. König, Observation of dipropenyl disulfide and other organic sulfur compounds in the atmosphere of a beech forest with *Allium ursinum* ground cover, *Atmos. Environ.*, 1997, **31**, 291–294.

R. Qi, F. Liao, K. Inoue, Y. Yatomi, K. Sato and Y. Ozaki, Inhibition by diallyl trisulfide, a garlic component, of intracellular Ca^{2+} mobilization without affecting inositol-1,4,5-trisphosphate (IP^3) formation in activated platelets, *Biochem. Pharmacol.*, 2000, **60**, 1475–1483.

D. J. Quer, *Flora Española, o historia de las plantas que se crian en España*, Tomo Segundo, Madrid, Joachin Ibarra, 1762.

A. Rabinkov, T. Miron, L. Konstantinovski, M. Wilchek, D. Mirelman and L. Weiner, The mode of action of allicin: trapping of radicals and interaction with thiol containing proteins, *Biochim. Biophys. Acta*, 1998, **1379**, 233–244.

A. Rabinkov, T. Miron, D. Mirelman, M. Wilchek, S. Glozman, E. Yavin and L. Weiner, *S*-Allylmercaptoglutathione: the reaction product of allicin with glutathione possesses SH-modifying and antioxidant properties, *Biochim. Biophys. Acta*, 2000, **1499**, 144–153.

H. D. Rabinowitch and L. Currah, *Allium Crop Science: Recent Advances*, CABI, Wallingford, UK, 2002.

H. Rackham, *English Translation of Pliny the Elder's Natural History*, Harvard University Press, Cambridge, MA, 1971.

M. Rafaat and A. K. Leung, Garlic burns, *Pediatr. Dermatol.*, 2000, **17**, 475–476.

K. Rahman, Effects of garlic on platelet biochemistry and physiology, *Mol. Nutr. Food Res.*, 2007, **51**, 1335–1344.

K. Rahman and G. M. Lowe, Garlic and cardiovascular disease: a critical review, *J. Nutr.*, 2006, **136**(3 Suppl), 736S–740S.

M. S. Rahman, Q. H. Al-Shamsi, G. B. Bengtsson, S. S. Sablani and A. Al-Alawi, Drying kinetics and allicin potential in garlic slices during different methods of drying, *Drying Technology*, 2009, **27**, 467–477.

T. V. Rajan, M. Hein, P. Porte and S. Wikel, A double-blinded, placebo-controlled trial of garlic as a mosquito repellant: a preliminary study, *Med. Veterinary Entomol.*, 2005, **19**, 84–89.

V. Ramakrishnan, G. J. Chintalwar and A. Banerji, Environmental persistence of diallyl disulfide, an insecticidal principle of garlic and its metabolism in mosquito, *Culex pipiens quinquifasciatus Say, Chemosphere*, 1989, **18**, 1525–1529.

R. R. Ramoutar and J. L. Brumaghim, Effects of inorganic selenium compounds on oxidative DNA damage, *J. Inorg. Biochem.*, 2007, **101**, 1028–1035.

S. V. Rana, R. Pal, K. Vaiphei and K. Singh, Garlic heptatotoxicity: safe dose of garlic, *Tropical Gastroenterol.*, 2006, **27**, 26–30.

D. M. Randel, *The Harvard Biographical Dictionary of Music*, Harvard University Press, Cambridge, MA, 1998 [entry for Karen Khachaturian, p. 445].

L. V. Ratcliffe, F. J. M. Rutten, D. A. Barrett, T. Whitmore, D. Seymour, C. Greenwood, Y. Aranda-Gonzalvo, S. Robinson and M. McCoustra, Surface analysis under ambient conditions using plasma-assisted desorption/ionization mass spectrometry, *Anal. Chem.*, 2007, **79**, 6094–6101.

P. Rattanachaikunsopon and P. Phumkhachorn, Diallyl sulfide content and antimicrobial activity against food-borne pathogenic bacteria of chives (*Allium schoenoprasum*), *Biosci. Biotechnol. Biochem.*, 2008, **72**, 2987–2991.

M. P. Rayman, Selenium in cancer prevention: a review of the evidence and mechanism of action, *Proc. Nutr. Soc.*, 2005, **64**, 527–542.

E. Regel, *Alliorum adhuc cognitorum monographia*, Petropolis, St. Petersburg, Russia, 1875.

E. Regel, *Allii species Asiae centralis in Asia media à Turcomania desertisque aralensibus et caspicis usque ad Mongoliam crescentes*, Petropoli, St. Petersburg, Russia, 1887.

M. E. Reid, M. S. Stratton, A. J. Lillicoc, M. Fakih, R. Natarajan, L. C. Clark and J. R. Marshall, A report of high-dose selenium supplementation: response and toxicities, *J. Trace Elements Med. Biol.*, 2004, **18**, 69–74.

K. M. Reinhart, C. I. Coleman, C. Teevan, P. Vachhani and C. M. White, Effects of garlic on blood pressure in patients with and without systolic hypertension: a meta-analysis, *Ann. Pharmacotherapy*, 2008, **42**, 1766–1771.

F. Rendu, B. Brohard-Bohn, S. Pain, C. Bachelot-Loza and J. Auger, Thiosulfinates inhibit platelet aggregation and microparticle shedding at a calpain-dependent step, *Thrombosis Haemostatis*, 2001, **86**, 1284–1291.

J. Resemann, B. Maier and R. Carle, Investigations on the conversion of onion aroma precursors *S*-alk(en)yl-L-cysteine sulphoxides in onion juice production, *J. Sci. Food Agric.*, 2004, **84**, 1945–1950.

H. D. Reuter, H. P. Koch and L. D. Lawson, Therapeutic effects and applications of garlic and its preparations, in *Garlic The Science and Therapeutic Applications of Allium sativum L. and Related Species*, ed. H. P. Koch and L. D. Lawson, Williams and Wilkins, Baltimore, MD, 1996, pp. 135–212.

K. Ried, O. R. Frank, N. P. Stocks, P. Fakler and T. Sullivan, Effect of garlic on blood pressure: a systematic review and meta-analysis, *BMC Cardiovascular Disorders*, 2008, **8**, 1: http://www.biomedcentral.com/1471-2261/8/13.

G. R. Rik, T. E. Pashchenko, A. F. Burtsev and O. V. Redman, Effect of volatile organic isolates of macerated onion bulb on the germination of cucumber seeds, *Dokl. Vses. Akad. S-kh. Nauk im. V. I. Lenina*, 1974, 14–15 (*Chem. Abstr.*, 1975, **82**, 165793).

R. S. Rivlin, Can garlic reduce risk of cancer? *Am. J. Clin. Nutr.*, 2009, **89**, 17–18.

R. J. Roberge, R. Leckey, R. Spence and E. J. Krenzelok, Garlic burns of the breast, *Am. J. Emergency Med.*, 1997, **15**, 548.

J. E. Robertson, M. M. Christopher and Q. R. Rogers, Heinz body formation in cats fed baby food containing onion powder, *J. Am. Veterinary Med. Assoc.*, 1998, **212**, 1260–1266.

G. Rodari, *Il romanzo di Cipollino*, ed. di Cultura Sociale, Rome, Italy, 1951 [subsequently published as *Le avventure di Cipollino* (*The Adventures of the Little Onion*), 1959; Editori riuniti, Rome, with multiple reprintings through 2000 as well as translations into German, Russian, Chinese, *etc*.].

I. S. Rombauer and M. R. Becker, *Joy of Cooking*, Bobbs-Merrill, Indianapolis, 1975.

K. D. Rose, P. D. Croissant, C. F. Parliament and M. B. Levin, Spontaneous spinal epidural hematoma with associated platelet dysfunction from excessive garlic ingestion: a case report, *Neurosurgery*, 1990, **26**, 880–882.

R. T. Rosen, R. D. Hiserodt, E. K. Fukuda, R. J. Ruiz, Z. Zhou, J. Lech, S. L. Rosen and T. G. Hartman, The determination of metabolites of garlic preparations in breath and human plasma, *Biofactors*, 2000, **13**, 241–249.

R. T. Rosen, R. D. Hiserodt, E. K. Fukuda, R. J. Ruiz, Z. Zhou, J. Lech, S. L. Rosen and T. G. Hartman, Determination of allicin, *S*-allylcysteine and volatile metabolites of garlic in breath, plasma or simulated gastric fluids, *J. Nutr.*, 2001, **131**, 968S–971S.

J. Rosso and S. Lukins, *The Silver Palate Cookbook*, Workman Publishing, New York, 1982.

P. S. Ruddock, M. Liao, B. C. Foster, L. Lawson, J. T. Arnason and J. A. Dillon, Garlic natural health products exhibit variable constituent levels and antimicrobial activity against *Neisseria gonorrhoeae, Staphylococcus aureus* and *Enterococcus faecalis*, *Phytotherapy Res.*, 2005, **19**, 327–334.

L. Rudkin, *Natural Dyes*, A and C Black Publishers, London, UK, 2007.

C. Rundqvist, Pharmacological investigation of *Allium* bulbs, *Pharm. Notisbl.*, 1909, **18**, 323–333.

P. E. Russell and A. E. A. Mussa, The use of garlic (*Allium sativum*) extracts to control foot rot of *Phaseolus vulgaris* caused by *Fusarium solani* f.sp. *phaseoli*, *Ann. Appl. Biol.*, 1977, **86**, 369–372.

Russian Federation, 1992, 30 Kopek Chipollino stamp: Scott Catalogue number 6077: ITC "Marka" 16; Michel 235; Stanley Gibbons 6355, Yvert et Tellier 5995.

Russian Federation, 2004, 4 Ruple "Chipollino on a car" series of six stamps: ITC "Marka" 963; Michel 1195, block 72; Stanley Gibbons MS7287, Yvert et Tellier 6826.

L. Ruzicka, Life and work of Arthur Stoll, *Helv. Chim. Acta*, 1971, **54**, 2601–2615.

M. E. Rybak, E. M. Calvey and J. M. Harnly, Quantitative determination of allicin in garlic: supercritical fluid extraction and standard addition of alliin, *J. Agric. Food Chem.*, 2004, **52**, 682–687.

A. Z. Rys and D. N. Harpp, Insertion of a two-sulfur unit into the S-S bond. Tailor-made polysulfides, *Tetrahedron Lett.*, 2000, **4**, 7169–7172.

S. C. Sahu, Dual role of organosulfur compounds in foods: a review, *J. Environ. Sci. Health, C Environ. Carcinogenesis Ecotoxicol. Rev.*, 2002, **20**, 61–76.

H. Salazar, I. Llorente, A. Jara-Oseguera, R. García-Villegas, M. Munari, S. E. Gordon, L. D. Islas and T. Rosenbaum, A single *N*-terminal cysteine in TRPV1 determines activation by pungent compounds from onion and garlic, *Nature Neurosci.*, 2008, **11**, 255–261.

D. Saleheen, S. A. Ali and M. M. Yasinzai, Antileishmanial activity of aqueous onion extract *in vitro*, *FitoterapiaI*, 2004, **75**, 9–13.

N. F. Salivon, Y. E. Filinchuk and V. V. Olijnyk, The first complex of diallyl polysulfide: synthesis and crystal structure of $[Cu_3Br_3(CH_2{=}CHCH_2(S)_4CH_2CH{=}CH_2)]$, *Z. Anorg. Allg. Chem.*, 2006, **632**, 1610–1613.

N. F. Salivon, V. V. Olijnik and A. A. Shkurenko, Synthesis and crystal structure of π complex of copper(I) chloride with diallyl trisulfide 2CuCl · DATrS, *Russ. J. Coord. Chem.*, 2007, **33**, 908–913.

M. K. S. R. D. Samarasinghe, B. S. Chhillar and R. Singh, Insecticidal properties of methanolic extract of *Allium sativum* L. and its fractions against *Plutella xylostella* (L.), *Pestic. Res. J.*, 2007, **19**, 145–148.

M. N. Sani, H. R. Kianifar, A. Kianee and G. Khatami, Effect of oral garlic on arterial oxygen pressure in children with hepatopulmonary syndrome, *World J. Gastroenterol.*, 2006, **12**, 2427–2431.

A. S. Saulis, J. H. Mogford and T. A. Mustoe, Effect of Mederma on hypertrophic scarring in the rabbit ear model, *Plastic Reconstructive Surgery*, 2002, **110**, 177–183.

P. Savi, J. L. Zachayus, N. Delesque-Touchard, C. Labouret, C. Hervé, M. F. Uzabiaga, J. M. Pereillo, J. M. Culouscou, F. Bono, P. Ferrara and J. M. Herbert, The active metabolite of Clopidogrel disrupts P2Y12 receptor oligomers and partitions them out of lipid rafts, *Proc. Natl. Acad. Sci. USA.*, 2006, **103**, 11069–11074.

G. K. Scadding, R. Ayesh, J. Brostoff, S. C. Mitchell, R. H. Waring and R. L. Smith, Poor sulphoxidation ability in patients with food sensitivity, *Br. Med. J.*, 1988, **297**, 105–107.

G. Scharbert, M. L. Kalb, M. Duris, C. Marschalek and S. A. Kozek-Langenecker, Garlic at dietary doses does not impair platelet function, *Anesthetics Analgesics*, 2007, **105**, 1214–1218.

H. Schindler, Concerning the origin of the onion dome and onion spires in Central European architecture, *J. Soc. Architectural Historians*, 1981, **40**, 138–142.

N. E. Schmidt, L. M. Santiago, H. D. Eason, K. A. Dafford, C. A. Grooms, T. E. Link, D. T. Manning, S. D. Cooper, R. C. Keith, W. O. Chance III, M. D. Walla and W. E. Cotham, Rapid extraction method of quantitating the lachrymatory factor of onion using gas chromatography, *J. Agric. Food Chem.*, 1996, **44**, 2690–2693.

B. Schmitt, H. Schulz, J. Storsberg and M. Keusgen, Chemical characterization of *Allium ursinum* L. depending on harvesting time, *J. Agric. Food Chem.*, 2005, **53**, 7288–7294.

I. Schüder, G. Port and J. Bennison, Barriers, repellents and antifeedants for slug and snail control, *Crop Protection*, 2003, **22**, 1033–1038.

J. C. Schultz, How plants fight dirty, *Nature*, 2002, **416**, 267.

O. E. Schultz and H. L. Mohrmann, Analysis of constituents of garlic, *Allium sativum*. II. Gas chromatography of garlic oil, *Pharmazie*, 1965, **20**, 441–447.

P. Schumacker, Reactive oxygen species in cancer cells: live by the sword, die by the sword, *Cancer Cell*, 2006, **10**, 175–176.

H. O. Schwabe, Germanic coin-names III, *Modern Philol.*, 1917, **14**, 611–638.

S. Schwimmer, J. F. Carson, R. U. Makower, M. Mazelis and F. F. Wong, Demonstration of alliinase in a protein preparation from onion, *Experientia*, 1960, **16**, 449–50.

R. Sealy, M. R. Evans and C. Rothrock, The effect of a garlic extract and root substrate on soilborne fungal pathogens, *HortTechnology*, 2007, **17**, 169–173.

E. Seebeck, Crystalline derivative from allium plants, *US2642374*, 1953.

H. Seifried, S. McDonald, D. Anderson, P. Greenwald and J. Milner, The antioxidant conundrum in cancer, *Cancer Res.*, 2006, **63**, 4295–4298.

T. Seki, T. Hosono, T. Hosono-Fukao, K. Inada, R. Tanaka, J. Ogihara and T. Ariga, Anticancer effects of diallyl trisulfide derived from garlic, *Asia Pacific J. Clin. Nutr.*, 2008, **17**(Suppl 1), 249–252.

F. W. Semmler, Essential oil of garlic (*Allium sativum*), *Arch. Pharm.*, 1892a, **230**, 434–443 (*Chem. Abstr.* 1906, 80662).

F. W. Semmler, Essential oil of onion (*Allium cepa*, L.), *Arch. Pharm.*, 1892b, **230**, 443–448 (*Chem. Abstr.*, 1906, 80663).

S. K. Senapati, S. Dey, S. K. Dwivedi and D. Swarup, Effect of garlic (*Allium sativum* L.) extract on tissue lead level in rats, *J. Ethnopharmacol.*, 2001, **76**, 229–232.

A. Sendl and H. Wagner, Isolation and identification of homologues of ajoene and alliin from bulb-extracts of *Allium ursinum*, *Planta Med.*, 1991, **57**, 361–362.

A. Sendl, G. Elbl, B. Steinke, K. Redl, W. Breu and H. Wagner, Comparative pharmacological investigations of *Allium ursinum* and *Allium sativum*, *Planta Med.*, 1992, **58**, 1–7.

A. Sendl, M. Schliack, R. Löser, F. Stanislaus and H. Wagner, Inhibition of cholesterol synthesis *in vitro* by extracts and isolated compounds prepared from garlic and wild garlic, *Atherosclerosis*, 1992, **94**, 79–85.

A. Sengupta, S. Ghosh and S. Bhattacharjee, *Allium* vegetables in cancer prevention: an overview, *Asian Pacific J. Cancer Prevention*, 2004, **5**, 237–245.

N. A. Shaath and F. B. Flores, Egyptian onion oil, *Dev. Food Sci.*, 1998, **40** (Food Flavors: Formation, Analysis, and Packaging Influences), 443–453. .

Y. Shadkchan, E. Shemesh, D. Mirelman, T. Miron, A. Rabinkov, M. Wilchek and N. Osherov, Efficacy of allicin, the reactive molecule of garlic, in inhibiting Aspergillus spp. *in vitro*, and in a murine model of disseminated aspergillosis, *J. Antimicrob. Chemotherapy*, 2004, **53**, 832–836.

M. Shah, J. Meija and J. A. Caruso, Relative mass defect filtering of high-resolution mass spectra for exploring minor selenium volatiles in selenium-enriched green onions, *Anal. Chem.*, 2007, **79**, 846–853.

M. Shams-Ghahfarokhi, M. -R. Shokoohamiri, N. Amirrajab, B. Moghadasi, A. Ghajari, F. Zeini, G. Sadeghi and M. Razzaghi-Abyaneh, *In vitro* antifungal activities of *Allium cepa, Allium sativum* and ketoconazole against some pathogenic yeasts and dermatophytes, *Fitoterapia*, 2006, **77**, 321–323.

M. Shams-Ghahfarokhi, M. Goodarzi, M. R. Abyaneh, T. Al-Tiraihi and G. Seyedipour, Morphological evidences for onion-induced growth inhibition of *Trichophyton rubrum* and *Trichophyton mentagrophytes*, *Fitoterapia*, 2004, **75**, 645–655.

S. Shannon, M. Yamaguchi and F. D. Howard, Reactions involved in formation of a pink pigment in onion purees, *J. Agric. Food Chem.*, 1967a, **15**, 417–422.

S. Shannon, M. Yamaguchi and F. D. Howard, Precursors involved in the formation of pink pigments in onion purees, *J. Agric. Food Chem.*, 1967b, **15**, 423–426.

E. Shemesh, O. Scholten, H. D. Rabinowitch and R. Kamenetsky, Unlocking variability: inherent variation and developmental traits of garlic plants originated from sexual reproduction, *Planta*, 2008, **227**, 1013–1024.

C. Shen and K. L. Parkin, *In vitro* biogeneration of pure thiosulfinates and propanethial S-oxide, *J. Agric. Food Chem.*, 2000, **48**, 6254–6260.

C. Shen, Z. Hong and K. L. Parkin, Fate and kinetic modeling of reactivity of alkanesulfenic acids and thiosulfinates in model systems and onion homogenates, *J. Agric. Food Chem.*, 2002, **50**, 2652–2659.

J. Shen, L. E. Davis, J. M. Wallace, Y. Cai and L. D. Lawson, Enhanced diallyl trisulfide has *in vitro* synergy with amphotericin B against *Cryptococcus neoformans*, *Planta Med.*, 1996, **62**, 415–418.

W. A. Sheppard and J. Diekmann, Sulfines, *J. Am. Chem. Soc.*, 1964, **86**, 1891–1892.

L. J. W. Shimon, A. Rabinkov, I. Shin, T. Miron, D. Mirelman, M. Wilchek and F. Frolow, Two structures of alliinase from *Alliium sativum* L.: *Apo* form and ternary complex with aminoacrylate reaction intermediate covalently bound to the PLP cofactor, *J. Mol. Biol.*, 2007, **366**, 611–625.

Y. Shukla and N. Kalra, Cancer chemoprevention with garlic and its constituents, *Cancer Lett.*, 2007, **247**, 167–181.

M. R. Shulman, *Garlic Cookery*, Thorsons, London,UK, 1984.

M. H. Siess, A. M. Le Bon, C. Teyssier, C. Belloir, V. Singh and R. Bergès, Garlic and cancer, *Med. Aromatic Plant Sci. Biotech.*, 2007, **1**, 25–30.

E. J. Siff, Method of using lachrymatory agents for moisturing the eyes, *US Pat.*, US 6 251 952, June 26, 2001.

E. J. Siff, Product for moisturing an eye, *US Pat.*, US 6 297 289, October 2, 2001.

C. A. Silagy and H. A. Neil, A meta-analysis of the effect of garlic on blood pressure, *J. Hypertension*, 1994, **12**, 463–468.

M. Silano, M. De Vincenzi, A. De Vincenzi and V. Silano, The new European legislation on traditional herbal medicines: main features and perspectives, *Fitoterapia*, 2004, **75**, 107–116.

A. Simmons, *American Cookery* (facsimile of 1796 work), Oxford University Press, Toronto, Canada, 1958.

F. J. Simoons, *Food in China: A Cultural and Historical Inquiry*, CRC Press, Boca Raton, FL, 1991.

F. J. Simoons, *Plants of Life, Plants of Death*, University of Wisconsin Press, Madison, WI, 1998.

M. W. Sinclair, N. Fourikis, J. C. Ribes, B. J. Robinson, R. D. Brown and P. D. Godfrey, Detection of interstellar thioformaldehyde, *Australian J. Physics*, 1973, **26**, 85–91.

D. K. Singh and A. Singh, *Allium sativum* (Garlic), a potent new molluscicide, *Biol. Agric. Horticulture*, 1993, **9**, 121–124.

D. K. Singh and T. D. Porter, Inhibition of sterol 4 α-methyl oxidase is the principal mechanism by which garlic decreases cholesterol synthesis, *J. Nutr.*, 2006, **136** (3 Suppl), 759S–764S.

S. V. Singh, A. A. Powolny, S. D. Stan, D. Xiao, J. A. Arlotti, R. Warin, E. R. Hahm, S. W. Marynowski, A. Bommareddy, D. M. Potter and R. Dhir, Garlic constituent diallyl trisulfide prevents development of poorly differentiated prostate cancer and pulmonary metastasis multiplicity in TRAMP mice, *Cancer Res.*, 2008, **68**, 9503–9511.

U. P. Singh, K. K. Pathak, M. N. Khare and R. B. Singh, Effect of leaf extract of garlic on *Fusarium oxysporum* f. sp. *ciceri*, *Sclerotinia sclerotiorum* and on gram seeds, *Mycologia*, 1979, **71**, 556–564.

U. P. Singh, V. B. Chauhan, K. G. Wagner and A. Kumar, Effect of ajoene, a compound derived from garlic (*Allium sativum*), on *Phytophthora drechsleri* f. sp. *cajani*, *Mycologia*, 1992, **84**, 105–108.

V. K. Singh and D. K. Singh, Characterization of allicin as a molluscicidal agent in *Allium sativum* (Garlic), *Biol. Agric. Hort.*, 1995, **12**, 119–131.

V. K. Singh and D. K. Singh, Pharmacological effects of garlic (*Allium sativum L.*), *ARBS Ann. Rev. Biomed. Sci.*, 2008, **10**, 6–26.

N. K. Sinha, D. E. Guyer, D. A. Gage and C. T. Lira, Supercritical carbon dioxide extraction of onion flavors and their analysis by gas chromatography-mass spectrometry, *J. Agric. Food Chem.*, 1992, **40**, 842–845.

S. Sitprija, C. Plengvidhya, V. Kangkaya, S. Bhuvapanich and M. Tunkayoon, Garlic and diabetes mellitus phase II clinical trial, *J. Med. Assoc. Thailand*, 1987, **70** (Suppl. 2), 223–227.

G. P. Sivam, Protection against *Helicobacter pylori* and other bacterial infections by garlic, *J. Nutr.*, 2001, **131**, 1106S–1108S.

G. P. Sivam, J. W. Lampe, B. Ulness, S. R. Swanzy and J. D. Potter, *Helicobacter pylori – in vitro* susceptibility to garlic (*Allium sativum*) extract, *Nutr. Cancer*, 1997, **27**, 118–121.

R. Slimestad, T. Fossen and I. M. Vågen, Onions: a source of unique dietary flavonoids, *J Agric Food Chem.*, 2007, **55**, 10067–10080.

A. J. Slusarenko, A. Patel and D. Portz, Control of plant diseases by natural products: allicin from garlic as a case study, *Eur. J. Plant Pathol.*, 2008, **121**, 313–322.

L. D. Small, J. H. Bailey and C. J. Cavallito, Alkyl thiolsulfinates, *J. Am. Chem. Soc.*, 1947, **69**, 1710–1713.

L. D. Small, J. H. Bailey and C. J. Cavallito, Comparison of some properties of thiolsulfinates and thiolsulfonates, *J. Am. Chem. Soc.*, 1949, **71**, 3565–3566.

R. G. Smith, Determination of the country of origin of garlic (*Allium sativum*) using trace metal profiling, *J. Agric. Food Chem.I*, 2005, **53**, 4041–4045.

E. Y. Sneeden, H. H. Harris, I. J. Pickering, R. C. Prince, S. Johnson, X. Li, E. Block and G. N. George, The sulfur chemistry of shiitake mushroom, *J. Am. Chem. Soc.*, 2004, **126**, 458–459.

S. B. Snell, Garlic on the baby's breath, *Lancet*, 1973, **7819**, 43.

G. M. Solomon and J. Moodley, Acute chlorpyrifos poisoning in pregnancy: a case report, *Clin. Toxicol.*, 2007, **45**, 416–419.

K. Song and J. A. Milner, The influence of heating on the anticancer properties of garlic, *J. Nutr.*, 2001, **131**, 1054S–1057S.

S. K. Soni and S. Finch, Laboratory evaluation of sulfur-bearing chemicals as attractants for larvae of the onion fly, *Delia antiqua* (Meigen) (Diptera: Anthomyiidae), *Bull. Entomol. Res.*, 1979, **69**, 291–298.

Soyuzmultfilm, *Chipollino i Zakoldovannyiy Mal'chik*, 1961, Russia [animated film].

C.-G. Spåre and A. I. Virtanen, On the lachrymatory factor in onion (*Allium cepa*) vapours and its precursor, *Acta Chem. Scand.*, 1963, **17**, 641–650.

C.-G. Spåre and A. I. Virtanen, Occurrence of free selenium-containing amino acids in onion (*Allium cepa*), *Acta Chem. Scand.*, 1964, **18**, 280–282.

R. N. Spice, Hemolytic anemia associated with ingestion of onions in a dog, *Can. Veterinary J.*, 1976, **17**, 181–183.

N. Sriram, S. Kalayarasan, P. Ashokkumar, A. Sureshkumar and G. Sudhandiran, Diallyl sulfide induces apoptosis in Colo 320 DM human colon cancer cells: involvement of caspase-3, NF-kappaB, and ERK-2, *Mol. Cell Biochem.*, 2008, **311**, 157–165.

F. A. Stafleu and R. S. Cowan, *Taxonomic Literature: A Selective Guide to Botanical Publications and Collections with Dates, Commentaries and Types*, 2nd Edition, Vol. 4, Utrecht: Bohn, Scheltema & Holkema, 1983.

S. D. Stan, S. Kar, G. D. Stoner and S. V. Singh, Bioactive food components and cancer risk reduction, *J. Cell Biochem.*, 2008, **104**, 339–356.

C. Starkenmann, B. Le Calvé, Y. Niclass, I. Cayeux, S. Beccucci and M. Troccaz, Olfactory perception of cysteine *S*-conjugates from fruits and vegetables, *J. Agric. Food Chem.*, 2008, **56**, 9575–9580.

Sterling-Winthrop Research Institution, personal communication of archival data, 1984.

D. Stern, 2009, personal communication: http://www.garlicseedfoundation.info/.

W. Stobie, Medical news. Wasp stings and bee stings, *Br. Med. J.*, March 5, 1932, 455.

A. Stoll and E. Seebeck, Allium compounds. I. *Alliin*, the true mother compound of garlic oil, *Helv. Chim. Acta*, 1948, **31**, 189–210.

A. Stoll and E. Seebeck, Allium compounds. II. Enzymic degradation of *alliin* and the properties of alliinase, *Helv. Chim. Acta*, 1949a, **32**, 197–205.

A. Stoll and E. Seebeck, Allium compounds III. Specificity of alliinase and synthesis of compounds related to alliin, *Helv. Chim. Acta*, 1949b, **32**, 866–876.

A. Stoll and E. Seebeck, Specific constituents of garlic, *Sci. Pharm.*, 1950, **18**, 61–79.

A. Stoll and E. Seebeck, Allium compounds. V. The synthesis of natural alliin and its three optically active isomers, *Helv. Chim. Acta*, 1951a, **34**, 481–487.

A. Stoll and E. Seebeck, Chemical investigations on alliin, the specific principle of garlic, *Adv. Enzymol.*, 1951b, **11**, 377–400.

J. Storsberg, H. Schulz and E. R. J. Keller, Chemotaxonomic classification of some *Allium* wild species on the basis of their volatile sulphur compounds, *J. Appl. Botany*, 2003, **77**, 160–162.

J. Storsberg, H. Schulz, M. Keusgen, F. Tannous, K. J. Dehmer and E. R. J. Keller, Chemical characterization of interspecific hybrids between *Allium cepa* L. and *Allium kermesinum* Rchb., *J. Agric. Food Chem.*, 2004, **52**, 5499–5505.

S. Stranges, J. R Marshall, R. Natarajan, R. P. Donahue, M. Trevisan, G. F. Combs, F. P. Cappuccio, A. Ceriello and M. E. Reid, Effects of long-term selenium supplementation on the incidence of type 2 diabetes: a randomized trial, *Ann. Internal Med.*, 2007, **147**, 217–223.

J. Strating, L. Thijs and B. Zwanenburg, A thioaldehyde *S*-oxide, *Recl. Trav. Chim. Pays-Bas*, 1964, **83**, 631–636.

L. Sturtevant, History of garden vegetables (continued), *Am. Naturalist*, 1888, **22**, 420–433.

F. Suarez, J. Springfield, J. Furne and M. Levitt, Differentiation of mouth *versus* gut as site of origin of odoriferous breath gases after garlic ingestion, *Am. J. Physiol.*, 1999, **276**, G425–30.

N. G. Sukul, P. K. Das and G. C. De, Nematicidal action of some edible crops, *Nematologica*, 1974, **20**, 187–199.

H. R. Superko and R. M. Krauss, Garlic powder, effect on plasma lipids, postprandial lipemia, low-density lipoprotein particle size, high-density lipoprotein subclass distribution and lipoprotein (a), *J. Am. College Cardiol.*, 2000, **35**, 321–326.

T. Suzuki, M. Sugii and T. Kakimoto, New γ-glutamyl peptides in garlic, *Chem. Pharm. Bull.*, 1961, **9**, 77–78.

K. T. Suzuki, Y. Tsuji, Y. Ohta and N. Suzuki, Preferential organ distribution of methylselenol source *Se*-methylselenocysteine relative to methylseleninic acid, *Toxicol. Appl. Pharmacol.*, 2008, **227**, 76–83.

C. A. Swanson, Suggested guidelines for articles about botanical dietary supplements, *Am. J. Clin. Nutr.*, 2002, **75**, 8–10.

V. Täckholm and M. Drar, *Flora of Egypt*, Cairo University Press, Cairo, 1954, **vol. 3**.

M. Tada, Y. Hiroe, S. Kiyohara and S. Suzuki, Nematicidal and antimicrobial constituents from *Allium grayi* Regel and *Allium fistulosum* L. var. caespitosum, *Agric. Biol. Chem.*, 1988, **52**, 2383–2385.

U. Takahama and S. Hirota, Deglucosidation of quercetin glucosides to the aglycone and formation of antifungal agents by peroxidase-dependent oxidation of quercetin on browning of onion scales, *Plant Cell Physiol.*, 2000, **41**, 1021–1029.

Z. Takats, J. M. Wiseman, B. Gologan and R. G. Cooks, Mass spectrometry sampling under ambient conditions with desorption electrospray ionization, *Science*, 2004, **306**, 471–473.

I. Takougang, J. Meli, S. Lemlenn, P. N. Tatah and M. Ntep, Loiasis – a neglected and under-estimated affliction: endemicity, morbidity and perceptions in eastern Cameroon, *Ann. Tropical Med. Parasitol.*, 2007, **101**, 151–160.

T. Tamaki and S. Sonoki, Volatile sulfur compounds in human expiration after eating raw or heat-treated garlic, *J. Nutr. Sci. Vitaminol. (Tokyo)*, 1999, **45**, 213–222.

K. Tamaki, T. Tamaki and T. Yamazaki, Studies on the deodorization by mushroom (*Agaricus bisporus*) extract of garlic extract-induced oral malodor, *J. Nutr. Sci. Vitaminol. (Tokyo)*, 2007, **53**, 277–286.

K. Tamaki, S. Sonoki, T. Tamaki and K. Ehara, Measurement of odour after *in vitro* or *in vivo* ingestion of raw or heated garlic, using electronic nose, gas chromatography and sensory analysis, *Int. J. Food Sci. Technol.*, 2008, **43**, 130–139.

H. Tan, H. Ling, J. He, L. Yi, J. Zhou, M. Lin and Q. Su, Inhibition of ERK and activation of p38 are involved in diallyl disulfide induced apoptosis of leukemia HL-60 cells, *Arch. Pharm. Res.*, 2008, **31**, 786–793.

X. Tang, Z. Xia and J. Yu, An experimental study of hemolysis induced by onion (*Allium cepa*) poisoning in dogs, *J. Veterinary Pharmacol. Therapeutics*, 2008, **31**, 143–149.

R. Tannahill, *Food in History*, Penguin, New York, 1992.

C. R. Taormina, J. T. Baca, S. A. Asher, J. J. Grabowski and D. N. Finegold, Analysis of tear glucose concentration with electrospray ionization mass spectrometry, *J. Am. Soc. Mass Spectrom.*, 2007, **18**, 332–336.

E. Tattelman, Health effects of garlic, *Am. Family Physician*, 2005, **72**, 103–106.

J. Taucher, A. Hansel, A. Jordan and W. Lindinger, Analysis of compounds in human breath after ingestion of garlic using proton-transfer-reaction mass spectrometry, *J. Agric. Food Chem.*, 1996, **44**, 3778–3782.

P. Taylor, R. Noriega, C. Farah, M. J. Abad, M. Arsenak and R. Apitz, Ajoene inhibits both primary tumor growth and metastasis of B16/BL6 melanoma cells in C57BL/6 mice, *Cancer Lett.*, 2006, **239**, 298–304.

K. Teranishi, R. Apitz-Castro, S. C. Robson, E. Romano and D. K. C. Cooper, Inhibition of baboon platelet aggregation *in vitro* and *in vivo* by the garlic derivative, ajoene, *Xenotransplantation*, 2003, **10**, 374–379.

J. Terrasson, B. Xu, M. Li, S. Allart, J. L. Davignon, L. H. Zhang, K. Wang and C. Davrinche, Activities of Z-ajoene against tumour and viral spreading *in vitro*, *Fundam. Clin. Pharmacol.*, 2007, **21**, 281–289.

C. Teyssier, L. Guenot, M. Suschetet and M. H. Siess, Metabolism of diallyl disulfide by human liver microsomal cytochromes P-450 and flavin-containing monooxygenases, *Drug Metab. Dispos.*, 1999, **27**, 835–41.

C. Teyssier and M. H. Siess, Metabolism of dipropyl disulfide by rat liver phase I and phase II enzymes and by isolated perfused rat liver, *Drug Metab. Dispos.*, 2000, **28**, 648–654.

E. Thibout and J. Auger, Defensive role of *Allium* sulfur volatiles against the insects, *Acta Botanica Gallica*, 1997, **144**, 419–426.

C. J. Thomas and A. Callaghan, The use of garlic (*Allium sativa*) and lemon peel (*Citrus limon*) extracts as *Culex pipiens* larvacides: persistence and interaction with an organophosphate resistance mechanism, *Chemosphere*, 1999, **39**, 2489–2396.

J. Thomas and L. Parkin, Quantification of alk(en)yl-L-cysteine sulfoxides and related amino acids in alliums by high-performance liquid chromatography, *J. Agric. Food Chem.*, 1994, **42**, 1632–1638.

R. J. Thornton, *A Family Herbal*, R. and R. Crosby, London, 2nd edn, 1814.

C. M. L. J. Tilli, A. J. W. Stavast-Kooy, J. D. D. Vuerstaek, M. R. T. M. Thissen, G. A. M. Krekels, F. C. S. Ramaekers and H. A. M. Neumann, The garlic-derived organosulfur component ajoene decreases basal cell carcinoma tumor size by inducing apoptosis, *Arch. Dermatol. Res.*, 2003, **295**, 117–123.

D. Trachootham, Y. Zhou, H. Zhang, Y. Demizu, Z. Chen, H. Pelicano, P. J. Chiao, G. Achanta, R. B. Arlinghaus, J. Liu and P. Huang, Selective killing of oncogenically transformed cells through a ROS-mediated mechanism by beta-phenylethyl isothiocyanate, *Cancer Cell*, 2006, **10**, 241–252.

S.-M. Tsao and M.-C. Yin, *In-vitro* antimicrobial activity of four diallyl sulphides occurring naturally in garlic and Chinese leek oils, *J. Med. Microbiol.*, 2001a, **50**, 646–649.

S.-M. Tsao and M.-C. Yin, *In vitro* activity of garlic oil and four diallyl sulphides against antibiotic-resistant *Pseudomonas aeruginosa* and *Klebsiella pneumoniae*, *J. Antimicrob. Chemotherapy*, 2001, **47**, 665–670.

S. Tsuno, F. Murakami, K. Tazoe and S. Kikumoto, The nutritional value of *Allium* plants. XXX. Isolation of methiin, *Bitamin*, 1960, **20**, 93–96.

S. Tuntipopipat, C. Zeder, P. Siriprapa and S. Charoenkiatkul, Inhibitory effects of spices and herbs on iron availability, *Int. J. Food Sci. Nutr.*, 2008, **60**, 43–55.

F. Turecek, L. Brabec, T. Vondák, V. Hanus, J. Hájícer and Z. Havlas, Sulfenic acids in the gas phase. Preparation, ionization energies and heats of formation of methane-, ethane-, ethyne- and benzenesulfenic acid, *Coll. Czech. Chem. Commun.*, 1988, **53**, 2140–2158.

F. Turecek, F. W. McLafferty, B. J. Smith and L. Radom, Neutralization–reionization and *ab initio* study of the CH_2=CHSOH to CH_3CH=S=O rearrangement, *Int. J. Mass Spectrom. Ion Processes*, 1990, **101**, 283–300.

B. Turner, C. Mølgaard and P. Marckmann, Effect of garlic (*Allium sativum*) powder tablets on serum lipids, blood pressure and arterial stiffness in normolipidaemic volunteers: a randomised, double-blind, placebo-controlled trial, *Br. J. Nutr.*, 2004, **92**, 701–706.

T. G. Tutin, Biological flora of the British Isles, *Allium ursinum* L., *J. Ecol.*, 1957, **45**, 1003–1010.

P. C. Uden, R. Hafezi, M. Kotrebai, P. Nolibos, J. Tyson and E. Block, Anticarcinogenic organoselenium compounds – chromatographic, atomic and molecular mass spectral speciation, *Phosphorus, Sulfur Silicon Related Elements*, 2001, **172**, 31–56.

United Nations Food and Agriculture Organization, updated June 2008: www.fao.org/.

USDA (United States Department of Agriculture), New Pest Response Guidelines – Leek Moth, *Acrolepiopsis assectella* (Zeller), November 25, 2004, 83 pages [photographs and extensive references]: http://www.aphis.usda.gov/import_export/plants/manuals/emergency/downloads/nprg_leek_moth.pdf.

I. I. I. Uvah and T. H. Coaker, Effect of mixed cropping on some insect pests of carrots and onions, *Entomol. Exp. Appl.*, 1984, **36**, 159–167.

V. Vaidya, K. U. Ingold and D. A. Pratt, Garlic: source of the ultimate antioxidants – sulfenic acids, *Angew. Chem., Int. Edn.*, 2009, **48**, 157–160.

R. Valdivieso, J. Subiza, S. Varela-Losada, J. L. Subiza, M. J. Narganes, C. Martinez-Cocera and M. Cabrera, Bronchial asthma, rhinoconjunctivitis, and contact dermatitis caused by onion, *J. Allergy Clin. Immunol.*, 1994, **94**, 928–930.

S. Vale, Hepatitis A associated with green onions, *N. Engl. J. Med.*, 2005, **353**, 2300–2301.

L. Valerio and M. Maroli, Evaluation of repellent and anti-feeding effect of garlic oil (*Allium sativum*) against the bite of phlebotomine sandflies (Diptera: Psychodidae) [in Italian], *Ann. Ist. Super Sanita*, 2005, **41**, 253–256.

M. Valko, C. J. Rhodes, J. Moncol, M. Izakovic and M. Mazur, Free radicals, metals and antioxidants in oxidative stress-induced cancer, *Chem.-Biol. Interact.*, 2006, **160**, 1–40.

M. B. A. van Doorn, S. M. Espirito Santo, P. Meijer, I. M. Kamerling, R. C. Schoemaker, V. Dirsch, A. Vollmar, T. Haffner, R. Gebhardt, A. F. Cohen, H. M. Princen and J. Burggraaf, Effect of garlic powder on C-reactive protein and plasma lipids in overweight and smoking subjects, *Am. J. Clin. Nutr.*, 2006, **84**, 1324–1329.

H. D. VanEtten, J. W. Mansfield, J. A. Bailey and E. E. Farmer, Two classes of plant antibiotics: phytoalexins *versus* "phytoanticipins, *Plant Cell*, 1994, **6**, 1191–1192.

L. Vasseur and D. Gagnon, Survival and growth of *Allium tricoccum* AIT. Transplants in different habitats, *Biol. Conservation*, 1994, **68**, 107–114.

A. I. Virtanen and E. J. Matikkala, Structure and synthesis of cycloalliin isolated from *Allium cepa*, *Acta Chem. Scand.*, 1959a, **13**, 623–626.

A. I. Virtanen and E. J. Matikkala, Isolation of *S*-methyl- and *S*-propylcysteine sulfoxide from onion and the antibiotic activity of crushed onion, *Acta Chem. Scand.*, 1959b, **13**, 1898–1900.

A. I. Virtanen and E. J. Matikkala, Evidence for the presence of γ-glutamyl-*S*-(1-propenyl)-cysteine sulfoxide and cycloalliin as original compounds in onion, *Suomen Kemistil., B*, 1961, **34B**, 114.

A. I. Virtanen and I. Mattila, γ-L-Glutamyl-*S*-allyl-L-cysteine in garlic, *Suomen Kemistil., B*, 1961, **34B**, 44.

A. I. Virtanen and E. J. Matikkala, New γ-L-glutamyl peptides in onion (*Allium cepa*). III, *Suomen Kemistil., B*, 1961, **34B**, 53–54.

A. I. Virtanen and C. -G. Spåre, Isolation of the precursor of the lachrimatory factor in onion (*Allium cepa*), *Suomen Kemistil., B*, 1961, **34**, 72.

A. I. Virtanen and E. J. Matikkala, Structure of the γ-glutamyl peptide 4 isolated from onion (*Allium cepa*)-γ-L-glutamyl-*S*-(1-propenyl)cysteine sulfoxide, *Suomen Kemistil., B*, 1961, **34B**, 84.

A. I. Virtanen, M. Hatanaka and M. Berlin, γ-L-Glutamyl-*S*-propylcysteine in garlic, *Suomen Kemistil., B*, 1962a, **35b**, 52.

A. I. Virtanen and E. J. Matikkala, γ-L-Glutamyl-*S*-(prop-1-enyl)-L-cysteine in the seeds of chives, *Suomen Kemistil., B*, 1962b, **35B**, 245.

A. I. Virtanen, Some organic sulfur compounds in vegetables and fodder plants and their significance in human nutrition, *Angew. Chem., Int. Edn. Engl.*, 1962c, **1**, 299–306.

A. I. Virtanen, Studies on organic sulphur compounds and other labile substances in plants, *Phytochemistry*, 1965, **4**, 207–228.

G. M. Volk, A. D. Henk and C. M. Richards, Genetic diversity among U.S. garlic clones as detected using AFLP methods, *J. Amer. Soc. Horticultural Sci.*, 2004, **129**, 559–569.

G. M. Volk and D. Stern, Phenotypic characteristics of ten garlic cultivars grown at diverse North American locations, *HortScience*, 2009, **44**, 1238–1247.

H. Wagner, W. Dorsch, T. Bayer, W. Breu and F. Willer, Antiasthmatic effects of onions: inhibition of 5-lipoxygenase and cyclooxygenase *in vitro* by thiosulfinates and cepaenes, *Prostaglandins Leukotrienes Essent. Fatty Acids*, 1990, **39**, 59–62.

O. Wahlroos and A. I. Virtanen, Volatiles from chives (*Allium schoenoprasum*), *Acta Chem. Scand.*, 1965, **19**, 1327–1332.

J. C. Walker and M. A. Stahman, Chemical nature of disease resistance in plants, *Annu. Rev. Plant Physiol.*, 1955, **6**, 351–366.

T. B. Walker III, Garlic press, *US Pat.*, US 7117785, October 10, 2006.

L. Walton, M. Herbold and C. C. Lindegren, Bactericidal effects of vapors from crushed garlic, *J. Food Sci.*, 1936, **1**, 163–169.

D. Wang, H. Nanding, N. Han, F. Chen and G. Zhao, 2-(1*H*-Pyrrolyl)carboxylic acids as pigment precursors in garlic greening, *J. Agric. Food Chem.*, 2008, **56**, 1495–1500.

H. Wang, J. Li, Z. Wang, X. Zhang and Y. Ni, Modified method for rapid quantitation of *S*-alk(en)yl-L-cysteine sulfoxide in yellow onions (*Allium cepa* L.), *J. Agric. Food Chem.*, 2007, **55**, 5429–5435.

J. Wang, Z. A. Luthey-Schulten and K. S. Suslick, Is the olfactory receptor a metalloprotein? *Proc. Natl. Acad. Sci. U.S.A.*, 2003, **100**, 3035–3039.

W. Wang, J. Tang and A. Peng, The isolation, identification and bioactivities of selenoproteins in selenium-rich garlic, *Shengwu Huaxue Zazhi*, 1989, **5**, 229–234 (*Chem. Abstr.*, 1989, **111**, 95847d).

W. L. Wang, Y. Liu, X. L. Ji, G. Wang and H. B. Zhou, Effects of wheat-oilseed rape or wheat-garlic intercropping on the population dynamics of *Sitobion avenae* and its main natural enemies, *Ying Yong Sheng Tai Xue Bao*, 2008, **19**, 1331–1336 [Article in Chinese].

J. Ward, A description of the town of Silchester in its present state, *Philos. Trans. R. Soc. London*, 1748, **45**, 603–614.

S. Warshafsky, R. S. Kamer and S. L. Sivak, Effect of garlic on total serum cholesterol. A metaanalysis, *Ann. Internal Med.*, 1993, **119**, 599–605.

T. Watters, *On Yuan Chwang's travels in India, 629–645 A.D.*, 2 vols, 1904–5, Oriental Translation Fund, n.s. 14, Royal Asiatic Society, London, UK.

N. D. Weber, D. O. Andersen, J. A. North, B. K. Murray, L. D. Lawson and B. G. Hughes, *In vitro* virucidal effects of *Allium sativum* (garlic) extract and compounds, *Planta Med.*, 1992, **58**, 417–423.

F. Wehner, F. Musshoff, M. M. Schulz, D. D. Matin and H.-D. Wehner, Detection of colchicine by means of LC-MS/MS after mistaking meadow saffron for bear's garlic, *Forensic Sci. Med. Pathol.*, 2006, **2**, 193–197.

L. Weiner, I. Shin, L. J. Shimon, T. Miron, M. Wilchek, D. Mirelman, F. Frolow and A. Rabinkov, Thiol-disulfide organization in alliin lyase (alliinase) from garlic (*Allium sativum*), *Protein Sci.*, 2009, **18**, 196–205.

A. S. Weisberger and J. Pensky, Tumor inhibition by a sulfhydrylblocking agent related to an active principle of garlic (*Allium sativum*), *Cancer Res.*, 1958, **18**, 1301–1308.

E. Weise, USA Today, 2/27/2007: http://www.usatoday.com/news/health/2007-02-26-garlic-cholesterol_x.htm.

T. Wertheim, Investigations on garlic oil, *Ann. Chem. Pharm.*, 1844, **51**, 289–315.

D. West, *Horace. The Complete Odes and Epodes*, Oxford University Press, Oxford, 1997.

H. A. Wetli, R. Brenneisen, I. Tschudi, M. Langos, P. Bigler, T. Sprang, S. Schuerch and R. C. Muehlbauer, A γ-glutamyl peptide isolated from onion (*Allium cepa* L.) by bioassay-guided fractionation inhibits resorption activity of osteoclasts, *J. Agric. Food Chem.*, 2005, **53**, 3408–3414.

P. D. Whanger and J. A. Butler, Effects of various dietary levels of selenium as selenite or selenomethionine on tissue selenium levels and glutathione peroxidase activity in rats, *J. Nutr.*, 1988, **118**, 846–852.

P. D. Whanger, C. Ip, C. E. Polan, P. C. Uden and G. Welbaum, Tumorigenesis, metabolism, speciation, bioavailability, and tissue deposition of selenium in selenium-enriched ramps (*Allium tricoccum*), *J. Agric. Food Chem.*, 2000, **48**, 5723–5730.

C. Wheeler, T. M. Vogt, G. L. Armstrong, G. Vaughan, A. Weltman, O. V. Nainan, V. Dato, G. Xia, K. Waller, J. Amon, T. M. Lee, A. Highbaugh-Battle, C. Hembree, S. Evenson, M. A. Ruta, I. T. Williams, A. E. Fiore and B. P. Bell, An outbreak of hepatitis A associated with green onions, *N. Engl. J. Med.*, 2005, **353**, 890–897.

S. D. Whiting and M. L. Guinea, Treating stingray wounds with onions, *Med. J. Aust.*, 1998, **168**, 584.

S. Widder, C. Sabater Lüntzel, T. Dittner and W. Pickenhagen, 3-Mercapto-2-methylpentan-1-ol, a new powerful aroma compound, *J. Agric. Food. Chem.*, 2000, **48**, 418–423.

W. F. Wilkens, Isolation and identification of the lachrymogenic compound of onion, *Cornell Univ. Agr. Exp. Station Mem.*, 1964, No. 385.

J. G. Wilkinson, in *The Ancient Egyptians*, ed. S. Birch, John Murray, London, UK, 1878.

G. Williamson, G. W. Plumb, Y. Uda, K. R. Price and M. J. C. Rhodes, Dietary quercetin gycosides: antioxidant activity and induction of the anticarcinogenic phase II marker enzyme quinone reductase in Hepalclc7 cells, *Carcinogenesis*, 1996, **17**, 2385–2387.

G. H. Willital and H. Heine, Efficacy of Contractubex gel in the treatment of fresh scars after thoracic surgery in children and adolescents, *Int. J. Clin. Pharmacol. Res.*, 1994, **14**, 193–202.

E. W. Wilson, The onion in folk belief, *West. Folklore*, 1953, **12**, 94–104.

G. Winnewisser, F. Lewen, S. Thorwirth, M. Behnke, J. Hahn, J. Gauss and E. Herbst, Gas-phase detection of HSOH: Synthesis by flash vacuum pyrolysis of di-*tert*-butyl sulfoxide and rotational-torsional spectrum, *Chem.–Eur. J.*, 2003, **9**, 5501–5510.

K. Wojcikowski, S. Myers and L. Brooks, Effects of garlic oil on platelet aggregation: a double-blind placebo-controlled crossover study, *Platelets*, 2007, **18**, 29–34.

P. M. Wolsko, D. K. Solondz, R. S. Phillips, S. C. Schachter and D. M. Eisenberg, Lack of herbal supplement characterization in published randomized controlled trials, *Am. J. Med.*, 2005, **118**, 1087–1093.

W. Woodville, *Medical Botany*, James Phillips, London, UK, 1793.

R. B. Woodward and R. Hoffmann, *The Conservation of Orbital Symmetry*, Verlag Chemie, Weinhein/Bergstr, Germany, 1970.

World Cancer Research Fund/American Institute for Cancer Research, *Food, Nutrition, Physical Activity, and the Prevention of Cancer: a Global Perspective*, Washington DC, AICR, 2007.

World Health Organization Monographs on Selected Medicinal Plants, 1999, vol. 1, Geneva: http://whqlibdoc.who.int/publications/1999/9241545178.pdf.

C.-C. Wu, J. G. Chung, S. J. Tsai, J. H. Yang and L. Y. Sheen, Differential effects of allyl sulfides from garlic essential oil on cell cycle regulation in human liver tumor cells, *Food Chem. Toxicol.*, 2004, **42**, 1937–1947.

X. J. Wu, Y. Hu, E. Lamy and V. Mersch-Sundermann, Apoptosis induction in human lung adenocarcinoma cells by oil-soluble allyl sulfides: triggers, pathways, and modulators, *Environ. Mol. Mutagenesis*, 2009, **50**, 266–275.

D. Wujastyk, *The Roots of Ayurveda: Selections from Sanskrit Medical Writings*, Penguin Books, London, UK, 2003.

D. Xiao, Y. Zeng, E. R. Hahm, Y. A. Kim, S. Ramalingam and S. V. Singh, Diallyl trisulfide selectively causes Bax- and Bak-mediated apoptosis in human lung cancer cells, *Environ. Mol. Mutagenesis*, 2009, **50**, 201–212.

D. Xiao, K. L. Lew, Y. A. Kim, Y. Zeng, E. R. Hahm, R. Dhir and S. V. Singh, Diallyl trisulfide suppresses growth of PC-3 human prostate cancer xenograft *in vivo* in association with Bax and Bak induction, *Clin. Cancer Res.*, 2006a, **12**, 6836–6843.

D. Xiao, M. Li, A. Herman-Antosiewicz, J. Antosiewicz, H. Xiao, K. L. Lew, Y. Zeng, S. W. Marynowski and S. V. Singh, Diallyl trisulfide inhibits angiogenic features of human umbilical vein endothelial cells by causing Akt inactivation and down-regulation of VEGF and VEGF-R2, *Nutr. Cancer*, 2006b, **55**, 94–107.

D. Xiao and S. V. Singh, Diallyl trisulfide, a constituent of processed garlic, inactivates Akt to trigger mitochondrial translocation of BAD and caspase-mediated apoptosis in human prostate cancer cells, *Carcinogenesis*, 2006c, **27**, 533–540.

B. Xu, B. Monsarrat, J. E. Gairin and E. Girbal-Neuhauser, Effect of ajoene, a natural antitumor small molecule, on human 20S proteasome activity *in vitro* and in human leukemic HL60 cells, *Fundam. Clin. Pharmacol.*, 2004, **18**, 171–180.

M. Yagami, S. Kawakishi and M. Namiki, Identification of intermediates in the formation of onion flavor, *Agric, Biol. Chem.*, 1980, **44**, 2533–2538.

O. Yamato, M. Hayashi, M. Yamasaki and Y. Maede, Induction of onion-induced haemolytic anaemia in dogs with sodium *n*-propylthiosulphate, *Veterinary Rec.*, 1998, **142**, 216–219.

O. Yamato, M. Hayashi, E. Kasai, M. Tajima, M. Yamasaki and Y. Maede, Reduced glutathione accelerates the oxidative damage produced by sodium *n*-propylthiosulfate, one of the causative agents of onion-induced hemolytic anemia in dogs, *Biochim. Biophys. Acta*, 1999, **1427**, 175–182.

O. Yamato, Y. Sugiyama, H. Matsuura, K.-W. Lee, K. Goto, M. A. Hossain, Y. Maede and T. Yoshihara, Isolation and identification of sodium 2-propenyl thiosulfate from boiled garlic (*Allium sativum*) that oxidizes canine erythrocytes, *Biosci. Biotech. Biochem.*, 2003, **67**, 1594–1596.

O. Yamato, E. Kasai, T. Katsura, S. Takahashi, T. Shiota, M. Tajima, M. Yamasaki and Y. Maede, Heinz body hemolytic anemia with eccentrocytosis from ingestion of Chinese chive (*Allium tuberosum*) and garlic (*Allium sativum*) in a dog, *J. Am. Animal Hosp. Assoc.*, 2005, **41**, 68–73.

M. Yamazaki, M. Sugiyama and K. Saito, Intercellular localization of cysteine synthase and alliinase in bundle sheaths of *Allium* plants, *Plant Biotechnol.*, 2002, **19**, 7–10.

Y. Yamazaki, T. Tokunaga and T. Okuno, Quantitative determination of eleven flavor precursors (*S*-alk(en)yl cysteine derivatives) in garlic with an HPLC method, *Nippon Shokuhin Kagaku Kogaku Kaishi*, 2005, **52**, 160–166 (*Chem. Abstr.*, 2005, **143**, 448406).

J.-S. Yang, G.-W. Chen, T.-C. Hsia, H.-C. Ho, C.-C. Ho, M.-W. Lin, S.-S. Lin, R.-D. Yeh, S.-W. Ip, H.-F. Lu and J.-G. Chung, Diallyl disulfide induces apoptosis in human colon cancer cell line (COLO 205) through the induction of reactive oxygen species, endoplasmic reticulum stress, caspases casade and mitochondrial-dependent pathways, *Food Chem. Toxicol.*, 2009, **47**, 171–179.

M. Yang, K. Wang, L. Gao, Y. Han, J. Lu and T. Zou, Exploration for a natural selenium supplement – characterization and bioactivities of *Se*-containing polysaccharide from garlic, *J. Chin. Pharm. Sci.*, 1992, **1**, 28–32 (*Chem. Abstr.*, 1993, **118**, 77092u).

Q. Yang, Q. Hu, O. Yamato, K. W. Lee, Y. Maede and T. Yoshihara, Organosulfur compounds from garlic (*Allium sativum*) oxidizing canine erythrocytes, *Z. Naturforsch., C: Biosci.*, 2003, **58**, 408–412.

W. Yang, J. Chen, W. Li and X. Chen, Preventive effects of 4 *Se*-enriched plants on rat stomach cancer induced by MNNG-3. Se accumulation and distribution in rats of different selenium resources for prevention of stomach cancer, *Wei Sheng Yan Jiu* [*J. Hyg. Res.*], 2008, **37**, 435–437 (in Chinese; PubMed ID 18839527 AN 2008647083).

M. C. Yarema and S. C. Curry, Acute tellurium toxicity from ingestion of metal-oxidizing solutions, *Pediatrics*, 2005, **116**, 319–321.

E. Yildirim and I. Guvenc, Intercropping based on cauliflower: more productive, profitable and highly sustainable, *Eur. J. Agron.*, 2005, **22**, 11–18.

M. C. Yin and S. M. Tsao, Inhibitory effect of seven *Allium* plants upon three *Aspergillus* species, *Int. J. Food Microbiol.*, 1999, **49**, 49–56.

K. S. Yoo and L. M. Pike, Determination of flavor precursor compound *S*-alk(en)yl-L-cysteine sulfoxides by an HPLC method and their distribution in *Allium* species, *Sci. Horticulture*, 1998, **75**, 1–10.

H. Yoshida, H. Katsuzaki, R. Ohta, K. Ishikawa, H. Fukuda, T. Fujino and A. Suzuki, Antimicrobial activity of the thiosulfinates isolated from oil-macerated garlic extract, *Biosci. Biotechnol. Biochem.*, 1999, **63**, 591–594.

H. Yoshida, N. Iwata, H. Katsuzaki, R. Naganawa, K. Ishikawa, H. Fukuda, T. Fujino and A. Suzuki, Antimicrobial activity of a compound isolated from an oil-macerated garlic extract, *Biosci. Biotechnol. Biochem.*, 1998, **62**, 1014–1017.

W. C. You, L. M. Brown, L. Zhang, J. Y. Li, M. L. Jin, Y. S. Chang, J. L. Ma, K. F. Pan, W. D. Liu, Y. Hu, S. Crystal-Mansour, D. Pee, W. J. Blot, J. F. Fraumeni, Jr., G. W. Xu and M. H. Gail, Randomized double-blind factorial trial of three treatments to reduce the prevalence of precancerous gastric lesions, *J. Natl. Cancer Inst.*, 2006, **98**, 974–983.

W. C. You, L. Zhang, M. H. Gail, J. L Ma, Y. S. Chang, W. J. Blot, J. Y. Li, C. L. Zhao, W. D. Liu, H. Q. Li, Y. R. Hu, J. C. Bravo, P. Correa, G. W. Xu and J. F. Fraumeni Jr., *Helicobacter pylori* infection, garlic intake and precancerous lesions in a Chinese population at low risk of gastric cancer, *Int. J. Epidemiol.*, 1998, **27**, 941–944.

D. Young, "Chopping Garlic" from *At the White Window*, Ohio State University, Columbus OH, 2000.

J. Q. Yu, Allelopathic suppression of *Pseudomonas solanacearum* infection of tomato (*Lycopersicon esculentum*) in a tomato–Chinese chive (*Allium tuberosum*) intercropping system, *J. Chem. Ecol.*, 1999, **25**, 2409–2417.

T. -H. Yu, C. -M. Wu and Y. C. Liou, Volatile compounds from garlic, *J. Agric. Food Chem.*, 1989a, **37**, 725–730.

T.-H. Yu, C.-M. Wu and S. Y. Chen, Effects of pH adjustment and heat treatment on the stability and the formation of volatile compounds of garlic, *J. Agric. Food Chem.*, 1989b, **37**, 730–734.

T.-H. Yu, C.-M. Wu, R. T. Rosen, T. G. Hartman and C.-T. Ho, Volatile compounds generated from thermal degradation of alliin and deoxyalliin in an aqueous solution, *J. Agric. Food Chem.*, 1994a, **42**, 146–153.

T.-H. Yu, C.-M. Wu and C.-T. Ho, Meat-like flavor generated from thermal interactions of glucose and alliin or deoxyalliin, *J. Agric. Food Chem.*, 1994b, **42**, 1005–1009.

T.-H. Yu, L.-Y. Lin and C.-T. Ho, Volatile compounds of blanched, fried blanched, and baked blanched garlic slices, *J. Agric. Food Chem.*, 1994c, **42**, 1342–1347.

H. Zeng and G. F. Combs Jr, Selenium as an anticancer nutrient: roles in cell proliferation and tumor cell invasion, *J. Nutr. Biochem.*, 2008, **19**, 1–7.

R. S. Zeng, Allelopathy in Chinese ancient and modern agriculture, in *Allelopathy in Sustainable Agriculture and Forestry*, ed. R.S. Zeng, A.U. Mallik, S.M. Luo, Springer, 2008.

G. Zhang, H. Wu, B. Zhu, Y. Shimoishi, Y. Nakamura and Y. Murata, Effect of dimethyl sulfides on the induction of apoptosis in human leukemia Jurkat cells and HL-60 cells, *Biosci. Biotechnol. Biochem.*, 2008a, **72**, 2966–2972.

L. Zhang, M. H. Gail, Y. Q. Wang, L. M. Brown, K. F. Pan, J. L. Ma, H. Amagase, W. C. You and R. Moslehi, A randomized factorial study of the effects of long-term garlic and micronutrient supplementation and of 2-wk antibiotic treatment for *Helicobacter pylori* infection on serum cholesterol and lipoproteins, *Am. J. Clin. Nutr.*, 2006b, **84**, 912–919.

X. Zhang and R. A. Laursen, Development of mild extraction methods for the analysis of natural dyes in textiles of historical interest using LC-diode array detector-MS, *Anal. Chem.*, 2005, **77**, 2022–2025.

Y. W. Zhang, J. Wen, J. B. Xiao, S. G. Talbot, G. C. Li and M. Xu, Induction of apoptosis and transient increase of phosphorylated MAPKs by diallyl disulfide treatment in human nasopharyngeal carcinoma CNE2 cells, *Arch. Pharm. Res.*, 2006a, **29**, 1125–1131.

Z. D. Zhang, Y. Li, Z. K. Jiao, *et al.*, [Effect of local application of allicin *via* gastroscopy on cell proliferation and apoptosis of progressive gastric carcinoma] *Chin. J. Integrated Traditional Western Med.*, 2008, **28**, 108–110 [*Zhongguo Zhong Xi Yi Jie He Za Zhi*; article in Chinese].

Z. M. Zhang, X. Y. Yang, S. H. Deng, W. Xu and H. Q. Gao, Anti-tumor effects of polybutylcyanoacrylate nanoparticles of diallyl trisulfide on orthotopic transplantation tumor model of hepatocellular carcinoma in BALB/c nude mice, *Chin. Med. J. (Beijing, Engl. Ed.)*, 2007, **120**, 1336–1342.

Z. M. Zhang, N. Zhong, H. Q. Gao, S. Z. Zhang, Y. Wei, H. Xin, X. Mei, H. S. Hou, X. Y. Lin and Q. Shi, Inducing apoptosis and upregulation of Bax and Fas ligand expression by allicin in hepatocellular carcinoma in Balb/c nude mice, *Chin. Med. J. (Beijing, Engl. Ed.)*, 2006b, **119**, 422–425.

H. Zhen, F. Fang, D. Y. Ye, S. N. Shu, Y. F. Zhou, Y. S. Dong, X. C. Nie and G. Li, Experimental study on the action of allitridin against human cytomegalovirus *in vitro*: inhibitory effects on immediate-early genes, *Antiviral Res.*, 2006, **72**, 68–74.

J. Zhou, S. Yao, R. Qian, Z. Xu, Y. Wei and Y. Guo, Observation of allicin–cysteine complex by reactive desorption electrospray ionization mass spectrometry for garlic, *Rapid Commun. Mass Spectrom.*, 2008, **22**, 3334–3337.

S. Ziaei, S. Hantoshzadeh, P. Rezasoltani and M. Lamyian, The effect of garlic tablet on plasma lipids and platelet aggregation in nulliparous pregnants at high risk of preeclampsia, *Eur. J. Obstetrics Gynecol. Reproductive Biol.*, 2001, **99**, 201–206.

S. J. Ziegler and O. Sticher, Optimization of the mobile phase for HPLC separation of *S*-alk(en)yl-L-cysteine derivatives and their corresponding sulfoxide isomers, *J. Chromatogr.*, 1989a, **12**, 199–220.

S. J. Ziegler and O. Sticher, HPLC of *S*-alk(en)yl-L-cysteine derivatives in garlic including quantitative determination of (+)-*S*-allyl-L-cysteine sulfoxide (alliin), *Planta Med.*, 1989b, **55**, 372–378.

D. Zohary and M. Hopf, *Domestication of Plants in the Old World*, Oxford University Press, Oxford, 3rd edn, 2000.

A. N. Zohri, K. Abdel-Gawad and S. Saber, Antibacterial, antidermatophytic and antitoxigenic activities of onion (*Allium cepa* L.) oil, *Microbiol Res.*, 1995, **150**, 167–172.

J. M. Zurada, D. Kriegel and I. C. Davis, Topical treatments for hypertrophic scars, *J. Am. Acad. Dermatol.*, 2006, **55**, 1024–1031.

Appendix 1

Garlic and Other Alliums: The Lore and the Science
By Eric Block

Published by the Royal Society of Chemistry, www.rsc.org

Table A.1 Flavor precursor content of representative genus *Allium* plants by different techniques.

Species	*Method*	*Enz*[i]	*Total*[a]	*Methiin*[b]	*Alliin*[b]	*Isoallin*[b]	*Propiin*[b]	*Reference*
A. altyncolicum	HPLC		3.4	20	1	75	4	Keusgen, 2002
A. ampeloprasum	HPLC		1.5	40	43	3	15	Fritsch, 2006
(elephant garlic)	HPLC		5	17	63	20	0	Yoo, 1998
A. ampeloprasum	LC-MS		2.6	5	tr	95	nd	Lundegardh, 2008;
(Leek)	CE		1.5	12	tr	88	Tr	Kubec, 2008
A. angulosum	HPLC		5.6	96	1	2	2	Fritsch, 2006;
	HPLC[j]			(52.5)	(0)	(35.1)	(12.4)	Ferary,1998
A. carinatum	HPLC		4.0	82	0	18	0	Fritsch, 2006
A. cepa	HPLC		2.5	11	10	79	0	Fritsch, 2006
	GC		1.5	14	3	82	0	Wang, 2007
	CE		0.59	18	nd	82	nd	Kubec, 2008
A. cepa (dehydrator)	HPLC		2.5	7	0	93	0	Yoo, 1998
A. cepa (shallot)	CE		1.35	19	tr	81	nd	Kubec, 2008
	HPLC		2.27	5	nd	95	nd	Yoo, 1998
A. chevsuricum	HPLC		1.5	48	2	49	1	Keusgen, 2002
A. fistulosum	HPLC		1.7	10	0	90	–	Yoo, 1998
A. globosum	HPLC		1.7	69	26	6	0	Fritsch, 2006;
	HPLC[j]			(77)	(20.5)	(2.5)	(0)	Ferary, 1998
A. hymenorrhizum	HPLC		0.6	79	2	19	nd	Krest, 2000
A. jesdianum DS[h]	HPLC	2	6.2	97	nd	3	nd	Krest, 2000
A. karelinii	HPLC[j]			(33.2)	(0)	(51.8)	(15)	Ferary, 1998
A. kermesinum	HPLC		7	28	59	7	6	Storsberg, 2004

A. moly	HPLC		1	22	8	70	0	Fritsch, 2006
A. neapolitanum	HPLC		1.4	98	2	0	1	Fritsch, 2006
A. obliquum	HPLC	13	6.4	44	56	nd	nd	Krest, 2000;
	HPLC		13.4	30	58	3	9	Keusgen, 2002;
	HPLC[j]			(47.7)	(46.5)	(5.8)	(0)	Ferary, 1998
A. oleraceum W[e]	HPLC		2.1	89	3	7	0	Fritsch, 2006
A. paniculatum	HPLC		–	100	0	0	0	Boscher, 1995
A. paradoxum	HPLC		0.4	80	nd	20	nd	Krest, 2000
A. roseum	HPLC		2.3	80	17	2	0	Fritsch, 2006
A. sativum China	CE	178	19	9	91	nd	nd	Horie, 2006;
	CE		12.3	10	81	9	nd	Kubec, 2008
A. sativum Japan	CE		14	9	91	nd	nd	Horie, 2006
A. sativum Texas	HPLC		12	5	84	11	0	Yoo, 1998
A. sativum various	HPLC[k]		31.1	4	84	12	nd	Yamazaki, 2005
A. saxatile	HPLC	18	6.7	94	2	4	0	Fritsch, 2006;
	HPLC[j]			(78)	(17.6)	(4.4)	(0)	Ferary, 1998
A. schoenoprasum[g]	HPLC		0.6	31	5	63	0	Fritsch, 2006
	CE		2.45	22	tr	78	tr	Kubec, 2008
	HPLC		1.42	48	6	46	0	Yoo, 1998
A. scorodoprasum	HPLC		1.5	25	69	5	0	Fritsch, 2006
A. senescens	HPLC		0.6	31	1	48	21	Fritsch, 2006;
	HPLC[j]			(7.6)	(0)	(92)	(0.4)	Ferary, 1998
A. siculum[f]	GC		0.4[c]	50	–	–	–	Kubec, 2002; 2009
A. sphaerocephalon	HPLC	6	0.5	42	18	39	0	Fritsch, 2006
A. stipitatum DS[h]	HPLC	7	4.4	98	nd	2	nd	
A. subhirsutum	HPLC	7	0.3	79	21	0	0	Fritsch, 2006
A. triquetrum	HPLC		1.1	56	21	19	4	Fritsch, 2006
A. tuberosum	CE HPLC[j]		5	80	20	–	–	Horie, 2007;
				(80)	(19.1)	(0.9)	(0)	Ferary, 1998
	HPLC		2.6	72	23	5	–	Yoo, 1998

Table A.1 (*Continued*).

Species	*Method*	*Enz*[i]	*Total*[a]	*Methiin*[b]	*Alliin*[b]	*Isoallin*[b]	*Propiin*[b]	*Reference*
A. ursinum[g]	HPLC		2[c]	35	28	37	0	Fritsch, 2006
A. victorialis[g]	HPLC	3	1[d]	75	25	–	0	Fritsch, 2006
A. vineale W[e]	HPLC		1	29	31	41	0	Fritsch, 2006

[a] g kg^{-1} fresh weight.
[b] % of total cysteine sulfoxides.
[c] In bulbs.
[d] In leaves.
[e] W = a weed
[f] Also contains 23% butiin.
[g] Also contains traces of ethiin (Kubec, 2000).
[h] DS = "drumstick" *Allium*.
[i] "enz" indicates amount of alliinase enzyme measured as μmol min^{-1} mg^{-1}.
[j] Based on HPLC analysis of thiosulfinates enzymatically formed (Ferary, 1998).
[k] Cation-exchange column; average % dry weight: alliin, 2.62; methiin, 0.36; isoalliin, 0.13.

Table A.2 Antibiotic activity of *Allium*-derived compounds [minimum inhibitory concentration (MIC) in mg L^{-1}].

Organism	*Allicin*	*Ajoene [Z (E)]*	*All_2S*	*All_2S_2*	*All_2S_3*	*All_2S_4*	*Other*	*References*
Bacteria, Gram negative								
E. coli	15	100–116 (200)	> 1000	> 1000	> 1000			Ankri, 1999; Naganawa, 1996; Yoshida, 1998; Kim, 2004
Helicobacter pylori	6–30	15–20 (25)	> 1000	100	13–25	3–6	8–32[e], 0.025[h]	O'Gara, 2000; Ohta, 1999
Klebsiella pneumoniae	8	113–152 (200)	96	72	40	20	24[e]	Ankri, 1999; Naganawa, 1996; Tsao, 2001b; Yoshida, 1998
Leptotrichia buccalis	13.8							Bakri, 2005
Pophyromonas gingivalis	1.7							Bakri, 2005
P. intermedia	1.7							Bakri, 2005
P. nigrescens	0.4							Bakri, 2005
Pseudomonas aeruginosa	15	> 500 (> 500)	80	64	32	12	16[e]	Naganawa, 1996; Ankri, 1999; Tsao, 2001b; Yoshida, 1998
Campylobacter jejuni			56	12	2	1	1[g]	Rattanachaikunsopon, 2008
Salmonella enterica			54	12	2	0.5	1[g]	Rattanachaikunsopon, 2008
V. cholerae			72	24	12	4	1[g]	Rattanachaikunsopon, 2008
Bacteria, Gram positive								
B. cereus			64	14	4	1	0.5[g]	Rattanachaikunsopon, 2008
B. subtilis		5 (14)						Naganawa, 1996; Yoshida, 1998
E. faecalis	28							Bakri, 2005
C. botulinum			64	20	4	1	0.5[g]	Rattanachaikunsopon, 2008

Table A.2 (*Continued*).

Organism	*Allicin*	*Ajoene [Z (E)]*	*All_2S*	*All_2S_2*	*All_2S_3*	*All_2S_4*	*Other*	*References*
Mycobacterium phlei		10–14 (30)						Naganawa, 1996; Yoshida, 1998
M. smegmatis		4 (Z only)					2[a], 4–16[b]	O'Donnell, 2007, 2008
Proteus mirablis	15							Ankri, 1999
S. aureus	12–28	20 (40)	20.0	4.0	2.0	0.5	24.0[e], 0.5[f], 1.0[g]	Ankri, 1999; Naganaw, 1996; Tsao, 2001a; Bakri, 2005; Yoshida, 1998
S. pyogenes	3							Ankri, 1999
Fungi								
A. flavus	16		64.0	12.0	4.0	2.0	40.0[e]	Shadkchan, 2004; Tsao, 2001a
A. fumigatus	8		54.0	12.0	8.0	4.0	32.0[e]	Shadkchan, 2004; Tsao, 2001a
Aspergillus niger	32	<20	40.0	8.0	2.0	1.0	20.0[e]	Shadkchan, 2004; Yoshida, 1987; Tsao, 2001a
Candida albicans	0.3	13	32.0	4.0	1.0	0.5	16[e]	Ankri, 1999; Naganawa, 1996; Tsao, 2001a
C. glabrata	0.3		54.0	8.0	4.0	2.0	32.0[e]	Ankri, 1999; Tsao, 2001a
C. neoformans	0.3							Ankri, 1999
C. krusei	0.3		72.0	12.0	8.0	4.0	24.0[e]	Ankri, 1999; Tsao, 2001a
C. parapsilosis	0.15							Ankri, 1999
C. tropicalis	0.3							Ankri, 1999
C. valida		15 (50)						Yoshida, 1998

Cryptococcus neoformans	6–12				2.5–100			Davis, 1990, 1994; Shen, 1996
Saccharomyces cerevisiae	10	12–20 (50)	> 1000	100	5	2	15[d], 10[e]	Naganawa, 1996; Kim, 2004; Yoshida, 1998, 1999
Scedosporium prolificans		4.0			8.0			Davis, 2003
Parasitic protozoa								
Entamoeba histolytica	30				59[c]			Mirelman, 1987; Lun, 1994
Giardia lamblia	30				14[c]			Mirelman, 1987; Lun, 1994
G. intestinalis			> 1000[c]	100[c]			7[i]	Harris, 2000
Leishmanial strains	5–30						376[c,d]	Mirelman, 1987; Saleheen, 2004; Khalid, 2005
Trypanosoma b. brucei					2.5[c]			Lun, 1994
T. congolense					5.5[c]			Lun, 1994
T. equiperdum					1.2[c]			Lun, 1994
T. evansi					0.8[c]			Lun, 1994

[a]8-Hydroxycanthin-6-one from *A. neapolitanum* (O'Donnell, 2007).
[b]Pyridine *N*-oxide disulfides from *A. stipitatum* (O'Donnell, 2009).
[c]IC_{50} ($\mu g\,mL^{-1}$).
[d]Onion oil.
[e]Garlic oil.
[f]Penicillin.
[g]Tetracycline.
[h]Amoxicillin.
[i]Allyl alcohol.

Table A.3 Recent *in vitro* studies on anticarcinogenic properties of *Allium*-derived compounds.

Compound	*Cancer cell type*	*Mechanism*	*Reference*
All_2S_4	human acute myeloid leukemia U937	induce apoptosis, Bax and Bak, caspase activation	Cerella, 2009
R_2S_3 including R = All	human colon cancer HCT-15 and DLD-1	microtubule disassembly, cell cycle arrest, cysteine –SH to –SSR reaction	Seki, 2008; Hosono, 2008, 2005
All_2S_3	DU145 and PC3 human prostate cancer	Bax and Bak induction; angiogenesis inhibition; Akt inactivation, ROS	Xiao, 2006a,b,c; Kim, 2007
same	prostate cancer in TRAMP mice	inhibition of lung metathesis	Singh, 2008
same	BGC823 human gastric cancer	induce apoptosis	Li, 2006
same	human childhood pre-B acute lymphoblastic leukemia	apoptosis	Hodge, 2008
All_2S_3 PBCA nanoparticles	hepatocellular carcinoma in mice	induce apoptosis	Zhang, 2007
All_2S_2	human colon cancer COLO 205	apoptosis *via* induction of reactive oxygen compounds (ROS)	Yang, 2008
same	human leukemia HL-60	apoptosis *via* inhibition of ERK & induction of p38	Tan, 2008
same	human prostate PC-3	induce apoptosis	Arunkumar, 2007
same	human CNE2 nasopharyngeal carcinoma	induce apoptosis	Zhang, 2006
same	human leukemia HL-60	inhibit NAT	Lin, 2002
same	non small cell lung cancer	apoptosis, induce Bax, suppress Bcl-2	Hong, 2000
All_2S_n, n = 1–3	human lung H358	Bax- and Bak-mediated apoptosis	Xiao, 2008
same	human lung A549	apoptosis *via* induction of ROS, JNK, p53	Wu, 2009

same	human glioblastoma	apoptosis *via* induction of ROS	Das, 2007
same	human liver J5	increase cyclin B1	Wu, 2004
All_2S_n, $n = 1–5$	human leukemia HL-60	inhibits repair polymerases	Nishida, 2008
Me_2S_n, $n = 1–4$	human leukemia HL-60	apoptosis *via* induction of ROS and caspase-3 activation	Zhang, 2008
All_2S	K562 human leukemia	modulate MDR	Arora, 2004
same	non small cell lung cancer	apoptosis, induce Bax, suppress Bcl-2	Hong, 2000
same	human leukemia HL-60	inhibit NAT	Lin, 2002
***S*-Allylmercaptocysteine**	human gastric cancer SNU-1	Bax, p53, caspase-9 induction	Lee, 2008
same	human prostate cancer PCa	up-regulation of E-cadherin	Chu, 2006
same	human prostate cancer PC-3	antimetastatic effect	Howard, 2007
S-Allylcysteine	human breast cancer MDA-MB-231	induce E-cadherin,inhibit MMP-2	Gapter, 2008
same	human prostate cancer PCa	restoration of E-cadherin expression	Chu, 2006, 2007
same	human colon cancer COLO 320DM	caspase-3 and NFκB induction	Sriram, 2008
Allicin	human leukemia HL60, myelo-monocytic U937	induce apoptosis, depleted GSH	Miron, 2008
same	cultured fibroblasts	microtubule disassembly	Prager-Khoutorsky, 2007
same	hepatocellular carcinoma in mice	induce apoptosis, upregulate Bax & Fas	Zhang, 2006b
same	human gastric epithelial carcinoma	AIF & PFK release	Park, 2005
same	human B chronic lymphocytic leukemia	tumor cell apoptosis *via* site-directed therapy	Arditti, 2005
same	human cervical cancer	induce apoptosis	Oommen, 2004

Table A.3 (*Continued*).

Compound	*Cancer cell type*	*Mechanism*	*Reference*
same	human mammary, (MCF7),colon (HT29)	depletes intracellular GSH	Hirsch, 2000
RS(O)SR from *A. tuberosum*	human prostate and colon cancers	caspase-3, -8, -9 induction, apoptosis	Park, 2007; Kim, 2008a,b; Lee, 2009
MeS(O)SMe and **PrS(O)SPr**	human acute myeloid leukemia U937	inhibition of cell proliferation, LD_{50} 2 μM	Merhi, 2008
Ajoene	human T lymphoma, neuroblastoma, & fibroblast, HL60	activation of p53, p63, p73 gene products, induce apoptosis	Terrasson, 2007
same	human promyeloleukemic	microtubule-interaction	Li, 2002
same	HL60 human leukemia	affect proteasome	Xu, 2004
same	HL60 human leukemia	ROS, induce apoptosis, NFκB and caspase-8	Dirsch, 2002, 1998
same	human promyeloleukemic	induce apoptosis, inhibit ERK	Antlsperger, 2003
same	human myeloid leukemia	enhance caspase-3 induction	Ahmed, 2001
same	mouse melanoma	inhibit metathesis	Taylor, 2006; Nishikawa, 2002
same	human childhood pre-B acute lymphoblastic leukemia	apoptosis	Hodge, 2008
***Se*-Methylselenocysteine**	murine B16F10 melanoma	reduces metathesis	Kim, 2008d
same	human prostate cancer	inhibit growth, reduce AR and PSA expression	Lee, 2006
Thiacremonone (2,4-dihydroxy-2,5-dimethyl-thiophene-3-one)	human colon cancer HCT-116	inhibit growth by inactivation of NF-κB	Ban, 2007
Garlic oligosaccharide (MW: 1800)	human lymphoma, colon adenocarcinoma	stimulates interferon-gamma	Tsukamoto, 2008

APPENDIX 2

Historical Illustrations of Alliums from *Flora Germanica*, Vol. 10 (Ludwig Reichenbach, 1848)

Ludwig Heinrich Gottlieb Reichenbach (1793–1879), a German physician, botanist, botanical artist and professor of natural history, was director of the botanical garden at Dresden, Germany, from 1820 to 1879 (Stafleu, 1983). This volume is from the rare book collection of the Cambridge University Botanic Garden Library; the plates are copied with kind permission. (L. Reichenbach, *Icones Florae Germanicae et Helveticae*, F. Hofmeister, Leipzig, Germany, 1848, vol. 10).

ALPHABETICAL LISTING OF *ALLIUM* SPECIES SHOWN

acutangulum (A2.19)
acutiflorum (A2.10)
ampeloprasum (A2.8) (great-headed garlic)
arenarium (A2.9)
ascalonicum (A2.10)
asperum (A2.2)
atropurpureum (A2.24)
carinatum (A2.1) (keeled garlic)
carneum (A2.23)
cepa (A2.13) (onion)
chamaemoly (A2.20)

Garlic and Other Alliums: The Lore and the Science
By Eric Block

Published by the Royal Society of Chemistry, www.rsc.org

controversum (A2.7) (serpent garlic)
descendens (A2.12)
fistulosum (A2.14) (Japanese bunching onion)
flavescens (A2.18)
flavum (A2.4) (small yellow onion)
fuscum (A2.4)
globosum (A2.16)
intermedium (A2.5)
kermesinum (A2.17)
longispathum (A2.6)
margaritaceum (A2.10)
moly (A2.20)
montanum (A2.19)
moschatum (A2.17)
multibulbosum (A2.25)
neapolitanum (daffodil garlic) (A2.26)
nigrum (A2.24) (black garlic)
ochroleucum (A2.17)
oleraceum (field garlic) (A2.26)
pallens (A2.3)
paniculatum (A2.3)
pedemontanum (A2.3)
pendulinum (A2.22)
permixtum (A2.21)
porrum (A2.8) (leek)
praescissum (A2.5)
pulchellum (A2.2)
roseum (rosy-flowered garlic) (A2.23)
rotundum (A2.11)
sativum (A2.7) (garlic)
saxatile (A2.16)
schoenoprasum (chives) (A2.15)
scorodoprasum (rocambole, sand leek) (A2.9)
sibiricum (A2.15) (hairy garlic)
sphaerocephalum (round-head garlic) (A2.11)
strictum (A2.12)
suaveolens (A2.18)
subhirsutum (A2.21)
triquetrum (A2.22) (three-cornered garlic)
ursinum (bear's garlic, ramsons) (A2.26)
victorialis (A2.27) (victory onion)
vineale (crow, field or stag's garlic) (A2.9)
violaceum (A2.1)

Figure A2.1 1057, *A. carinatum*; 1058, *A. violaceum*

Figure A2.2 1059, *A. asperum*; 1060, *A. pulchellum*

Figure A2.3 1061, *A. paniculatum*; 1062, *A. pallens*

Figure A2.4 1063, *A. flavum*; 1064, *A. fuscum*

Figure A2.5 1065, *A. intermedium*; 1066, *A. praescissum*

Figure A2.6 1067, *A. oleraceum*; 1068, *A. longispathum*

Figure A2.7 1069, *A. sativum*; 1070, *A. controversum*

Figure A2.8 1071, *A. porrum*; 1072, *A. ampeloprasum*

Figure A2.9 1073, *A. scorodoprasum*; 1074, *A. arenarium*; 1075, *A. vineale*

Figure A2.10 1076, *A. ascalonicum*; 1077, *A. margaritaceum*; 1078, *A. acutiflorum*

Figure A2.11 1079, *A. rotundum*; 1080, *A. sphaerocephalum*

Figure A2.12 1081, *A. strictum*; 1082, *A. descendens*

Figure A2.13 1083, *A. cepa*

Figure A2.14 1084, *A. fistulosum*

Figure A2.15 1085, *A. schoenoprasum*; 1086, *A. sibiricum*

Figure A2.16 1087, *A. saxatile*; 1088, *A. globosum*

Figure A2.17 1089, *A. kermesinum*; 1090, *A. ochroleucum*; 1091, *A. moschatum*

Figure A2.18 1092, *A. flavescens*; 1093, *A. suaveolens*

Figure A2.19 1094, *A. montanum*; 1095, *A. acutangulum*

Figure A2.20 1096, *A. chamaemoly*; 1097, *A. moly*

Figure A2.21 1098, *A. permixtum*; 1099, *A. subhirsutum*

Figure A2.22 1100, *A. pendulinum*; 1101, *A. triquetrum*

Figure A2.23 1102, *A. roseum*; 1103, *A. carneum*; 1104, *A. pedemontanum*

Figure A2.24 1105, *A. atropurpureum*; 1106, *A. nigrum*

Figure A2.25 1107, *A. multibulbosum*

Figure A2.26 1108, *A. neapolitanum*; 1109, *A. ursinum*

Figure A2.27 1110, *A. victorialis*

Subject Index

Boldface denotes illustration or figure